METHODS IN MOLECULAR BIOLOGY

Series Editor
John M. Walker
School of Life and Medical Sciences
University of Hertfordshire
Hatfield, Hertfordshire, UK

For further volumes:
http://www.springer.com/series/7651

For over 35 years, biological scientists have come to rely on the research protocols and methodologies in the critically acclaimed Methods in Molecular Biology series. The series was the first to introduce the step-by-step protocols approach that has become the standard in all biomedical protocol publishing. Each protocol is provided in readily-reproducible step-by step fashion, opening with an introductory overview, a list of the materials and reagents needed to complete the experiment, and followed by a detailed procedure that is supported with a helpful notes section offering tips and tricks of the trade as well as troubleshooting advice. These hallmark features were introduced by series editor Dr. John Walker and constitute the key ingredient in each and every volume of the Methods in Molecular Biology series. Tested and trusted, comprehensive and reliable, all protocols from the series are indexed in PubMed.

Manuscript Instructions https://drive.google.com/file/d/1K4Az5vpAFVzYadJ3jKx2-QSo7Dqa_SNA/view?usp=sharing

Computational Neuroscience

Edited by

Antonio Di Ieva

Computational NeuroSurgery (CNS) Lab, Faculty of Medicine, Health and Human Sciences, Macquarie Medical School, Macquarie University, Sydney, NSW, Australia

Jennilee M. Davidson

Macquarie Medical School, Faculty of Medicine, Health and Human Sciences, Macquarie University, Sydney, NSW, Australia

Humana Press

Editors
Antonio Di Ieva
Computational NeuroSurgery (CNS) Lab
Faculty of Medicine
Health and Human Sciences
Macquarie Medical School
Macquarie University
Sydney, NSW, Australia

Jennilee M. Davidson
Macquarie Medical School, Faculty of Medicine
Health and Human Sciences
Macquarie University
Sydney, NSW, Australia

ISSN 1064-3745 ISSN 1940-6029 (electronic)
Methods in Molecular Biology
ISBN 978-1-0716-5439-2 ISBN 978-1-0716-5440-8 (eBook)
https://doi.org/10.1007/978-1-0716-5440-8

This Humana imprint is published by the registered company Springer Science+Business Media, LLC, part of Springer Nature.
The registered company address is: 1 New York Plaza, New York, NY 10004, U.S.A.

Foreword

Hippocrates' ode to the brain is as relevant today as it was 2,500 years ago: "Men ought to know that from the brain, and from the brain only, arise our pleasures, joys, laughter and jests, as well as our sorrows, pain, griefs and tears." Today, however, scientists have tools to record and influence brain activity for the prognosis, treatment, and cure of brain disorders, with AI being the latest tool.

The arc from Hippocrates to the present is long but coherent. We have moved from philosophical speculations to anatomical precision, from gross morphology to cellular architecture, and from descriptive atlases to quantitative, multiscale modeling. Each epoch has refined our capacity to see the brain more clearly. Yet vision alone is not understanding. Understanding requires structure, measurement, comparison, and, increasingly, computation.

Computational neuroscience is not a fashionable appendage to neuroscience—it is its inevitable maturation. The brain is a biological organ, but it is also a system—complex, hierarchical, dynamic, and governed by principles that can be formalized. To map it is one endeavor. To model it is another. To predict its functions, and to intervene with precision, demands integration across levels of organization, from genes and receptors to system networks.

This volume reflects that integration. It does not treat computation as an external instrument imposed upon biology. Rather, it presents computational methods as intrinsic to the study of neural organization and function. Neuroanatomy, once tethered to the microscope and the atlas, now interfaces with imaging pipelines, multimodal connectomic analyses, statistical modeling, and artificial intelligence. The structural blueprint of the brain, so long the foundation of neuroscience, becomes, in this framework, a substrate for simulation, prediction, and translation.

Artificial intelligence, in particular, occupies a paradoxical position. It is both a tool and a metaphor. As a tool, it enhances pattern recognition, accelerates data analysis, and supports clinical decision-making. As a metaphor, it challenges us to clarify what biological intelligence truly is. The distinction is essential. Machines compute; brains integrate, adapt, and generate meaning. The task before us is not to conflate the two but to use one to better understand the other.

The contributors to this book approach computational neuroscience with intellectual discipline and translational intent. They traverse theoretical modeling, neuroimaging, connectomics, digital neuropathology, computational fractal-based analyses, cognitive neurosciences, and clinical applications. Importantly, they recognize that computation without anatomical grounding risks abstraction detached from biology. Equally, anatomy without quantitative synthesis risks remaining descriptive. Their work stands at this intersection.

The maturation of neuroscience has always depended on methodological revolutions: the microscope, histochemical staining, electrophysiology, imaging, computation, and AI, the last representing the current (r-)evolution. Whether it will prove enduring depends not on its novelty but on its rigor. Sound models must rest upon sound data, algorithms must be interpretable, and conclusions must be anatomically and biologically plausible.

Hippocrates intuited the centrality of the brain to human experience. We now possess the instruments to interrogate that centrality in depth. The responsibility that accompanies such capacity is considerable. To measure is to approximate; to model is to simplify; to intervene is to assume consequence. Computational neuroscience must, therefore, advance with precision, humility, and ethical awareness, as stated by the 2026 Declaration of Sydney, led by Prof. Antonio Di Ieva and an international multidisciplinary task force.

This book, edited by Prof. Antonio Di Ieva and Dr. Jennilee M. Davidson, contributes meaningfully to that advance. It situates computation not as spectacle but as scholarship. It recognizes that the brain remains the most intricate structure known to science and that its understanding will demand both biological fidelity and mathematical clarity.

The millenary evolution from the cardiocentric theory to the encephalocentric foundation of modern science is seeing another paradigm shift, not toward a "silicocentric" foundation but rather toward a synergistic AI-human intelligence convergence, as stated multiple times by Antonio in his works and at his 1st World Conference of Computational Neurosurgery in Sydney.

If Hippocrates were to observe contemporary neuroscience, he might marvel at our technologies. Yet he would likely remind us that the object of our study is the seat of human joy and suffering. Computational tools may refine our insight, but the purpose remains unchanged: to understand the brain to alleviate disorder and preserve the qualities that define our humanity.

Sydney, Australia
February 2026

Scientia Professor George Paxinos, AO, DSc, FASSA, FAA (Dist)FRSN, FAHMS

Preface

The brain is the most complex system in the universe, and all neuroscientific discoveries have been related to or enabled by the refinement of experimental techniques: from clinical observation to experimental tests, from Golgi staining to MRI, and from electrophysiology to multi-omics approaches, each technological leap has expanded our capacity to understand and characterize the nervous system. The analysis of such a complex system requires modeling, quantification, and inference that go far beyond typical analytical techniques, thereby necessitating computational tools and advanced technologies.

Furthermore, the metacognitive problem of *brains studying and analyzing brains* also requires assistance, or enhancement, through artificial intelligence techniques and methodologies to overcome, or augment, human intelligence.

From the foundation of computational neurosciences as a revolutionary field to the most recent and trailblazing computational psychiatry, computational neurosurgery, and other computational fields in the neurosciences domain, the Cartesian *Cogito ergo sum* is slowly becoming an *Infero ergo sum*, where the convergence of natural and artificial intelligences can be synergized to shift the paradigms in the study of the brain and nervous system in general.

The exploitation of computational techniques over the last decades to analyze the supra-human complexity of multimodal and large datasets has created new challenges, above all in terms of academic curricula to teach and learn novel computational techniques applied to the neurosciences.

This volume was conceived in response to this emerging reality. It brings together leading contributors across molecular neurobiology, systems neuroscience, neuroimaging, neuropathology, clinical and experimental neurology, neurosurgery, genomics/proteomics, and artificial intelligence to provide practical and reproducible computational techniques and pipelines that span the translational neurosciences continuum.

This book is structured into two complementary sections: Wet-Lab Computational Neuroscience (Chaps. 1–9) and Computational Clinical Neurosciences (Chaps. 10–15).

The first section focuses on experimental platforms that generate high-dimensional biological data and the computational methodologies required to transform raw signals into mechanistic insights. The section covers a broad variety of computational analytical techniques, including mass spectrometry-based neuroproteomics techniques and computational analyses to characterize complex spectral fingerprints; structural phylogenetics to deepen our understanding of neural function and diseases; computational fractal-based analysis of microglia; neural dynamics analysis to infer attractor structures underlying brain-state transitions across micro-, meso-, and macro-scales; computational modeling of extrachromosomal DNA in brain oncogenesis; network-level inference from multielectrode recordings in 2D cultures and 3D organoids to analyze functional connectivity and neuro-topology; and finally, whole-brain histology imaging pipelines to confront one of neuroscience's most formidable data challenges: terabyte-scale volumetric datasets.

The second section of the volume transitions from bench to bedside, illustrating how computational methodologies reshape diagnosis, prognosis, and mechanistic inference across the whole pathophysiological spectrum of the nervous system, from normal brain function to neurological diseases.

The section covers: neuroradiomics pipelines to provide a comprehensive, reproducible framework for extracting quantitative imaging biomarkers from neuroimaging datasets; clinical connectomics and computational morphometric approaches to map and quantify structural brain connections as well as disconnections in neurological diseases; digital neuropathology pipelines and federated machine learning techniques to improve and accelerate diagnosis and discovery in neuropathology; integration of large language models with electrophysiological recordings to bridge computational linguistics and human neurophysiology; and, finally, computational genomics pipelines for precision medicine in neuroscience.

This volume does not present computational neuroscience as a monolithic discipline but as an ecosystem of interoperable methodologies, aimed at teaching naïve researchers and more expert scholars the methodological foundations of the computational neurosciences in a unifying perspective.

Importantly, this book is intentionally protocol-driven. Each chapter provides step-by-step workflows, code snippets, tool recommendations, and practical guidance. The objective is not only conceptual understanding but also operational capability. Researchers, clinicians, students, and data scientists should emerge equipped not merely to appreciate computational methods but to implement them.

The future of neuroscience will be shaped by our ability to integrate scales—molecular to systems, experimental to clinical, biological to artificial. Computational methods provide the language of that integration.

As neuroscience continues to generate unprecedented volumes of multimodal data, the distinction between "wet lab" and "clinical neuroscience" will increasingly dissolve, thanks to computational techniques merging them into a more comprehensive and unified translational field. Computational literacy will become as fundamental as anatomical knowledge or electrophysiological techniques.

This volume aspires to serve as both a reference and a catalyst: a reference for reproducible pipelines across diverse subfields and a catalyst for interdisciplinary collaboration. It reflects a collective conviction shared by its contributors that understanding the brain in the twenty-first century requires not only observation but also algorithmic processing.

The brain is a system of extraordinary structural and functional complexity. It demands methods that match that complexity, and it deserves complex techniques and sophisticated methods explained in a simple way, for reproducibility and advancement.

Progress in neuroscience has always required precision—precision of dissection, of staining, of recording, of description. It now requires precision of code, of modeling assumptions, and of statistical inference.

We hope that this book will be a step in the translational neurosciences direction, contributing to this endeavor.

Sydney, Australia *Antonio Di Ieva and Jennilee M. Davidson*

Contents

Editors and Contributors

About the Editors

DR.ANTONIO DI IEVA MD, PHD, FRACS, IFAANS is a professor of neurosurgery and associate professor of neuroanatomy at Macquarie University in Sydney, specializing in neuro-oncology, skull base surgery, microneurosurgery, pain neuromodulation, peripheral nerve surgery, spine surgery, and computational neurosurgery. He leads the Computational NeuroSurgery (CNS) Lab, founded by him in 2018. Prof. Di Ieva has published more than 200 peer-reviewed articles and five textbooks, including the *Handbook of Skull Base Surgery* (Thieme, 2015), *The Fractal Geometry of the Brain* (Springer, 1st Ed. in 2016 and 2nd Ed. in 2024), *Computational Neurosurgery* (Springer, 2024), and *Computational Neuroscience* (Springer, 2026). He has received more than $9Mil in research grants, including a prestigious Australian Research Council (ARC) Future Fellowship, several major grants from the National Health and Medical Research Council (NHMRC), and the John Mitchell Crouch Fellowship from the Royal Australasian College of Surgeons (RACS). A global pioneer in computational modeling, fractal analysis, and AI in neurosurgery, Prof. Di Ieva has also developed the world's first Computational Neurosurgery Fellowship and led the 1st World Conference of Computational Neurosurgery (Sydney, February 13–15, 2026), where the Declaration of Sydney on the Ethical Use of Artificial Intelligence in Neurosurgery was pledged, too. Furthermore, he leads workshops on neuroanatomy and connectomics. Prof. Antonio Di Ieva has been named the 2026 Eccles Lecturer—one of the highest honors in the neurosciences, awarded by the Australasian Neuroscience Society and the Neurosurgical Society of Australasia, in tribute to Nobel Laureate Sir John Eccles.

DR.JENNILEE M. DAVIDSON PHD, MSC, BSCH is a Macquarie University Research Fellow (Sydney, Australia) and an international ALS Scholar in Therapeutics with the Healey & AMG Center for ALS (Massachusetts General Hospital, USA, in partnership with FightMND and ALS Finding a Cure). Her research lies at the intersection of cellular and molecular neurobiology and translational biomedicine, receiving major grants, including from Australia's National Health and Medical Research Council (NHMRC) and the National Science and Engineering Research Council of Canada (NSERC). Dr Davidson's research on pre-clinical therapeutic development has been licensed to a neurodegenerative disease therapeutics company. She has led the neuroproteomics discovery of neuronal disease biomarkers in collaboration with the Computational NeuroSurgery (CNS) Lab, of which she is an associate member, at Macquarie University. She is a certified omics analyst for a specialized biomedical software company and a scientific consultant to multinational medical device companies.

Contributors

ION ANDRONACHE • *Advanced Digital Archaeological Historical Network (ADAHN), Alma Mater Europaea, Maribor, Slovenia*

GHASEM AZEMI • *Computational NeuroSurgery (CNS) Lab, Faculty of Medicine, Health and Human Sciences, Macquarie Medical School, Macquarie University, Sydney, NSW, Australia*

SARA BERTAGNOLI • *Social and Cognitive Neuroscience Laboratory, Department of Psychology, University La Sapienza, Roma, Italy; IRCCS Sacro Cuore Don Calabria, Neurorehabilitation Unit, Negrar, Verona, Italy*

JING CAI • *Department of Neurosurgery, Massachusetts General Hospital, Harvard Medical School, Boston, MA, USA*

TRANG CAO • *School of Psychological Sciences, Turner Institute for Brain and Mental Health, and Monash Biomedical Imaging, Monash University, Melbourne, Australia*

LI-AN CHU • *Department of Biomedical Engineering, National Taiwan University, Hsinchu City, Taiwan; Brain Research Center, National Tsing Hua University, Hsinchu City, Taiwan*

CONG CONG • *Centre for Health Informatics, Macquarie University, Sydney, Australia*

NICHOLAS J. DEBONO • *Institute for Biomedicine and Glycomics, Griffith University, Southport, QLD, Australia*

ANTONIO DI IEVA • *Computational NeuroSurgery (CNS) Lab, Faculty of Medicine, Health and Human Sciences, Macquarie Medical School, Macquarie University, Sydney, NSW, Australia*

JORDAN DOUGLAS • *Centre for Computational Evolution, The University of Auckland, Auckland, New Zealand; Department of Physics, The University of Auckland, Auckland, New Zealand; Macroevolution and Macroecology, Research School of Biology, Australian National University, Canberra, Australia*

ALEX FORNITO • *School of Psychological Sciences, Turner Institute for Brain and Mental Health, and Monash Biomedical Imaging, Monash University, Melbourne, Australia*

ANNA GULLER • *Computational NeuroSurgery (CNS) Lab, Faculty of Medicine, Health and Human Sciences, Macquarie Medical School, Macquarie University, Sydney, NSW, Australia; Macquarie Medical School, Faculty of Medicine, Health, and Human Sciences, Macquarie University, Sydney, NSW, Australia; Translational NanoMedicine Group, Children's Cancer Institute at Minderoo Children's Comprehensive Cancer Centre, Sydney, NSW, Australia; School of Clinical Medicine, UNSW Medicine & Health, Sydney, NSW, Australia*

MAGNUS HAUGHEY • *Evolutionary Dynamics Group, Centre for Cancer Evolution, Barts Cancer Institute, Queen Mary University of London, London, UK*

LYNDAL HENDEN • *Macquarie Medical School, Faculty of Medicine Health and Human Sciences, Macquarie University, Sydney, Australia*

BENJAMIN HENG • *Macquarie Medical School, Faculty of Medicine, Health, and Human Sciences, Macquarie University, Sydney, NSW, Australia*

CHAO-YING HUANG • *Department of Institute of Information Systems and Applications, National Tsing Hua University, Hsinchu City, Taiwan*

WILLIAM P. KLARE • *Australian Proteomics Analysis Facility (APAF), Macquarie University, Sydney, NSW, Australia*

LUCAS CARDOSO LAZARI • *GlycoProteomics Laboratory, Department of Parasitology, ICB, University of São Paulo, São Paulo, Brazil*

SIDONG LIU • *Centre for Health Informatics, Macquarie University, Sydney, Australia*
SUELY K.N. MARIE • *Laboratory of Molecular and Cellular Biology, Lim 15, Department of Neurology, Faculdade de Medicina USP, São Paulo, Brazil*
NICHOLAS MATZKE • *School of Biological Sciences, The University of Auckland, Auckland, New Zealand; Centre for Computational Evolution, The University of Auckland, Auckland, New Zealand*
SUSANNA B. MIERAU • *Division of Cognitive and Behavioral Neurology, Brigham & Women's Hospital, Boston, MA, USA*
BRANDON R. MUNN • *Neuroscience, School of Medical Sciences, Faculty of Medicine and Health, University of Sydney, Sydney, Australia; School of Physics, Faculty of Science, University of Sydney, Sydney, Australia*
IMRAN NOORANI • *Institute of Neurology, University College London, London, UK*
VALENTINA PACELLA • *Groupe d'Imagerie Neurofonctionnelle (GIN), Institut des Maladies Neurodégénératives-UMR 5293, University of Bordeaux, Bordeaux, France; Brain Connectivity and Behaviour Laboratory, Paris, France*
GIUSEPPE PALMISANO • *GlycoProteomics Laboratory, Department of Parasitology, ICB, University of São Paulo, São Paulo, Brazil*
CHI NAM IGNATIUS PANG • *Australian Proteomics Analysis Facility (APAF), Macquarie University, Sydney, NSW, Australia*
JAMES C. PANG • *School of Psychological Sciences, Turner Institute for Brain and Mental Health, and Monash Biomedical Imaging, Monash University, Melbourne, Australia*
CAROLINE PUENTE-LELIEVRE • *School of Biological Sciences, The University of Auckland, Auckland, New Zealand; Centre for Computational Evolution, The University of Auckland, Auckland, New Zealand*
LIVIA ROSA-FERNANDES • *GlycoProteomics Laboratory, Department of Parasitology, ICB, University of São Paulo, São Paulo, Brazil*
PENGLI SHAO • *Department of Neurology, Massachusetts General Hospital, Harvard Medical School, Boston, MA, USA*
GRIFFIN A. SHIMAMOTO • *Department of Neurology, Massachusetts General Hospital, Harvard Medical School, Boston, MA, USA*
JAMES M. SHINE • *Neuroscience, School of Medical Sciences, Faculty of Medicine and Health, University of Sydney, Sydney, Australia*
TIMOTHY P.H. SIT • *Sainsbury Wellcome Centre, University College London, London, UK*
MILLIJOY D. VILLANUEVA • *Computational NeuroSurgery (CNS) Lab, Faculty of Medicine, Health and Human Sciences, Macquarie Medical School, Macquarie University, Sydney, NSW, Australia; Macquarie Medical School, Faculty of Medicine, Health, and Human Sciences, Macquarie University, Sydney, NSW, Australia*

Part I

Wet-Lab Computational Neuroscience

Chapter 1

Data Processing and Machine Learning Methods to Analyze MALDI-TOF MS Data

Lucas Cardoso Lazari, Livia Rosa-Fernandes, Suely K.N. Marie, Antonio Di Ieva and Giuseppe Palmisano

Abstract

MALDI-TOF MS is a rapid, sensitive method for generating complex protein profiles from minimal sample volumes. Although its spectra lack direct peptide sequence information, combining MALDI-TOF output with machine learning enables discrimination of diseases based on intensity patterns. Accurate machine learning model development depends on standardized preprocessing to reduce noise, dimensionality, and bias. This protocol demonstrates two key preprocessing approaches—peak picking and spectral binning—and demonstrates their application in training and testing a machine learning model for the diagnosis of glioblastoma.

Keywords MALDI-TOF MS, Machine learning, Preprocessing, Diagnosis, Prognosis

1 Introduction

Matrix-assisted laser desorption/ionization time-of-flight mass spectrometry (MALDI-TOF MS) is a soft-ionization technique used for the rapid analysis of large biomolecules. This technique is capable of generating the protein profiles of complex sample mixtures in seconds, consuming very little sample volume and with simplified sample preparation protocols [1]. Due to its speed and sensitivity, MALDI-TOF MS has been extensively used in proteomics workflows for microbial identification, biomarker discovery, diagnosis, and prognosis of diseases [2–7].

The complex protein profile generated by this technique can be combined with machine learning (ML) to build models capable of identifying patterns in the spectra and distinguishing groups based on the intensity patterns of the protein profile.

The use of the MALDI-TOF MS spectral data requires the application of preprocessing steps for feature extraction and filtration prior to model training. Raw mass spectra are noisy, high-dimensional, and

Antonio Di Ieva and Jennilee M. Davidson (eds.), *Computational Neuroscience*, Methods in Molecular Biology, vol. 3062,
https://doi.org/10.1007/978-1-0716-5440-8_1,

biased; therefore, standardized preprocessing is essential to prevent ML models from learning artifacts and to ensure it is being trained with relevant biological patterns [8].

This protocol describes a MALDI-TOF MS data preprocessing pipeline to select and filter features for machine learning applications, demonstrating how these features can be used to train and test a machine learning model for disease diagnosis and prognosis, focusing on glioblastoma.

2 Materials

This protocol uses MALDI-TOF MS files from glioblastoma patients, and will be accessible in the PRIDE database under accession number PXD060922 [7], or upon request. This dataset includes serum protein profiles from samples categorized into glioblastoma and healthy control patients. Preprocessing is done using R (version 4.5.0) and Python (version 13.2.8). Spectral processing (including normalization, transformation, and baseline correction) is performed in R using the MALDIquant and MALDIQuantForeign library [9, 10]. Python is primarily used for model training and other data preprocessing steps, such as feature selection, oversampling, and scaling.

3 Methods

3.1 Initial Preprocessing

Regardless of the method used for feature extraction, there are a few steps that are important for both approaches. These steps consist of: transformation, smoothing, baseline correction, and intensity normalization (*see* **Note 1**). Many of the steps applied in this section were adapted from the MALDIquant documentation and introduction material [11]; therefore, for more details on parameters and other functions, please refer to the original documentation. Two approaches of feature extraction will be demonstrated: (1) peak picking and (2) spectral binning. The spectral binning approach was adapted from Weis et al. [12].

1. First, use MALDIquantForeign to import the mzML files (*see* **Note 2**). Each step performed here is done within each spectrum. This is important since later, during machine learning implementation for classification, we must ensure that there is no data leakage between the train and test datasets. Details on the choice of the parameters can be found in **Note 3**.
2. The parameters are dataset dependent; therefore, the choice for the optimal parameters depends on the inspection of the

spectra and the evaluation of the effect of these parameters on the data. The following script shows how to use one sample to perform these preprocessing steps and visually inspect the spectrum in all steps. Figure 1 contains the result of this analysis.

```
# You can select any random sample, or work with
the average spectra
spectra_raw <- gbm[[16]]
spectra_sqrt <- transformIntensity(spectra_raw,
method="sqrt")
spectra_smooth <- smoothIntensity(spectra_sqrt,
method="SavitzkyGolay", halfWindowSize=10)
baseline <- estimateBaseline(spectra_smooth,
method="TopHat")
spectra_corr <- removeBaseline(spectra_smooth,
method="TopHat")


# Set up plotting layout
```

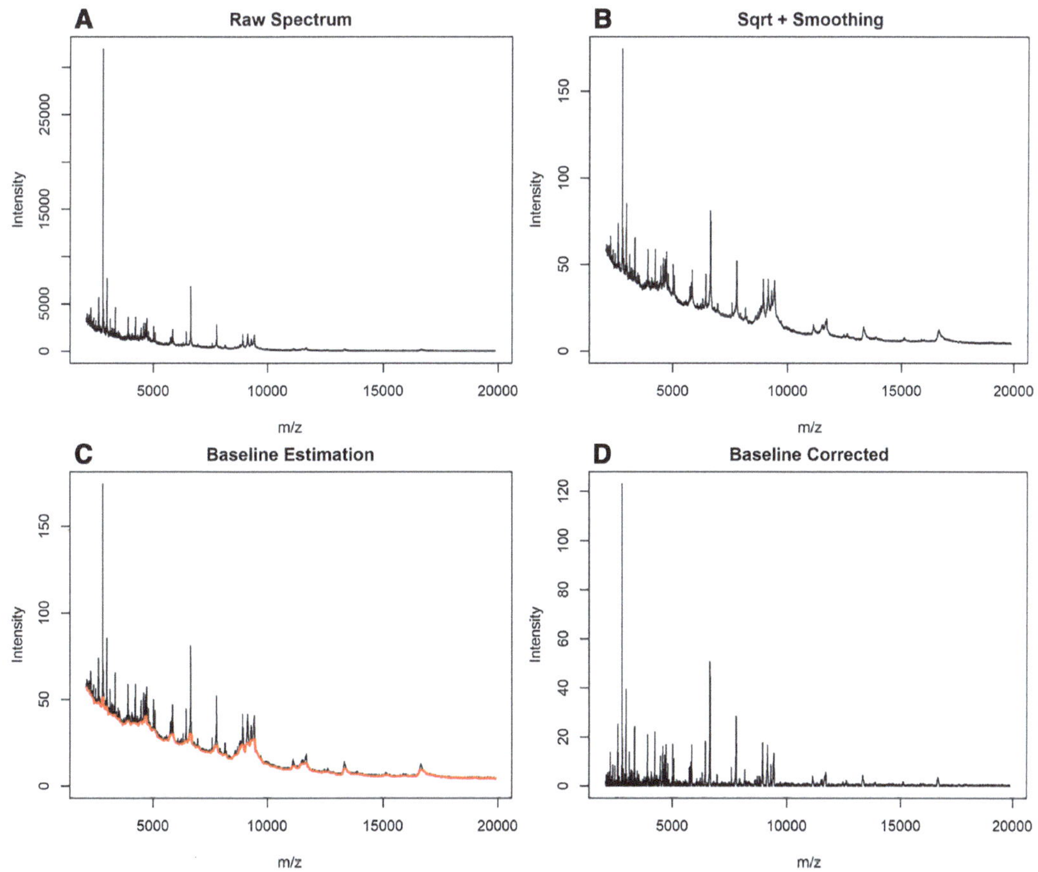

Fig. 1 (**A**) Raw unprocessed MALDI-TOF spectra. (**B**) Spectra after square-root transformation and smoothing. (**C**) Estimated baseline. (**D**) Spectra after baseline correction using the estimation

```
par(mfrow=c(2,2), mar=c(4,4,2,1))

# Raw spectrum
plot(spectra_raw, main="Raw Spectrum", xlab="m/z",
ylab="Intensity")

# After sqrt + smoothing
plot(spectra_smooth, main="Sqrt + Smoothing",
xlab="m/z", ylab="Intensity")

# Smoothed + Baseline
plot(spectra_smooth, main="Baseline Estimation",
xlab="m/z", ylab="Intensity")
lines(baseline, col="red", lwd=2)

# Baseline corrected
plot(spectra_corr, main="Baseline Corrected", xlab=
"m/z", ylab="Intensity")
```

3. Then, setup the function to process the raw spectrum and return a normalized spectrum. The raw spectra must be transformed, then smoothed, followed by base correction and finally normalization.

```
# Importing libraries
packages <- c("MALDIquant","MALDIquantForeign")
lapply(packages, require, character.only = TRUE)

# First, import all mzML files using the MALDIquant
library
gbm <-  import("<path_to_gbm_files>", type="mzML")
controls  <- import("<path_to_controls_files>",
type="mzML")

# Now we create a function to process the spectra.
processData <- function(data_list) {
  samp_transf <- transformIntensity(data_list,
method="sqrt")
  samp_smooth <- smoothIntensity(samp_transf,
method="SavitzkyGolay", halfWindowSize = 10)
  base_corr <- removeBaseline(samp_smooth, method=
"TopHat")
  int_norm <- calibrateIntensity(base_corr, method=
"TIC")
  return(int_norm)}

# Then we apply the function to both datasets
processed_gbm <- processData(gbm)
processed_ct <- processData(controls)
```

4. After the initial processing, the spectral data is ready to be stored in a text file to be used in the binning approach. For this step, create a file where each line is composed of the sample ID, the intensities, and the corresponding m/z, all separated by a semicolon. This can be done with the following:

```
# Loop through each list and write to the same text
file
for (i in 1:length(processed_gbm)) {
  # Convert the list to a character vector and then
to a comma-separated string
  list_content <- paste(unlist(processed_gbm[[i]]@
intensity), collapse = ",")
  mass_charge <- paste(unlist(processed_gbm[[i]]@
mass), collapse = ",")


  # Create the formatted line with ID and list
content. The string split to get the IDs differs
depending on your file path. Thus, it is advisable
to manually inspect this variable to ensure the ID
is being retrieved correctly
  line <- paste(paste(str_split(str_
split(processed_gbm[[i]]@metaData$file, '\\\\')[[1]]
[8], "_")[[1]][1], list_content, sep = ";"), mass_
charge, sep = ";")
  # Write to file (append = TRUE to add to the
existing file)
  write(line, file = tumor, append = TRUE)
}
# Loop through each list and write to the same text
file
for (i in 1:length(processed_ct)) {
  # Convert the list to a character vector and then
to a comma-separated string
  list_content <- paste(unlist(processed_ct[[i]]@
intensity), collapse = ",")
  mass_charge <- paste(unlist(processed_ct[[i]]@
mass), collapse = ",")
  # Create the formatted line with ID and list
content
  line <- paste(paste(str_split(str_
split(processed_ct[[i]]@metaData$file, '\\\\')[[1]]
[8], "_")[[1]][1], list_content, sep = ";"), mass_
charge, sep = ";")


  # Write to file (append = TRUE to add to the
existing file)
  write(line, file = control, append = TRUE)
```

3.2 Peak Picking

The peak picking step consists of separating what a signal is and what is noise from the spectra, then assigning a m/z and an intensity to the signals found. This approach yields highly informative features, since the biological signals are being directly extracted from the spectra. However, due to mass shifts, it can be challenging to generate comparable features when separating the dataset into train and test sets (*see* **Note 4**).

1. Before proceeding to peak picking, first, it is necessary to estimate the signal-to-noise ratio (SNR) in order to filter out most of the noise in the spectra (Fig. 2). This can be done with the following (*see* **Note 5**):

```
# We group all samples and calculate a  mean
spectrum to get a general noise estimation
all_samples <- c(processed_gbm, processed_ct)
avg_spec<- averageMassSpectra(all_samples, method=
"mean")
# Then we estimate the noise using the median
absolute deviation
noise_intensity <- estimateNoise(avg_spec)
```

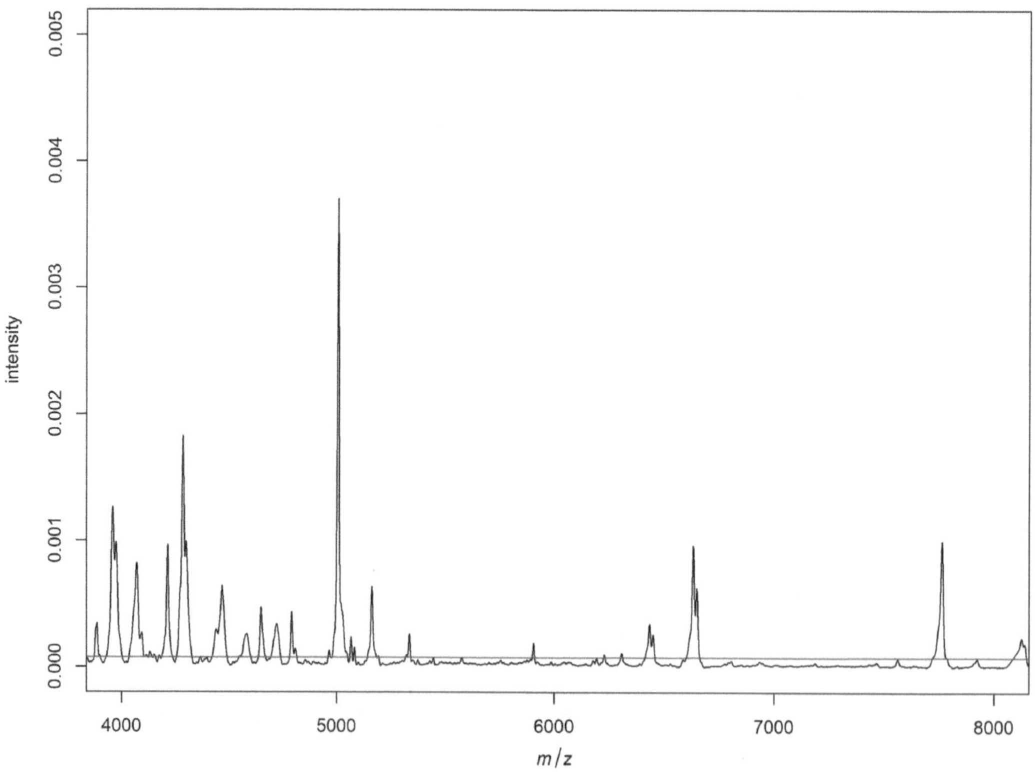

Fig. 2 A segment of a processed and normalized mean MALDI-TOF spectra (average of all samples), showing an estimated noise intensity (red horizontal line)

```
# noise_intensity[,1] contains the m/z from the
average spectra, while noise_intensity[,2] contains
the estimated noise intensity. The multiplication
value ('6' in this case) is our selected SNR (see
Note 5).
plot(avg_spec, xlim=c(4000, 8000), ylim=c(0,
0.005))
lines(noise_intensity[,1], noise_intensity[, 2]*6,
col="blue")
```

2. With the estimation of the SNR, proceed to detect the peaks and align them into discrete bins (*see* **Note 6**). The resulting plot can be observed in Fig. 3.

```
# halfWindowSize is the window size that the local
maxima will be searched on.
detected_peaks<- detectPeaks(all_samples,
halfWindowSize = 10, SNR = 8)
# We can evaluate the peak detection by taking a
look at the spectra
par(mfrow=c(2,1), mar=c(4,4,2,1))
plot(all_samples[[1]], xlim=c(5000,7000), ylim=
c(0,0.002),
main="Spectrum", xlab="m/z", ylab="Intensity")
plot(detected_peaks[[1]], xlim=c(5000,7000), ylim=
c(0,0.002),
main="Detected Peaks", xlab="m/z", ylab=
"Intensity", col="red")
```

```
# A tolerance of 0.003 captures a shift of 3000 ppm
(see Note 6)
bin_peaks <- binPeaks(detected_peaks, method =
"strict", tolerance=0.003)
```

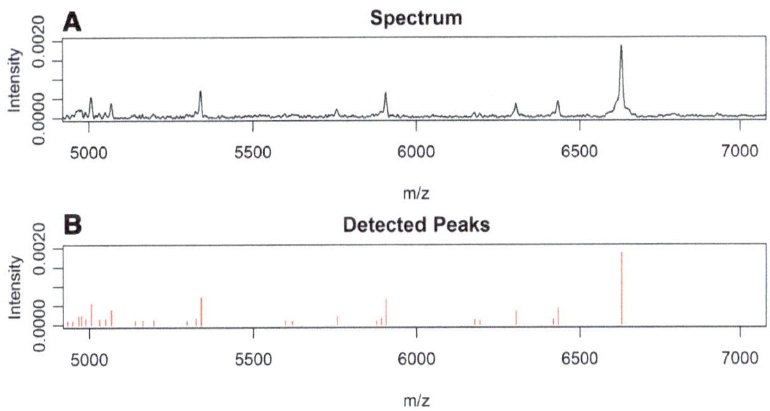

Fig. 3 (**A**) Segment of a processed spectrum before peak detection. (**B**) Detected peaks from the segment

3. Then, the identified peaks can be used to create the intensity matrix and then passed to a dataframe (*see* **Note 7**). This can be done with the following:

```
matrix_all <- intensityMatrix(bin_peaks,
all_samples)
df <- as.data.frame(as.table(matrix_all))
df <- reshape(df,direction = "wide",idvar =
"Var1",timevar = "Var2")
df$Var1 <- NULL
df$Group <- rep(c(1, 0), c(length(processed_gbm),
length(processed_ct)))
# Saving the dataframe in csv format
write.csv(df, "<your_path>.csv", row.names =
FALSE)
```

3.3 Machine Learning for Classification

With both datasets loaded in Python, *sklearn* library can be used to train the machine learning models. In this protocol, one model will be trained and tested as an example.

1. Start by importing all necessary libraries for this task.

```
import os
import numpy as np
import pandas as pd
from sklearn.model_selection import StratifiedKFold,
GridSearchCV
from sklearn.preprocessing import StandardScaler
from sklearn.svm import SVC
from sklearn.metrics import roc_auc_score,
confusion_matrix, precision_recall_curve, auc
from sklearn.decomposition import PCA
```

2. Then, create a function to load the data that was acquired using the R pipeline, and apply this function to obtain the numpy arrays with the sample IDs, mass-to-charge ratios, and intensity measurements. Also, load the *csv* table with the identified peaks, which was saved in the previous code snippet.

```
def load_data_from_file(filepath):
"""
Loads and parses data from a semicolon-separated
file.


Args:
filepath (str): The path to the input file.


Returns:
tuple: A tuple containing numpy arrays for IDs,
values, and masses.
```

```
Returns (None, None, None) if the file cannot be
processed.


"""
try:
# Using pandas for robust file reading
df = pd.read_csv(filepath, sep=';', header=None,
names=['id', 'values', 'masses'])
# Process the string data into numpy arrays
ids = df['id'].to_numpy()
values = np.array(df['values'].str.split(',').
tolist(), dtype=float)
masses = np.array(df['masses'].str.split(',').
tolist(), dtype=float)
return ids, values, masses
except (FileNotFoundError, Exception) as e:
print(f"Error processing file {filepath}: {e}")
return None, None, None


t_id, t_intensity, t_masses =
load_data_from_file("tumor_processed_spectra_
chapter.txt")
ct_id, ct_intensity, ct_masses =
load_data_from_file("control_processed_spectra_
chapter.txt")


labels_t = np.repeat(1, t_id.shape[0])
labels_ct = np.repeat(0, ct_id.shape[0])
peaks_df = pd.read_csv('peak_detection_dataset.
csv')
```

3. Now, create the binned spectra from the *txt* files. For that, it is necessary to read all spectra, and for each window of 3 Da (*see* **Note 6**), sum all intensity values. By doing this, the spectra dimensionality is reduced significantly, since all data points within a range of 3 Da are combined to a single value. Also, this binning serves as a way to reduce intersample variability.
4. Start by creating a function to perform the spectra binning. It is advisable to set a lower and upper bound for the mass-to-charge ratio so the same number of bins with the same feature name are generated for all samples. This can be done as follows:

```
def bin_mass_spectra(mz_list, intensity_list,
sample_ids, lower_bound=2000, upper_bound=15000,
bin_size=3):
"""
Trims and bins the intensity values of mass spectra
into fixed-width m/z bins.
```

```
Parameters:
mz_list (List[np.ndarray]): List of m/z arrays (one
per sample)
intensity_list (List[np.ndarray]): Corresponding
list of intensity arrays
sample_ids (List[str] or List[int]): Sample
identifiers
lower_bound (float): Minimum m/z value for binning
upper_bound (float): Maximum m/z value for binning
bin_size (float): Width of each m/z bin


Returns:
pd.DataFrame: DataFrame with binned intensity
values for each sample
"""
bins = np.arange(start=lower_bound, stop=upper_
bound + bin_size, step=bin_size)
df = pd.DataFrame()


for sample_id, (mz, intensity) in enumerate(zip(mz_
list, intensity_list)):
# Trim the spectrum
valid_idx = (mz >= lower_bound) & (mz <=
upper_bound)
mz_trimmed = mz[valid_idx]
intensity_trimmed = intensity[valid_idx]


# Bin the intensities
indices = np.digitize(mz_trimmed, bins)
sum_intensities = [intensity_trimmed[indices ==
i].sum() for i in range(1, len(bins))]


# Create row and append
row_df = pd.DataFrame([sum_intensities],
columns=[f"bin_({bins[i]:.2f}-{bins[i+
1]-0.01:.2f})"
for i in range(len(bins) - 1)],
index=[sample_ids[sample_id]])
df = pd.concat([df, row_df])


df.index.name = "ID"
return df.reset_index()
```

5. Next, apply this function to both groups of arrays to generate the dataframes for model training. Also, since the peaks dataframe does not contain the *ID* column, use the binning dataframe to add that information (*see* **Note 8**).

```
tumor_df = bin_mass_spectra(t_masses, t_intensity,
t_id)
ct_df = bin_mass_spectra(ct_masses, ct_intensity,
ct_id)


# Add group labels
tumor_df['Group'] = 1
ct_df['Group'] = 0


# Merge both
binned_df = pd.concat([tumor_df, ct_df], axis=0)


# Prepare the group and ids columns of the peaks
dataframe
# Change the group names to 0 or 1 for model
training
peaks_df['Group'] = peaks_df['Group'].map({'GBM':
1, 'CTR': 0})
peaks_df['ID'] = binned_df['ID'].values
```

6. Then, create functions to train the model and evaluate it. Start by setting the desired model (SVC will be used in this example) and setting a grid for hyperparameter tuning (*see* **Note 9**).
7. Next, use a stratified split to separate the data into train and test using a total of fivefold (meaning a split of 80%/20%, this can change based on how many samples are available).
8. With the training set, fit a scaler using the *StandardScaler* function from *sklearn* and apply the scaler to the test set.
9. For feature reduction, a Principal Component Analysis can be done to reduce the number of features to a desired number of components.

```
# --- Define the model and hyperparameters to test
---
svc_model = SVC(probability=True, random_state=42)
svc_param_grid = {
'C': [0.1, 1, 10],
'kernel': ['rbf', 'linear']
}


def run_pipeline(dataframe, model, param_grid, n_
splits=5, n_components=10):
X = dataframe.drop(columns=["ID", "Group"])
y = dataframe["Group"]
```

```
# Setup for cross-validation
skf = StratifiedKFold(n_splits=n_splits, shuffle=
True, random_state=42)
results_log = []


# Loop through each fold
for fold, (train_idx, test_idx) in enumerate(skf.
split(X, y)):
fold_name = f"Fold {fold + 1}/{n_splits}"
print(f"--- {fold_name} ---")


# Split data for the current fold
X_train, X_test = X.iloc[train_idx],
X.iloc[test_idx]
y_train, y_test = y.iloc[train_idx],
y.iloc[test_idx]


# 1. Scale the data
scaler = StandardScaler()
X_train_scaled = scaler.fit_transform(X_train)
X_test_scaled = scaler.transform(X_test)
# 2. Reduce dimensionality with PCA
# Ensure n_components is not greater than the
number of features
actual_n_components = min(n_components, X_train_
scaled.shape[1])
print(f"Running PCA, reducing to {actual_n_
components} components...")
pca = PCA(n_components=actual_n_components, random_
state=42)
X_train_pca = pca.fit_transform(X_train_scaled)
X_test_pca = pca.transform(X_test_scaled)


# 3. Tune hyperparameters
print("Tuning hyperparameters with
GridSearchCV...")
search = GridSearchCV(model, param_grid, cv=3,
scoring='accuracy', n_jobs=-1)
search.fit(X_train_pca, y_train)
best_model = search.best_estimator_
print(f"Best parameters found: {search.
best_params_}")


# 4. Evaluate the best model on the test set
y_pred = best_model.predict(X_test_pca)
y_prob = best_model.predict_proba(X_test_pca)[:, 1]


# Calculate Precision-Recall curve and AUC
```

```
precision, recall, _ = precision_recall_curve(y_
test, y_prob)
pr_auc = auc(recall, precision)


# Store results for this fold
fold_results = {
"fold": fold + 1,
"best_params": search.best_params_,
"roc_auc": roc_auc_score(y_test, y_prob),
"pr_auc": pr_auc,
"confusion_matrix": confusion_matrix(y_test,
y_pred)
}


results_log.append(fold_results)
print(f"Fold ROC AUC: {fold_
results['roc_auc']:.4f}, PR AUC:
{fold_results['pr_auc']:.4f}\n")


print("--- Pipeline Execution Complete ---")
return results_log
```

10. Apply the function to both datasets.

```
pipeline_results_peaks = run_pipeline(
dataframe=peaks_df,
model=svc_model,
param_grid=svc_param_grid,
n_splits=5,
n_components=20
)


pipeline_results_binned = run_pipeline(
dataframe=binned_df,
model=svc_model,
param_grid=svc_param_grid,
n_splits=5,
n_components=20
)
```

11. Create the function to summarize the results and apply it to both datasets. The results are combined in Table 1.

```
def print_summary(results):
"""Prints a summary of the pipeline results."""
roc_aucs = [r['roc_auc'] for r in results]
pr_aucs = [r['pr_auc'] for r in results]
sensitivities = []
```

```
specificities = []
balanced_accuracies = []


print("\n--- Fold-by-Fold Performance ---")
for res in results:
cm = res['confusion_matrix']
tn, fp, fn, tp = cm.ravel()


# Sensitivity (Recall)
sensitivity = tp / (tp + fn) if (tp + fn) > 0 else
0
sensitivities.append(sensitivity)


# Specificity
specificity = tn / (tn + fp) if (tn + fp) > 0 else
0
specificities.append(specificity)


# Balanced Accuracy
balanced_accuracy = (sensitivity + specificity) / 2
balanced_accuracies.append(balanced_accuracy)

print(f"Fold {res['fold']}: Bal. Accuracy=
{balanced_accuracy:.4f}, Sensitivity=
{sensitivity:.4f}, Specificity={specificity:.4f}")


print("\n--- Average Performance Across All Folds
---")
print(f"Mean Balanced Accuracy: {np.
mean(balanced_accuracies):.4f} (+/- {np.
std(balanced_accuracies):.4f})")
```

Table 1

Average metrics considering all five folds for each approach tested. The model performance is higher for the binning approach with a lower standard deviation as well

Metric	Peak picking	Fixed-size binning
Mean balanced accuracy	0.8909±0.0547	0.9131±0.0449
Mean sensitivity	0.9484±0.0269	0.9671±0.0187
Mean specificity	0.8333±0.0936	0.8590±0.0919
Mean ROC AUC	0.9612±0.0137	0.9865±0.0092
Mean PR AUC	0.9855±0.0056	0.9955±0.0028

```
print(f"Mean Sensitivity (Recall): {np.
mean(sensitivities):.4f} (+/- {np.
std(sensitivities):.4f})")
print(f"Mean Specificity:       {np.
mean(specificities):.4f} (+/- {np.
std(specificities):.4f})")
print(f"Mean ROC AUC:           {np.mean(roc_
aucs):.4f} (+/- {np.std(roc_aucs):.4f})")
print(f"Mean PR AUC:            {np.mean(pr_
aucs):.4f} (+/- {np.std(pr_aucs):.4f})")


print_summary(pipeline_results_peaks)
print_summary(pipeline_results_binned)
```

4 Notes

1. The steps included here were obtained from the MALDIquant library. It should be noted that there are other functions and methods available for these tasks.
2. MALDI-TOF data are typically uploaded in the mzML format. However, raw data may also be found, requiring conversion to the mzML format. This conversion can be accomplished using the MSconvert application within the ProteoWizard suite [13].
3. A square-root transformation of the data helps to stabilize the variance. This is because variance typically increases with higher intensity peaks (heteroscedasticity), and the transformation makes the variance more consistent (homoscedastic) across all peak intensity levels [14]. Other methods for transformation can be used depending on your purpose. Smoothing is used to remove instrumental and chemical noise from the spectrum, improving the signal-to-noise ratio of true peaks [15]. However, over-smoothing the spectrum leads to the elimination of low-intensity biological signals. The purpose of baseline correction is to remove unwanted background signals, known as baseline drift, from analytical data. This step is crucial because the presence of this drift can distort the true chemical information within a signal. Not correcting it can negatively impact the accuracy of subsequent data analysis [16]. The selection of the baseline correction algorithm and its parameters can be done by visually inspecting the curve and evaluating if it correctly identifies the baseline of the spectrum.
4. It is necessary to align the spectra and generate bins before picking the peaks. This is one of the main problems when training machine learning models with this approach. This means that every time a new set of samples needs to be classified in

the model, it would be required to align them with the dataset used for training in order for their features to be comparable.

5. MALDIquant will consider a signal any peak whose intensity is higher than *SNR*noise_intensity*. Therefore, we must search in the spectra for a value of SNR that, when multiplied by the noise intensity, eliminates most of the very low intensity peaks. The choice of the value of "8" as the SNR came with manual inspection of multiple values in the search for one that minimizes the loss of low-intensity protein peaks. It is important to consider that the average spectra tend to present less noise than the samples spectra, therefore, it might help to increase in one or two points the chosen SNR (e.g., if we plot a SNR of 6 in the average spectra, it will be enough to eliminate most of the noise, but when applied to the samples individually it may not filter most of the noise, leading us to increase the SNR to 8).
6. The peak location can often shift between 0.15% and 0.3% of the m/z value [17]. Therefore, a conservative tolerance value of 0.003 accounts for the high end of a 0.3% mass shift, which translates to 3000 ppm or 3 Da (e.g., for a value of 1000 m/z). The binning procedure ensures that the peaks match their positions across all samples (i.e., if a peak has the position at 1250 m/z in one sample and 1251 m/z in another sample, they will be corrected to have the same m/z). The main downside of using the peak picking approach is that every time a new set of samples needs to be tested, peak binning must be performed again. This can generate peaks at slightly different positions, which requires retraining the model. In contrast, dividing the spectra into fixed-size windows—3 Da in this case—helps account for this variability without needing peak detection. Each feature in this approach represents a region of the spectrum with summed intensities. Since the features are always in the same positions, this approach allows new samples to be tested using the existing model, without the need for retraining. The result is a spectrum that closely resembles the original but has a much lower dimensionality.
7. The intensity matrix function is used to fill the intensity values from the peaks that are identified in one sample but not the other. For instance, if a peak is identified at the m/z 500 in sample A, but not in sample B, the function will search for the intensity value of the m/z 500 in sample B normalized spectra.
8. It is also possible to add that information directly in the R pipeline before generating the peaks *csv* file. If the binned dataframe information is used to fill those columns, ensure that the samples are aligned. The pipeline always starts with

the tumor samples, followed by the controls, always keeping the same ID order.

9. This machine learning pipeline is a simplified suggestion. To increase model performance, consider testing different feature selection approaches, adding oversampling/undersampling to account for class imbalance, and increasing the grid search for parameters.

Acknowledgments

This work was supported by the São Paulo Research Foundation (FAPESP, grants 2018/15549-1 and 2020/04923-0 to G.P., and 2025/00049-7 to L.C.L.), the Brazilian National Council for Scientific and Technological Development (CNPq, grants 307854/2018-3 and 317353/2021-7 to G.P.), and the Coordination for the Improvement of Higher Education Personnel (CAPES, PIPD 88887.111997/2025-00 to L.C.L.). The funding agencies had no involvement in the study design, data collection and analysis, decision to publish, or paper preparation.

References

1. Singhal N, Kumar M, Kanaujia PK, Virdi JS (2015) MALDI-TOF mass spectrometry: an emerging technology for microbial identification and diagnosis. Front Microbiol 6:1–16. https://doi.org/10.3389/fmicb.2015.00791
2. Brégeon F, Brioude G, De Dominicis F, Atieh T, D'Journo XB, Flaudrops C, Rolain JM, Raoult D, Thomas PA (2014) MALDI-ToF mass spectrometry for the rapid diagnosis of cancerous lung nodules. PLoS One 9:1–7. https://doi.org/10.1371/journal.pone.0097511
3. Hou TY, Chiang-Ni C, Teng SH (2019) Current status of MALDI-TOF mass spectrometry in clinical microbiology. J Food Drug Anal 27:404–414. https://doi.org/10.1016/j.jfda.2019.01.001
4. Lazari LC, de Rose Ghilardi F, Rosa-Fernandes L, Assis DM, Nicolau JC, Santiago VF, Dalçóquio TF, Angeli CB, Bertolin AJ, Marinho CRF, Wrenger C, Durigon EL, Siciliano RF, Palmisano G (2021) Prognostic accuracy of MALDI-TOF mass spectrometric analysis of plasma in COVID-19. Life Sci Alliance 4. https://doi.org/10.26508/LSA.202000946
5. Lazari LC, Zerbinati RM, Rosa-Fernandes L, Santiago VF, Rosa KF, Angeli CB, Schwab G, Palmieri M, Sarmento DJS, Marinho CRF, Almeida JD, To K, Giannecchini S, Wrenger C, Sabino EC, Martinho H, Lindoso JAL, Durigon EL, Braz-Silva PH, Palmisano G (2022) MALDI-TOF mass spectrometry of saliva samples as a prognostic tool for COVID-19. J Oral Microbiol 14. https://doi.org/10.1080/20002297.2022.2043651
6. Nachtigall FM, Pereira A, Trofymchuk OS, Santos LS (2020) Detection of SARS-CoV-2 in nasal swabs using MALDI-MS. Nat Biotechnol 38:1168–1173. https://doi.org/10.1038/s41587-020-0644-7
7. Lazari LC, Silva JM, Donado PRS, Shinjo SMO, Rosa-Fernandes L, Di Ieva, A., Marie SKN, Palmisano G (2025) Identification of astrocytomas through serum protein fingerprint using MALDI-TOF MS and machine learning. bioRxiv. https://doi.org/10.1101/2025.02.21.639567
8. Tong DL, Boocock DJ, Coveney C, Saif J, Gomez SG, Querol S, Rees R, Ball GR (2011) A simpler method of preprocessing MALDI-TOF MS data for differential biomarker analysis: stem cell and melanoma cancer studies. Clin Proteomics 8:1–18. https://doi.org/10.1186/1559-0275-8-14
9. Gibb S, Strimmer K (2012) MALDIquant: a versatile R package for the analysis of mass spectrometry data. Bioinformatics 28:2270–2271. https://doi.org/10.1093/BIOINFORMATICS/BTS447
10. Gibb S (2024) MALDIquantForeign: Import/Export routines for MALDIquant. Reference manual, version 0.14.1. Comprehensive R Archive Network (CRAN). https://

doi.org/10.32614/CRAN.package.MALDIquantForeign
11. Gibb S (2024) MALDIquant: Quantitative Analysis of Mass Spectrometry Data. MALDIquant package vignette, version 1.22.3. Comprehensive R Archive Network (CRAN). https://cran.r-project.org/web/packages/MALDIquant/vignettes/MALDIquant-intro.pdf
12. Weis C, Cuénod A, Rieck B, Dubuis O, Graf S, Lang C, Oberle M, Brackmann M, Søgaard KK, Osthoff M, Borgwardt K, Egli A (2022) Direct antimicrobial resistance prediction from clinical MALDI-TOF mass spectra using machine learning. Nat Med 28:164–174. https://doi.org/10.1038/s41591-021-01619-9
13. Matthew CC, Brendan M, Robert B, Daria A, Daniel LR, Steffen N, Laurent G, Bernd F, Brian P, Jarrett E, Katherine H, Darren K, Natalie T, Nicholas S, Barbara F, Tahmina AB, Mi-youn B, Christopher P, David C, Lisa F, Parag M (2012) A cross-platform toolkit for mass spectrometry and proteomics. Nat Biotechnol 30:1–8. https://doi.org/10.1038/nbt.2377.A
14. Deininger SO, Cornett DS, Paape R, Becker M, Pineau C, Rauser S, Walch A, Wolski E (2011) Normalization in MALDI-TOF imaging datasets of proteins: practical considerations. Anal Bioanal Chem 401:167–181. https://doi.org/10.1007/s00216-011-4929-z
15. Savitzky A, Golay MJE (1964) Smoothing and differentiation of data by simplified least squares procedures. 36:1639–1643. https://doi.org/10.1021/ac60214a048
16. Liu X, Zhang Z, Liang Y, Sousa PFM, Yun Y, Yu L (2014) Baseline correction of high resolution spectral profile data based on exponential smoothing. Chemom Intell Lab Syst 139:97–108. https://doi.org/10.1016/j.chemolab.2014.09.018
17. Lin SM, Haney RP, Campa MJ, Fitzgerald MC, Patz EF (2007) Characterising phase variations in MALDI-TOF data and correcting them by peak alignment. Cancer Inform 1:32–40. https://doi.org/10.1177/117693510500100107

Chapter 2

An R-Based Workflow for Analysis of Data Independent Acquisition (DIA) Neuroproteomics Data

William P. Klare, Nicholas J. DeBono and Chi Nam Ignatius Pang

Abstract

Proteomic studies are highly informative for identifying biological signatures. The data generated from such experiments are highly complex and require substantial computational expertise to analyze appropriately. Rigorous analysis is further confounded by the presence of missing values and multiple sources of technical variance. Here, we present a methodical walkthrough of a workflow for the analysis of Data Independent Acquisition (DIA) proteomics data. The workflow is derived from the MultiScholaR framework, which aims to make best-in-class tools and practices available to researchers of all skill levels. Users will analyze a publicly available neuroproteomics dataset, searching data using DIA-NN, filtering and normalizing resultant data, before imputing missing values and removing technical variance using relevant tools in the field. Users will be able to generate easy to understand results on the individual protein and functional annotation level, with results and plots automatically formatted to be publication-ready. Designed for researchers of all skill levels, MultiScholaR emphasizes modularity, transparency, and learning through tunable parameters, extensive documentation, and automated reporting. Upon completion, users should feel equipped to analyze their own datasets confidently, bridging the gap between data generation and biological interpretation.

Keywords Proteomics, Data independent acquisition, Neuroscience, R

1 Introduction

Modern omics datasets require substantial programming and statistical expertise to analyze to an analytical standard, from which useful biological insight can be derived. Our ability to generate omics data far exceeds our capabilities to rigorously analyze it to an acceptable and, importantly, reproducible standard. The ability to both initiate and complete an analysis is heavily skill-gated, which makes it challenging for already time-deficient scientists to be able to analyze the data they generate.

Proteomics, particularly in the context of neuroscience (neuroproteomics), stands as a cornerstone of modern biomedical research and biomarker discovery. The complexity of the nervous system, with

Antonio Di Ieva and Jennilee M. Davidson (eds.), *Computational Neuroscience*, Methods in Molecular Biology, vol. 3062, https://doi.org/10.1007/978-1-0716-5440-8_2,

its vast array of cell types and intricate signaling pathways, presents a unique analytical challenge. Data-Independent Acquisition (DIA) mass spectrometry has become a key technology in proteomics, offering the depth and reproducibility required to probe the neural proteome. However, the path from raw DIA data to meaningful biological insight is fraught with analytical hurdles. These include the systematic, non-random nature of missing values, which are often intensity-dependent; the presence of batch effects and other unwanted technical variation that can obscure true biological signals; and the need for a robust and reproducible framework for experimental design and data standardization.

This chapter presents a comprehensive, end-to-end workflow designed to address these challenges. We provide a step-by-step guide that takes the user from raw data processing with the popular search engine, DIA-NN [1] to sophisticated downstream analysis, including quality control, normalization, imputation, differential expression, and functional enrichment. The workflow is facilitated by the MultiScholaR Github (https://github.com/APAF-BIOINFORMATICS/MultiScholaR).

The workflow presented integrates *limpa*-based imputation [2] and RUV-III-C normalization within a modular framework, emphasizing transparency, parameter tunability, and educational value. The modified workflow implements enhanced peptide-level processing, filtering, normalization, *limpa*-based imputation, and visualization steps, ensuring higher data quality prior to protein rollup. The workflow is designed to foster learning by doing with easy-to-understand architecture, extensive technical markup, further educational reading and parameter manipulation made easy for data quality control.

This chapter will guide users step-by-step through processing the Neurolincs dataset [3], with each section detailing rationale, followed by numbered code chunks. All function names and their corresponding parameters will be highlighted in italics to clearly link the protocol to the underlying R code. Users are highly encouraged to read all the commentary above code chunks, including additional reading, to fully understand the technicalities of the workflow.

2 Materials

MultiScholaR is implemented on the open-source statistical programming language R [4]. This Materials section will detail the steps required to setup a programming environment from scratch on either Windows or Mac operating systems.

2.1 Hardware

Proteomics datasets require substantial computational resources to analyze, with the most crucial component being sufficient memory (RAM) due to the size of peptide and protein matrices commonly encountered in DIA experiments.

2.1.1 For Smaller Datasets (<100 Samples)

1. CPU: Minimum four cores.
2. RAM: Minimum 16 GB; recommended 32+ GB.
3. Storage: Minimum 500 GB SSD; recommended 1 TB SSD.

2.1.2 For Large Datasets (>100 Samples)

1. *CPU: 16+* cores.
2. *RAM: 64+* GB (essential for large matrix operations).
3. *Storage: 2+* TB NVMe SSD for fast I/O operations.

2.2 Software

MultiScholaR leverages Bioconductor packages and a number of other CRAN packages for statistical analysis and visualization. The following software components are required to utilize the framework.

2.2.1 Core Environment

1. R version 4.3 or higher.
2. RStudio (or other R compatible IDE, however RStudio is recommended).
3. RTools (version matched to R, required for compiling packages on Windows).

2.2.2 R Packages

1. MultiScholaR (latest version from GitHub—BookChapter branch, github.com/APAF-bioinformatics/MultiScholaR/tree/BookChapter).
2. devtools (CRAN): For installing packages from GitHub.

2.3 Workflow and Project Setup

Setting up a proteomics data project with MultiScholaR requires downloading the setup script from the MultiScholaR GitHub repository (https://github.com/APAF-BIOINFORMATICS/MultiScholaR). The script automates workflow retrieval, project setup, and directory management for users; creating a standardized directory structure for project management. A typical file directory setup is depicted below.

```
my_analysis/  # Project root
├── config.ini  # Central configuration file
├── my_analysis.Rproj  # RStudio project file
├── .Rprofile  # Startup script for auto-opening workflows
├── scripts/  # Scripts and workflows
|   ├── proteomics/  # Proteomics-specific
|   |   ├── DIA_workflow_experienced.rmd  # Workflow R Markdown
(example for experienced user)
|   |   └── DIANN_report.rmd  # Report template
|   ├── metabolomics/  # Metabolomics-specific (similar
structure)
|   |   ├── metabolomics_workflow_experienced.rmd
|   |   └── metabolomics_report.rmd
|   ├── transcriptomics/  # And so on for each omic_type
|   └── integration/  # Integration workflows if selected
├── data/  # Data directories
|   ├── proteomics/
|   ├── UniProt/  # UniProt annotations (for proteomics)
|   ├── metabolomics/
|   ├── transcriptomics/
|   └── integration/  # If multi-omics integration is selected
└── results/  # Results (created during analysis, not by script)
    └── proteomics/  # Example: peptide_qc/, protein_qc/, etc.
```

2.4 Sourcing Demonstrative Neuroproteomics Data

Demonstration DIA data from the Neurolincs project [4] (searched with DIA-NN v2.2) provides beginner-friendly complexity for workflow testing, including Amyotrophic Lateral Sclerosis patient samples with appropriate controls. Much of programming requires referencing objects, file locations, directory structures, file in/out operations—in this chapter, we have created an easy-to-use setup script that fetches the latest workflow template documents, as well as the searched output files and associated FASTA files for *Homo sapiens*, which are provided in a Google Drive repository. Links to this repository are available on the MultiScholaR github (https://github.com/APAF-bioinformatics/MultiScholaR/), and files can be directly downloaded there. The setup script can be found by clicking the link indicated in the red box shown in Fig. 1, with installation instructions details on the GitHub "README.md (https://github.com/APAF-bioinformatics/MultiScholaR/tree/BookChapter)." If users wish to use the workflow for their own analysis—please set the following variable to NULL as follows: *tutorial_type <- NULL*. Users conducting their own analysis should ensure they place their DIA-NN search results report.parquet file in the /project/data/proteomics directory, and their organism FASTA file in the /project/scripts/proteomics directory.

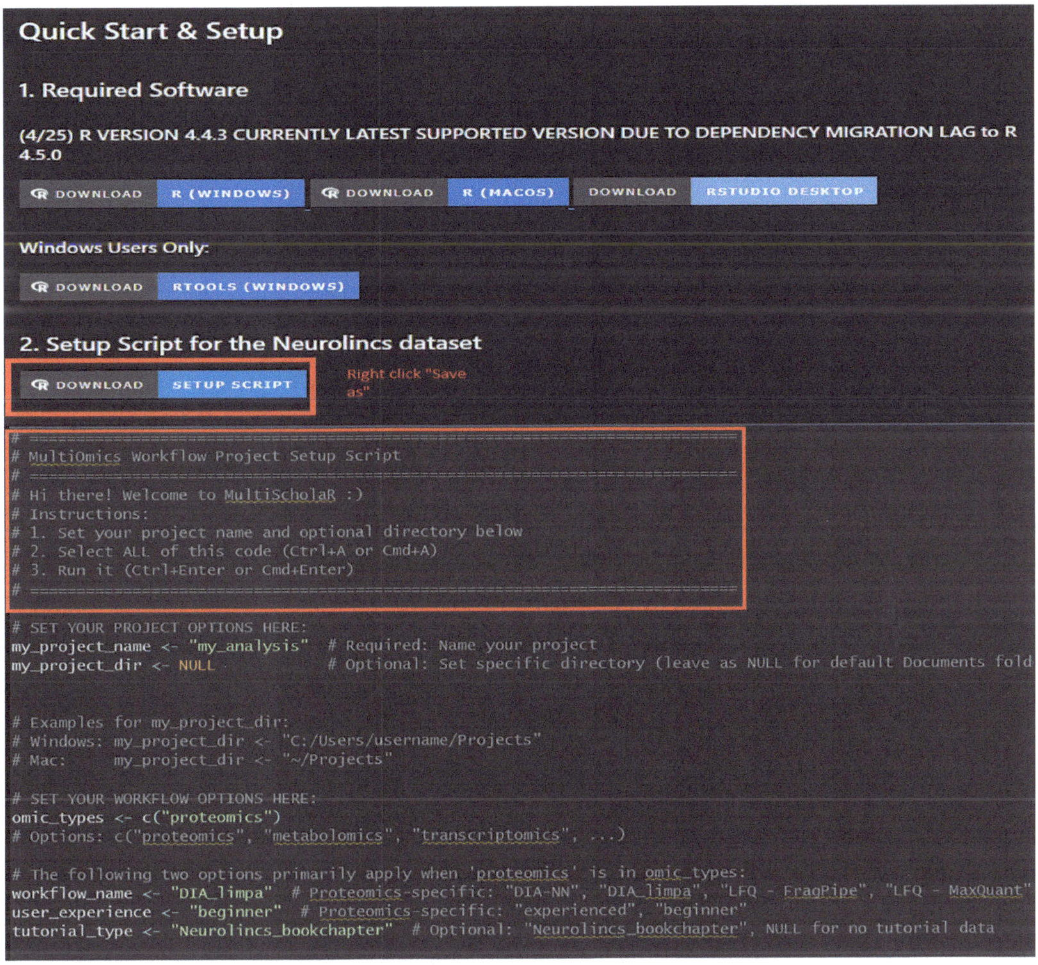

Fig. 1 Setup Script download button and directions for using the script in RStudio

3 Methods

The following section is a guide for users wanting to run DIA-NN on their own datasets and modify the workflow template to accommodate their data. It is not necessary to complete analysis of the Neurolincs dataset in this tutorial.

Users wanting to utilize the workflow on their own data can modify the variables of any code chunk to accommodate their desired inputs or parameters to suit the needs of their own data.

3.1 Running DIA-NN for Data Processing

Raw mass spectrometry data must be processed using a search engine to identify and quantify peptides. This workflow is optimized for use with DIA-NN, a powerful, open-source software that uses deep neural networks for improved signal processing. DIA-NN can be run in a "library-free" mode directly from a protein sequence database (FASTA file), or with a user-supplied spectral library.

Retrieving DIA-NN: Download the latest version of DIA-NN from the official GitHub repository (https://github.com/vdemichev/diann). It is available for Windows and Linux. Installation instructions are details on the GitHub.

3.1.1 Prepare Inputs

1. Raw Data:
   Depending on spectra filetype, additional resources may need to be installed (e.g., additional.dll for sciex files)
2. FASTA File:
   Provide a protein sequence database for your organism. This should be in FASTA format, and Swiss-Prot reviewed databases are always preferred (*see* **Note 4.1**).

3.1.2 Configuring DIA-NN

DIA-NN can be configured with the following steps:

1. Add your raw data files to the search queue.
2. Specify the path to your FASTA file.
3. Set the output file name (e.g., report). DIA-NN will generate several output files, with *report.tsv* or *report.parquet* being the primary input for the MultiScholaR workflow.
4. Adjust key parameters as needed. Important considerations include:
   - *Precursor FDR*: Set the false discovery rate for precursor identification (typically 1%).
   - *Modifications*: Specify any expected variable modifications (e.g., methionine oxidation).
   - *Missed cleavages*: Two missed cleavages is considered the sweet spot between permissible and restrictive.

3.1.3 Run Analysis

This step is time intensive. The software will perform chromatogram extraction, peak scoring, interference correction, and statistical validation.

3.1.4 Locate Output

Once the run is complete, locate the main output file (report.parquet). This file contains the quantified peptide and protein information and will be the starting point for the R-based workflow. Move this file into the data/proteomics directory created by the project_setup.R script in the following section.

3.2 Installing MultiScholaR and Dependencies

When users have run setup_project.R—a new window will open in RStudio with the downloaded workbook to complete the analysis of the Neurolincs dataset. The initial setup involves installing the core *MultiScholaR* package from GitHub and all its dependencies from CRAN and Bioconductor (*see* **Note 4.2**). This process is automated to ensure a smooth start.

3.2.1 Execute Installation Function

Run the first code chunk *{r 1 MultiScholaR FIRST INSTALL, message=TRUE, warning=TRUE}*. This contains the *installMultiScholaR()* and *loadDependencies()* functions as can be seen in Fig. 2. This function first checks if *devtools* is installed (a package necessary for installing from GitHub) and installs it if not. It then proceeds to download and install *MultiScholaR* and its extensive list of dependencies. Be aware that this step can be time-consuming, especially on a fresh R installation.

3.3 Setup Project Environment

A standardized directory structure is essential for reproducibility and organization. This step *{r 2 Project Environment Management}* creates a set of nested folders for a specific project, ensuring that data, scripts, and results are stored in a consistent and predictable manner. The code chunk and its contents can be seen below.

```
"'{r 2 Project Environment Management}
# Directory Management
## Set up the project directory structure
## This section sets up the project directory
structure for MultiScholaR
## Directory management can be challenging,
particularly when managing objects
## across multiple chunks within a single R
Markdown document.
workflow_name <- "DIA_limpa"
experiment_label <- "iMNS_bookchapter_analysis"
omic_type <- "proteomics" # Set this to the
type of analysis you are doing eg "proteomics",
"metabolomics", "transcriptomics"
# Setup for the proteomics only for this workflow
project_dirs <- setupDirectories(
#omic_types = "metabolomics"
# Or:
omic_types = c("proteomics"), #, "metabolomics",
"transcriptomics", "lipidomics", "integration"
label = experiment_label,
force = FALSE # Set to TRUE to skip prompts if dirs
exist
)
"'
```

3.3.1 Define Project Variables

The variables necessary to direct function logic *workflow_name*, *experiment_label* and *omic_type* are set here. For this workflow, *workflow_name* should be set to "*DIA_limpa*" and *omic_type* should be set to *"proteomics"*.

3.3.2 Create Directories

setupDirectories() creates a root project folder named after your *experiment_label* with the appropriate subdirectory structure, critical for the workflow in terms of input/output functionality.

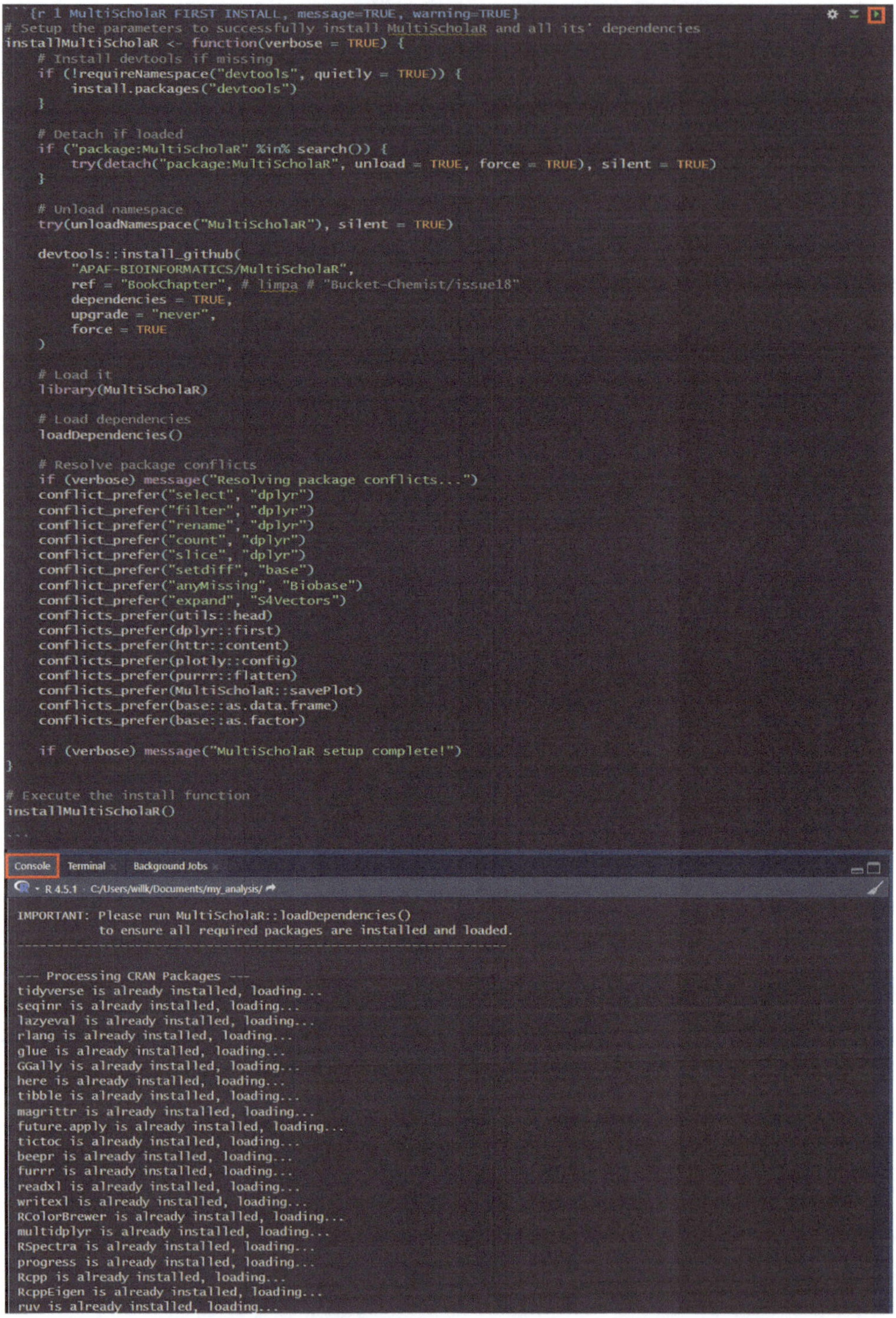

Fig. 2 Installing the MultiScholaR package and required dependencies. The constructed functions will fetch the most recent build of MultiScholaR from GitHub, after which all associated dependencies will be installed from various sources. Users are directed to either hit ctrl+shift+return or ctrl+cmd+enter, or click the red arrow highlighted in green to run the chunk. Users will be notified in the console tab of RStudio when the setup is done. As a general note, all functions will print extensive logging detail that users are able to peruse in the event of errors, or to inspect what functions are doing

3.3.3 Place Data Files

Data will be automatically populated by the setup_project.R script, taken from a Google Drive repository.

3.4 Data Import, Annotation, and Design Matrix Specification

This section covers the initial steps of bringing the processed DIA-NN data into the R environment, enriching it with protein annotations, and defining the experimental structure using *{r 3 Data Management}*. The code chunk for Sects. 3.4.1, 3.4.2, and 3.4.3 are directly below each subsection, and an example of the GUI interface through Shiny can be seen in Fig. 3.

3.4.1 Load Configuration

The workflow uses a centralized parameter configuration file config. ini file to manage parameters. *readConfigFile()* parses the.ini structure to make the parameters available in RStudio in an R-friendly format for the workflow functions (see code chunk below).

```
"'{r 3 Load configuration file}
## Input Parameters for Quality Control
## Parameters in this section are experiment-
specific. Their default parameters
## are intended as a guide only - every source of
variance is different just as
## every set of proteins going through a mass
spectrometer is different!
## One size does not fit all and you *will* most
likely need to fine tune these
## to get the most out of your data.
config_list <- readConfigFile(file = file.path(project_
dirs$proteomics$base_dir, "config.ini"))
"'
```

3.4.2 Import DIA-NN Data

DIA-NN data are read into the RStudio environment by *readDIANN()*, a *limpa* package that imports the searched *report.parquet* output file (*see* **Note 4.3**). Searched data are filtered based on *q*-value columns (*q.columns*) and cutoffs (*q.cutoffs*), in this case, stringently set to 0.01. The imported data are formatted with *formatDIANN()* into the long format for downstream processing (see code chunk below).

```
"'{r 4 Import DIA-NN Data}
# Convert your EList to DIA-NN format
data_tbl_parquet_filt <- limpa::readDIANN(file.
path(project_dirs[[paste0(omic_type, "_",
experiment_label)]]$data_dir, "report.parquet"),
format="parquet", q.columns = c("Q.
Value","Lib.Q.Value","Lib.PG.Q.Value"), q.cutoffs =
0.01)
data_tbl <- formatDIANN(data_tbl_parquet_filt)
"'
```

3.4.3 Process FASTA and Annotate Proteins

Protein identifiers from the search engine need to be mapped to up to date annotations in *{r 5 Process FASTA and Annotate Proteins}*

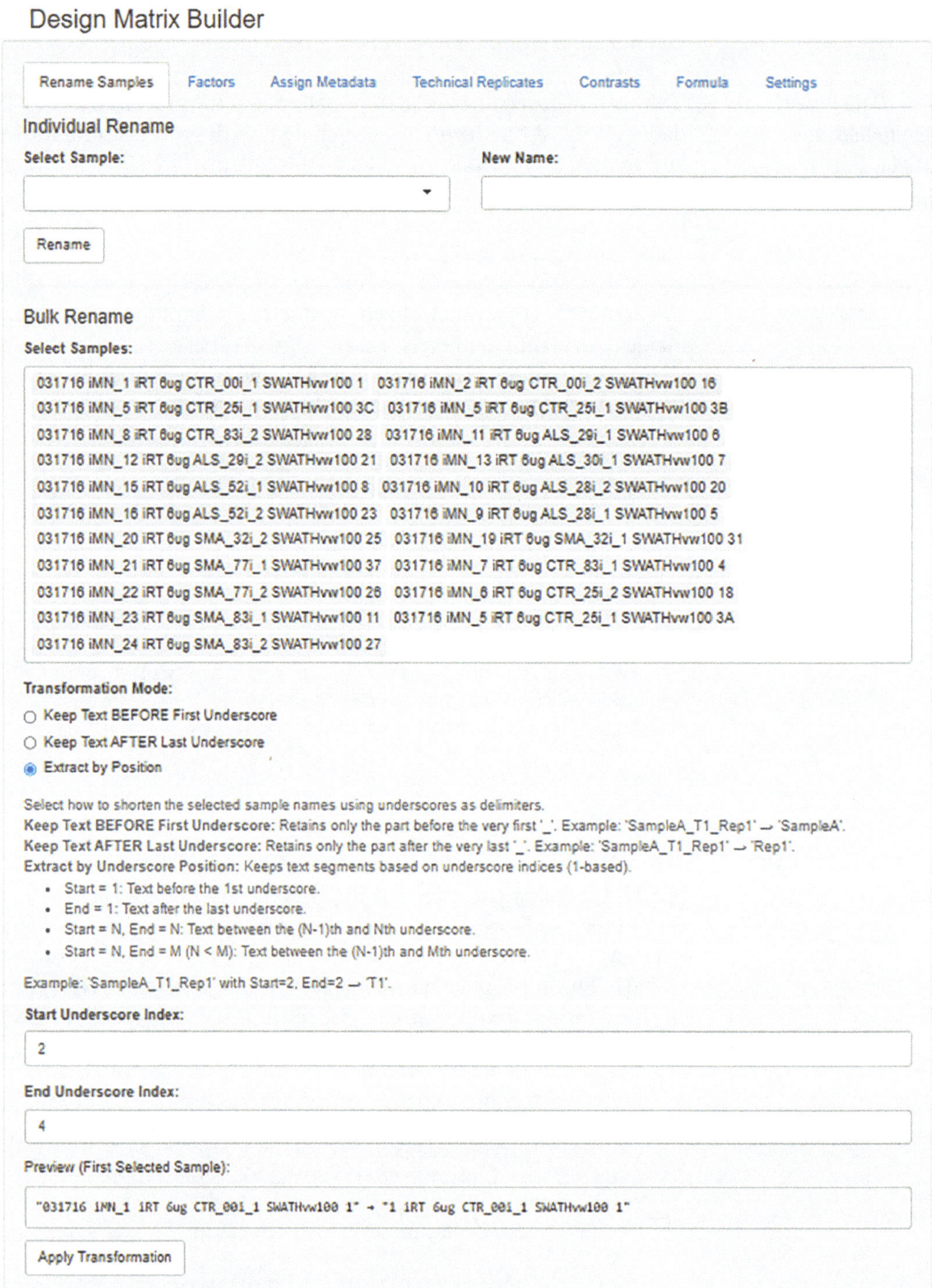

Fig. 3 A subsection of the Design Matrix Builder shiny application. Users are able to assign all features of experimental design such as grouping factors and statistical contrasts. Clear instructions and information are given throughout to guide the user through the experience

and *{r 6 Protein ID Conversion and Uniprot Lookup}*. *processFastaFile()* parses the provided FASTA file to extract protein accessions and associated metadata. It can also integrate user-supplied mapping files from UniProt/UniParc, if available; however, these functions are for poorly annotated organisms and will not be covered in this walkthrough. *updateProteinIDs()* aligns the protein IDs in your main data table with the cleaned accessions from the step preceding. *getUniprotAnnotationsFull()* fetches extensive annotations from the UniProt database through API calls for all identified proteins, using the specified *taxon_id*. This function queries UniProt for details like gene names, functional descriptions, and GO terms. This step can take some time depending on the number of proteins.

```
"'{r 5 Process FASTA and Annotate Proteins}
# Annotation Management
## Please download the organism fasta file from
UniProt. If UniProt is not
## available, the program will extract the relevant
identifiers from the fasta
## provided and attempt to match them to user
supplied UniProt / UniParc
## conversions
## Please set the name of your fasta file here in
the root directory if you
## already have it
fasta_filename <- "CanonicalIsoform uniprotkb_
taxonomy_id_9606_AND_reviewed_2025_05_06.fasta" ##
copy to /data/UniProt in your project directory
uniprot_search_results <- NULL ## copy to /data/
UniProt in your project directory eg "idmapping.
tsv"
uniparc_search_results <- NULL ## copy to /data/
UniProt in your project directory eg "idmapping.
tsv"
## Please supply your organism's taxon ID here
taxon_id <- 9606 # You can find this at the links
above.
## Please supply your organism's name here
organism_name <- "Homo sapiens" # If you don't know
this by now, you have bigger problems than this
workflow!
fasta_file_path <- file.path(project_
dirs[[paste0(omic_type, "_", experiment_
label)]]$uniprot_annotation_dir, fasta_filename)
config_list[["globalParameters"]][["fasta_file"]] <
- fasta_filename
# Load search results if files exist
if (!is.null(uniprot_search_results) &&!is.
null(uniparc_search_results)) {
uniprot_search_results <- vroom::vroom(
file.path(project_dirs$proteomics$uniprot_
annotation_dir, uniprot_search_results))
```

```
uniparc_search_results <- vroom::vroom(
file.path(project_dirs$proteomics$uniprot_
annotation_dir, uniparc_search_results))
}
"'
"'{r 6 Protein ID Conversion and Uniprot Lookup}
fasta_meta_file <- "parsed_fasta_data.rds"
aa_seq_tbl_final <- processFastaFile(
fasta_file_path,
uniprot_search_results,
uniparc_search_results,
fasta_meta_file,
organism_name)
data_tbl <- updateProteinIDs(data_tbl,
aa_seq_tbl_final)
uniprot_cache_file <- file.path(
project_dirs[[paste0(omic_type, "_",
experiment_label)]]$uniprot_annotation_dir,
"uniprot_annotations.RDS")
if (file.exists(uniprot_cache_file)) {
uniprot_dat_cln <- readRDS(uniprot_cache_file)
} else {
uniprot_dat_cln <- getUniprotAnnotationsFull(
data_tbl = data_tbl,
protein_id_column = "Protein.Ids",
cache_dir = file.path(project_
dirs[[paste0(omic_type, "_",
experiment_label)]]$uniprot_annotation_dir),
taxon_id = taxon_id)
saveRDS(uniprot_dat_cln, uniprot_cache_file)
}
"'
```

3.4.4 (Optional for This Tutorial) Specify Experimental Design

Curating sample names from instrument runs and assigning metadata can be tedious. Moreover, many user groups and facilities have their own standard conventions of what these look like. In chunk `{r 7 Design Matrix Setup}` MultiScholaR provides an easy to use Shiny app to create and manipulate their design matrix such that it is compatible with the rest of the workflow (*see* **Note 4.4**). The app will create a design matrix and a contrast table for statistical analysis downstream and modify sample names as directed to create a new data matrix *dat_cln*. This step is commented out by default as the authors have done this in the interest of avoiding monotony for the user.

3.4.5 Launch the Interactive Design Matrix Creator

Do this by running `RunApplet(applet_type = "designMatrix",...)`. This Shiny applet provides a guided interface to create the design matrix, ensuring sample names match the data and the design is balanced. For detailed notes on how to use this applet please visit this documentation webpage: https://github.com/

APAF-bioinformatics/MultiScholaR/blob/BookChapter/HowTo/how_to_use_the_design_matrix_builder.md.

3.4.6 Read in Previously Run Design Matrix Outputs

In *{r 8 Design Matrix Output Read In - optional}*, users are able to import data saved by the Design matrix applet, as seen in code chunk below. To begin, we load a previously saved design matrix with the *read.table()* function. For those interested in simple data cleaning operations—we highlight an often overlooked source of errors in programming, the trailing whitespace. The *mutate()* function from the *dplyr* package is used to remove the empty space at the end of the sample name. The cleaned data table, *data_cln*, is read in with the use of *vroom()* from the *vroom* package, optimized for larger data objects. A sanity check is performed using *setdiff()* on the sample names (*Run*) in both the *design_matrix* and *data_cln* to ensure that they are identical. Finally, the contrast table is read into *contrasts_tbl* into a character vector, filtered to remove the header row, then converted into a single-column tibble (with the default column name value) using *as_tibble()*, and finally, that column is renamed to contrasts.

```
"'{r 8 Design Matrix Output Read In - optional}
design_matrix <- read.table(
file = file.path(project_dirs[[paste0(omic_type, "_",
experiment_label)]]$source_dir, "design_matrix.
tab"),
sep = "\t",
header = TRUE,
stringsAsFactors = FALSE) |>
mutate( Run = trimws(Run))
data_cln <- vroom::vroom(
file = file.path(project_dirs[[paste0(omic_type, "_",
experiment_label)]]$source_dir, "data_cln.tab"),
delim = "\t",
col_types = cols(.default = col_guess()),
na = c("", "NA"))
contrasts_tbl <- file.path(project_
dirs[[paste0(omic_type, "_", experiment_
label)]]$source_dir, "contrast_strings.tab") |>
readLines() |>
{
\(x) x[!grepl("^contrasts", x)]
}() |>
as_tibble() |>
dplyr::rename(contrasts = value)
"'
```

3.5 Data Containerization and Peptide-Level QC

This section marks the beginning of the core analysis. The imported and annotated data are firstly containerized into an S4 object. This object-oriented approach ensures data integrity and provides a structured container for the multistep quality control (QC) and filtering pipeline that follows.

3.5.1 Create PeptideQuantitativeData Object

In chunk *{r 9 Peptide Data S4 Object Creation & Annotation API Lookup}*, the S4 class is the central data structure for peptide-level analysis. When creating the object, users must map the columns from your input data to the object's slots, such as *peptide_data*, *protein_id_column*, *peptide_sequence_column*, *q_value_column*, and *raw_quantity_column*. Users must also provide the *design_matrix* created in the previous step and specify the column names for *sample_id*, *group_id*, and *technical_replicate_id*. This function and parameters can be seen in the code chunk below.

```
"'{r 9 Peptide Data S4 Object Creation & Annotation
API Lookup}
peptide_data <- new(
"PeptideQuantitativeData",
# Peptide vs Sample quantitative data
peptide_data = data_cln,
protein_id_column = "Protein.Ids",
peptide_sequence_column = "Stripped.Sequence",
q_value_column = "Q.Value",
global_q_value_column = "Global.Q.Value",
proteotypic_peptide_sequence_column =
"Proteotypic",
raw_quantity_column = "Precursor.Quantity",
norm_quantity_column = "Precursor.Normalised",
is_logged_data = FALSE,
# Design Matrix Information
design_matrix = design_matrix,
sample_id = "Run",
group_id = "group",
technical_replicate_id = "replicates",
args = config_list)
"'
```

3.5.2 Initial Raw Data QC

Raw data should always be interrogated to establish a reference baseline before performing data QC downstream. Here, we use the *updateProteinFiltering()* function in *{r 10 Raw Data QC}* with *step_name* set to "1_Raw Data". This function generates a panel of plots, including peptide/protein counts per sample and missing value distributions, providing a snapshot of the initial data quality seen in the code chunk below.

```
"'{r 10 Raw Data QC}
```

updateProteinFiltering(# Please note that these images don't render properly in RStudio, please see the *publication_graphs_dir* for the images

```
data = data_cln,
step_name = "1_Raw Data",
omic_type = omic_type, # e.g., "proteomics"
```

```
experiment_label = experiment_label, # e.g.,
"workshop_data"
return_grid = TRUE,
overwrite = TRUE)
"'
```

3.5.3 Filter by Q-value and Proteotypic Status

This step in {r 11 Filter Peptides on q Value and Proteotypic Peptide Match} enhances the quality of the dataset by removing low-confidence and ambiguous peptide identifications. The *srlQvalueProteotypicPeptideClean()* function applies filters based on three key parameters from the configuration file. These are:

1. *qvalue_threshold*: Filters peptides based on their individual identification confidence (e.g., <0.01).
2. *global_qvalue_threshold*: Filters based on the protein group-level confidence (e.g., <0.01).
3. *choose_only_proteotypic_peptide*: A binary flag gating whether to keep only unique peptides.

After filtering, call *updateProteinFiltering()* again with *step_name* set to *"2_qval Filtered"* to visualize the impact of this cleaning step on peptide and protein counts. This function and parameters can be seen in the code chunk below.

```
"'{r 11 Filter Peptides on q Value and Proteotypic
Peptide Match}
search_srl_quant_cln <-
srlQvalueProteotypicPeptideClean(theObject =
peptide_data)
qval_protein_count <- search_srl_quant_cln@peptide_
data |>
distinct(Protein.Ids) |>
nrow()
message(paste("Number of distinct proteins
remaining after q-value and proteotypic filtering:",
qval_protein_count))
# Example: Interactively change q-value thresholds
and re-run this chunk
# peptide_data <- updateConfigParameter(
# theObject = peptide_data, # Use the input object
from the previous step
# function_name =
"srlQvalueProteotypicPeptideClean",
# parameter_name = "qvalue_threshold", # Effect:
Lower = stricter peptide ID confidence
# new_value = NEW_VALUE_HERE)
# peptide_data <- updateConfigParameter(
# theObject = peptide_data,
# function_name =
"srlQvalueProteotypicPeptideClean",
```

```
# parameter_name = "global_qvalue_threshold", #
Effect: Lower = stricter protein group ID confidence
# new_value = NEW_VALUE_HERE)
# peptide_data <- updateConfigParameter(
# theObject = peptide_data,
# function_name =
"srlQvalueProteotypicPeptideClean",
# parameter_name = "choose_only_proteotypic_
peptide",# Effect: 1 = Keep only unique peptides, 0
= Keep shared peptides
# new_value = NEW_VALUE_HERE)
# search_srl_quant_cln <-
srlQvalueProteotypicPeptideClean(theObject =
peptide_data)
# Re-run after update
updateProteinFiltering(
data = search_srl_quant_cln@peptide_data,
step_name = "2_qval Filtered",
omic_type = omic_type, # e.g., "proteomics"
experiment_label = experiment_label, # e.g.,
"workshop_data"
return_grid = TRUE,
overwrite = TRUE)
"'
```

3.5.4 Roll-Up Precursors to Peptides

Aggregate intensity data from the precursor level to the peptide level. Figure 4 depicts a typical QC figure at this stage. In {r 12 Roll-up Precursor to Peptide}, the *rollUpPrecursorToPeptide()* function sums the intensities of all precursors (i.e., different charge states and modified forms) that map to the same stripped peptide sequence. This reduces data dimensionality and provides an aggregated measure of peptide abundance for quantification downstream.

3.6 Peptide-Level Normalization and Quality Control

This section details the multistage process of data transformation, normalization, and batch correction at the peptide level, with the aim of removing technical variation and preparing the data for robust statistical analysis. The entire process is monitored with a suite of QC plots.

3.6.1 Data Transformation and Initial QC

Before normalization, the peptide-level data are transformed to be suitable for normalization in code chunk *{r 13 QC of peptide matrix}*.

1. Run *calcPeptideMatrix()* to pivot the data from a long to a wide format, creating a matrix of peptides versus samples.
2. Apply a log2 transformation using *log2TransformPeptideMatrix()* to stabilize variance and make the data more suitable for linear modeling.

Generate a panel of prenormalization QC plots. This includes Relative Log Expression (RLE) plots (*plotRle*), Principal Component Analysis (PCA) plots (*plotPca*), density plots (*plotDensity*), and

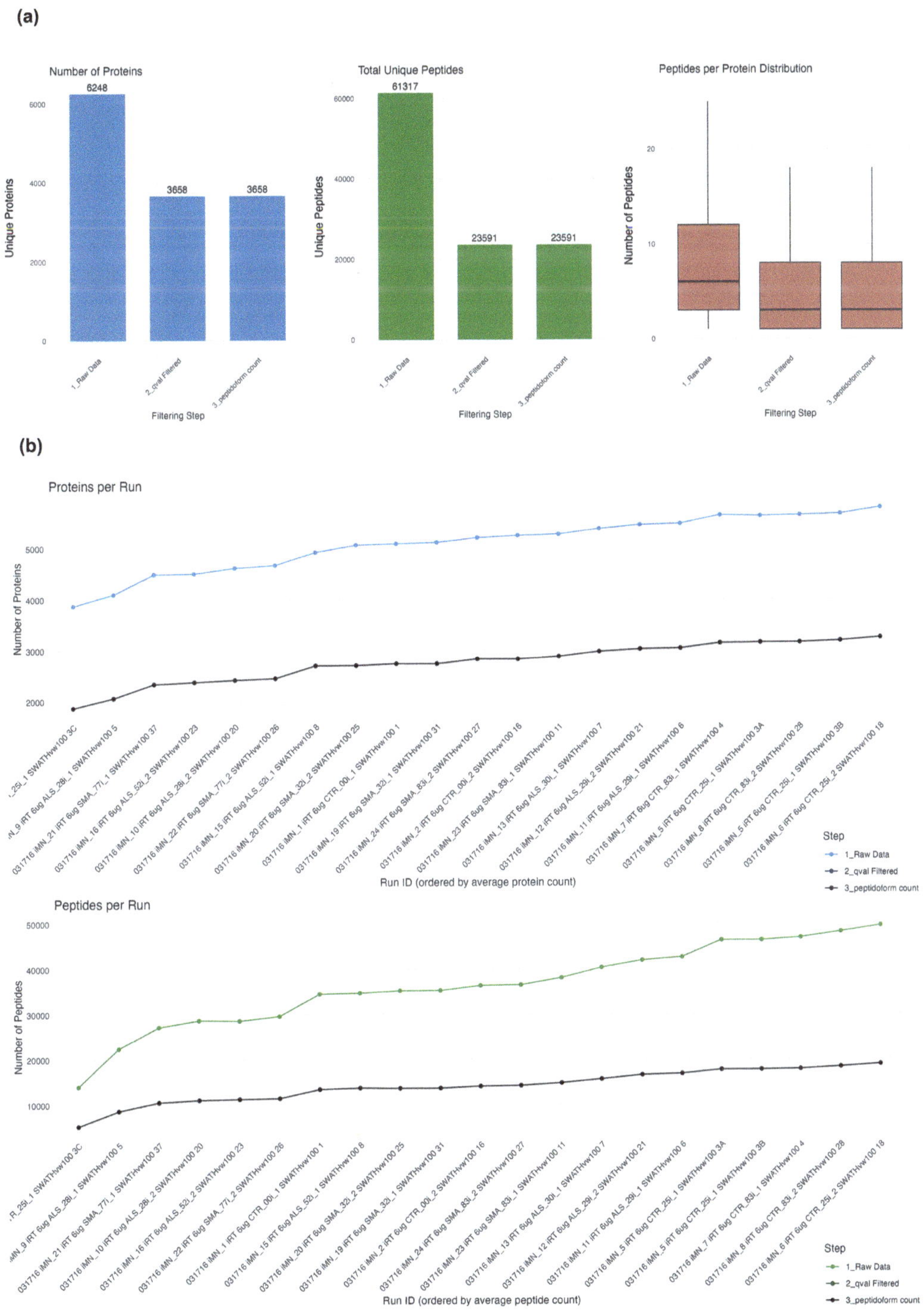

Fig. 4 Rolling QC metrics facilitated by *updateProteinFiltering()*. Users are able to track the distribution of proteins and peptides mapping to samples across all stages of filtering and QC

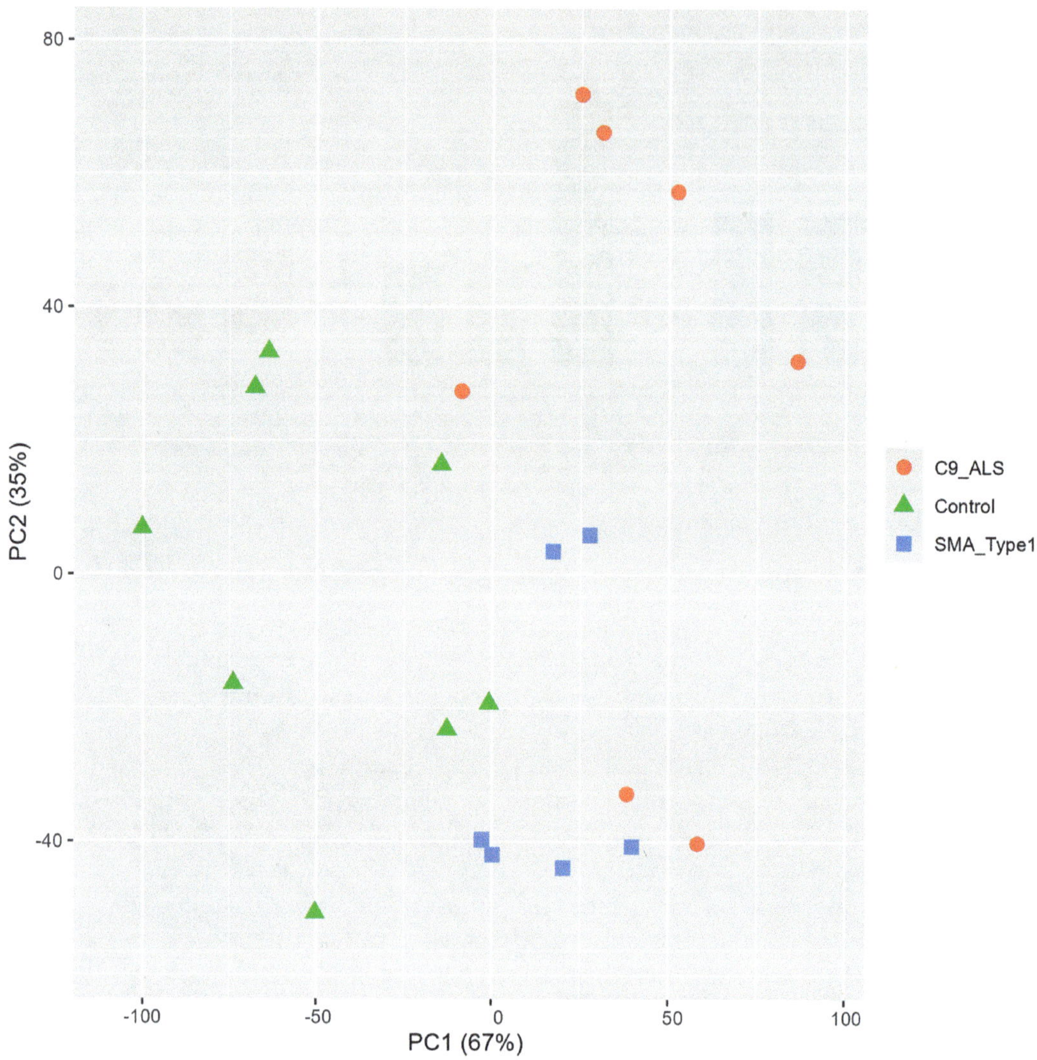

Fig. 5 Principal component analysis plot. Successive rounds of normalization and batch effect removal through the workflow should see groups separate from each other, but cluster more closely within each grouping

Pearson correlation heatmaps (*plotPearson*). These plots provide a comprehensive baseline of the data's structure, sample relationships, and the extent of technical variation before correction. A typical PCA plot from this step can be seen in Fig. 5.

3.6.2 Cyclic Loess Normalization

Apply an intensity-dependent normalization to correct for non-linear biases between samples in code chunk *{r 14 Cyclic loessnormalisation}*.

1. Execute *normaliseBetweenSamples()* with normalization_method set to "*cyclicloess*".

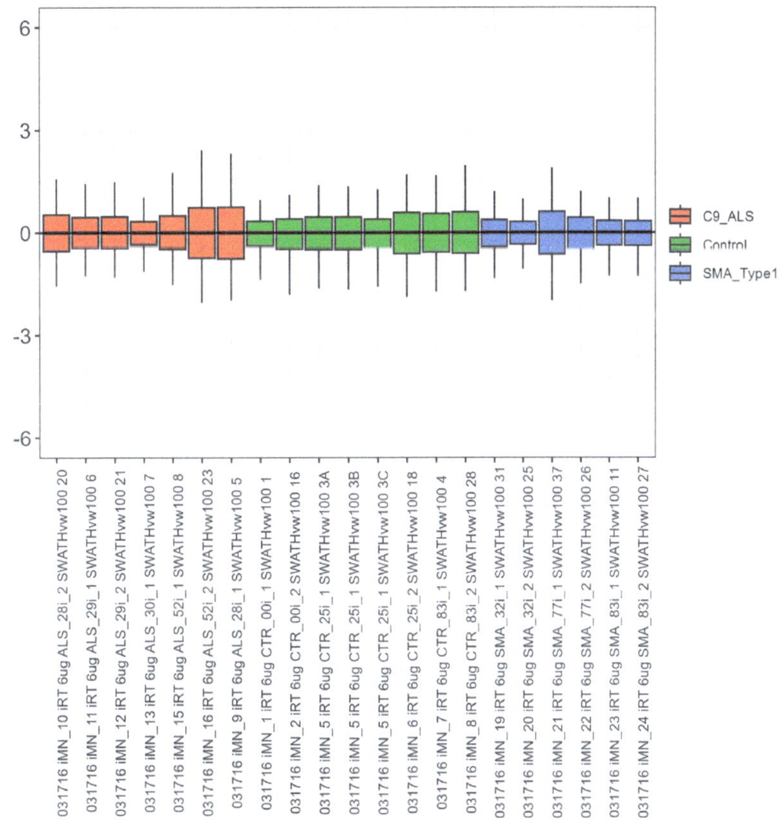

Fig. 6 Relative log expression plot. This plot shows the distribution of log-transformed expression values relative to the median across samples. Successive rounds of normalization should sequentially center medians about zero and shrink the variance of boxplots

This method fits local regression models iteratively across pairs of samples to adjust for systematic trends in intensity, which is a common artifact in mass spectrometry data.

2. After normalization, regenerate the full panel of QC plots (RLE, PCA, etc.).

This is to visually confirm that the distributions in the RLE plot are similar, and sample correlations have improved. Figure 6 depicts an RLE plot where samples have been normalized, with distributions centered around 0 and displaying similarly even interquartile ranges.

3.6.3 Protein Accession Cleanup, Imputation, and Protein Rollup

Remove Unwanted Variance (RUV) is a powerful tool used to identify and remove sources of technical variation, such as batch effects, while preserving the biological variation of interest. We perform several complex operations in code chunk *{r 15 RUV-III-c parameter tuning and normalisation}* in this final round of normalization.

1. Automated Parameter Tuning:
   First, determine the optimal parameters for RUV using *findBestNegCtrlPercentagePeptides()*. This function automates the selection of two key parameters to the tool's use. The first is the percentage of peptides to be used as *negative controls* (i.e., peptides assumed to be non-differentially expressed across conditions). The function tests the model across a *percentage_range* (e.g., 1%–30%) and selects the value that maximizes the separation between biological signal and unwanted variation based on a *separation_metric* (e.g., *"max_difference"*). Secondly, the number of unwanted factors (*k*) to remove. The canonical correlation between the data and the experimental design is systematically evaluated to identify the optimal *k* that balances variance removal with signal preservation, applying a *k_penalty_weight* to avoid over-correction.
2. Apply RUV Correction:
   Use the optimized parameters (*best_percentage* and *best_k*) from the previous step to run the final normalization with *ruvIII_C_Varying()*. The QC step for this is assessment of the canonical correlation plot as per Fig. 7. The number *k* removed should correspond to separation metric chosen.
3. Final QC Assessment:
   The QC plot panel is regenerated for the third time to assess the final state of the data. At this stage, PCA plots should ideally show clear clustering by the main biological *grouping_variable*, and RLE plots should be tightly centered around a median of zero, indicating that technical variation has been successfully minimized (*see* **Note 4.5**).

3.7 RUV-III-C Normalisation for Batch Correction

After peptide-level processing, peptides are grouped to protein-level information. This involves standardizing protein accessions, imputing missing values using the *limpa* method, and summarizing peptide data into robust protein-level quantities.

3.7.1 Select Best Protein Accession

DIA-NN often reports multiple protein accessions for a single peptide group (demarked with a; between accessions). The *chooseBestProteinAccession()* function in *{r 16 Annotation Rollup}* resolves this ambiguity by selecting a single, representative protein ID for each group. It uses the preprocessed FASTA file (*seqinr_obj*) to prioritize accessions based on evidence criteria (e.g., preferring Swiss-Prot over TrEMBL entries). An *aggregation_method* (e.g., *"mean"*) is used to handle any duplicate peptide measurements that arise after collapsing protein groups. Users can modify the method if they so wish, but the default setting is recommended.

3.7.2 Impute Missing Values and Rollup to Protein Level

Missing values are a persistent challenge in proteomics and contribute to a loss of potential signal as proteins with missing values are

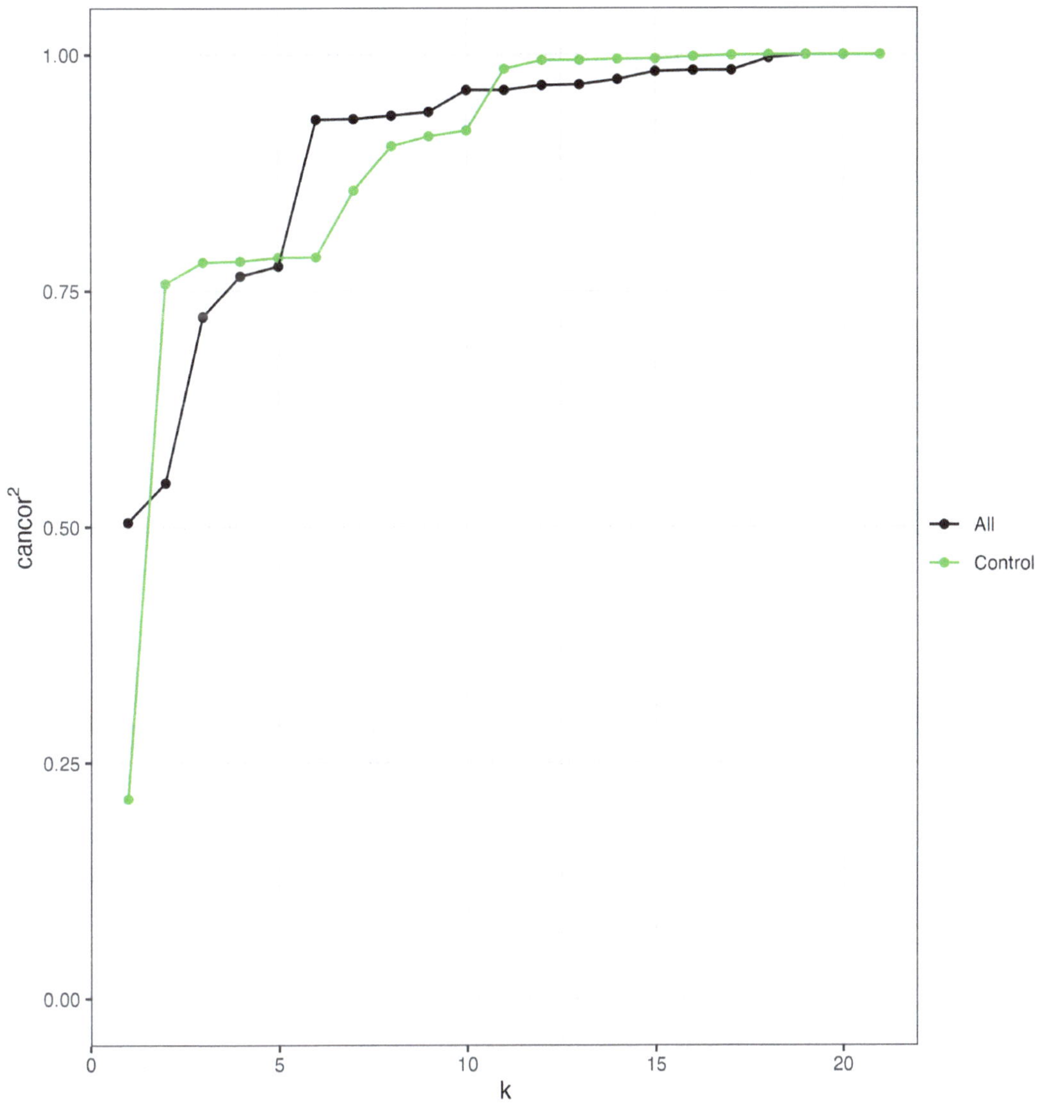

Fig. 7 RUV-III-C Canonical Correlation (Cancor) Plot for Optimal Normalization Factor Selection. The plot displays the correlations between estimated factors of unwanted variation and experimental design groups across number of factors (*k*). The "elbow" inflection point typically marks the optimal number *k* to remove from the dataset

typically removed for all samples before statistics can be performed. This workflow uses the *limpa* package to impute missing values, which models the probability of a peptide being detected based on its intensity. The function *proteinMissingValueImputationLimpa()* is performed in *{r 17 Limpa imputation and protein rollup}* on the peptide-level data object. This function performs two key operations:

1. Imputation:
   Missing peptide intensities are imputed based on the Detection Probability Curve (DPC), which is more appropriate for mass spectrometric data rather than other traditional machine learning methods or simple replacement methods.
2. Protein Rollup:
   Peptide data are aggregated to the protein level using a robust summarization method, producing a protein-by-sample quantification matrix contained in a new S4 object.

3.7.3 Visualize Imputation QC

generateLimpaQCPlots() produces a set of diagnostic plots, including the DPC itself and comparisons of data distributions before and after imputation as can be seen in Fig. 8 (*see* **Note 4.6**).

3.8 Protein-Level Filtering and Final QC

Final round of QC and filtering on the normalized protein-level data based on the number of unique and prototypic peptides and removing outlier samples is necessary before statistical analysis.

3.8.1 Filter by Peptide Evidence (Two-Peptide Rule)

In *{r 18 Filter Proteins on Peptide Number}*, *filterMinNumPeptidesPerProtein()* filters the dataset to proteins identified by at least 2 peptides quantified that match to it (*see* **Note 4.7**). The thresholds can be set for both unique peptides (*num_peptides_per_protein_thresh*, typically 1) and total peptidoforms (*num_peptidoforms_per_protein_thresh*, typically 2).

3.8.2 Filter Samples by Correlation

Poorly correlating samples within a replicate group can indicate outlier samples. In *{r 19 Sample Pearson Correlation Filtering}* pairwise correlations between all samples within each group are calculated using *pearsonCorForSamplePairs()*, after which *filterSamplesByProteinCorrelationThreshold()* removes any samples that do not meet a *min_pearson_correlation_threshold* (e.g., 0.5) with other samples in their group.

3.8.3 Save Intermediate Results

The normalized, filtered protein quantification data are saved in multiple formats for audit trail and downstream analysis in *{r 20 Output Files For Audit Trail}*. The data are exported as both.tsv (human-readable) and RDS (R object) formats, with the latter allowing easy import for manipulation at a later date.

3.9 Composite QC Figure Generation and Data Export

The final QC plots provide a final checkpoint for interrogating on the data quality and structure before proceeding to statistical testing. This code chunk *{r 21Composite QC Figure Generation}* assembles them into a single, multipanel figure, as can be seen in Fig. 9. The *createGridQC()* function takes the object containing all previously generated plots and arranges them into a structured grid. For the *"DIA_limpa" workflow_name*, it creates a specifically formatted layout that compares the data at each major stage (log2 transformed, cyclic

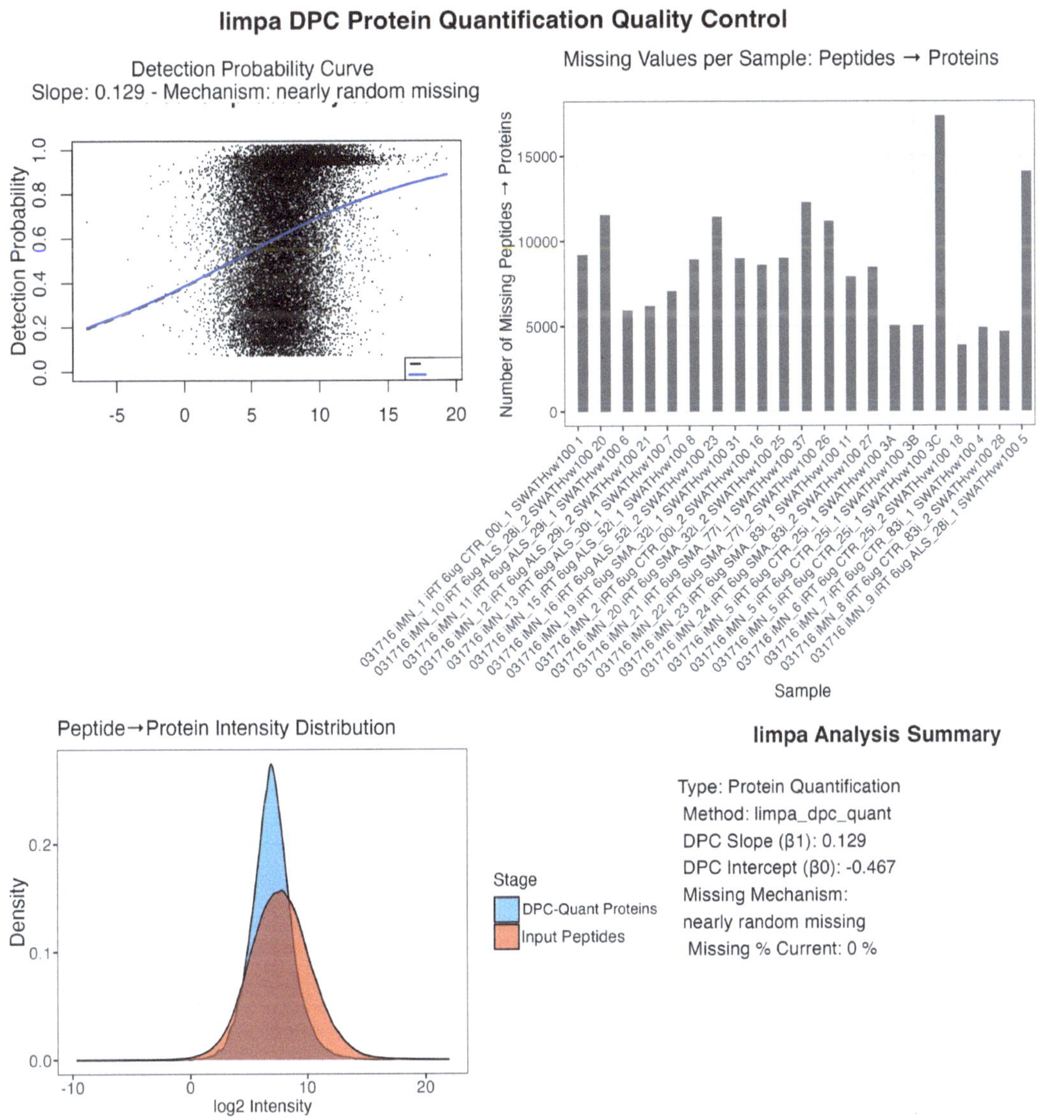

Fig. 8 QC metrics from *limpa* DPC imputation. This composite figure allows users to evaluate the effectiveness of the DPC method for imputing missing values. Users are able to visualize the shape of their data in the DPC, the reduction of missing values across samples, histogram shifts of imputed compared to original data and model parameters and outputs

loess normalized, RUV corrected, and final protein-level), making it easy to visualize the impact of each normalization step.

3.10 Differential Expression Analysis

With a clean, normalized, and filtered protein matrix, statistical analysis is needed to identify proteins that are differentially expressed between experimental conditions. The workflow uses the *limma* package, a robust and widely used tool for analyzing omics data (*see* **Note 4.8**).

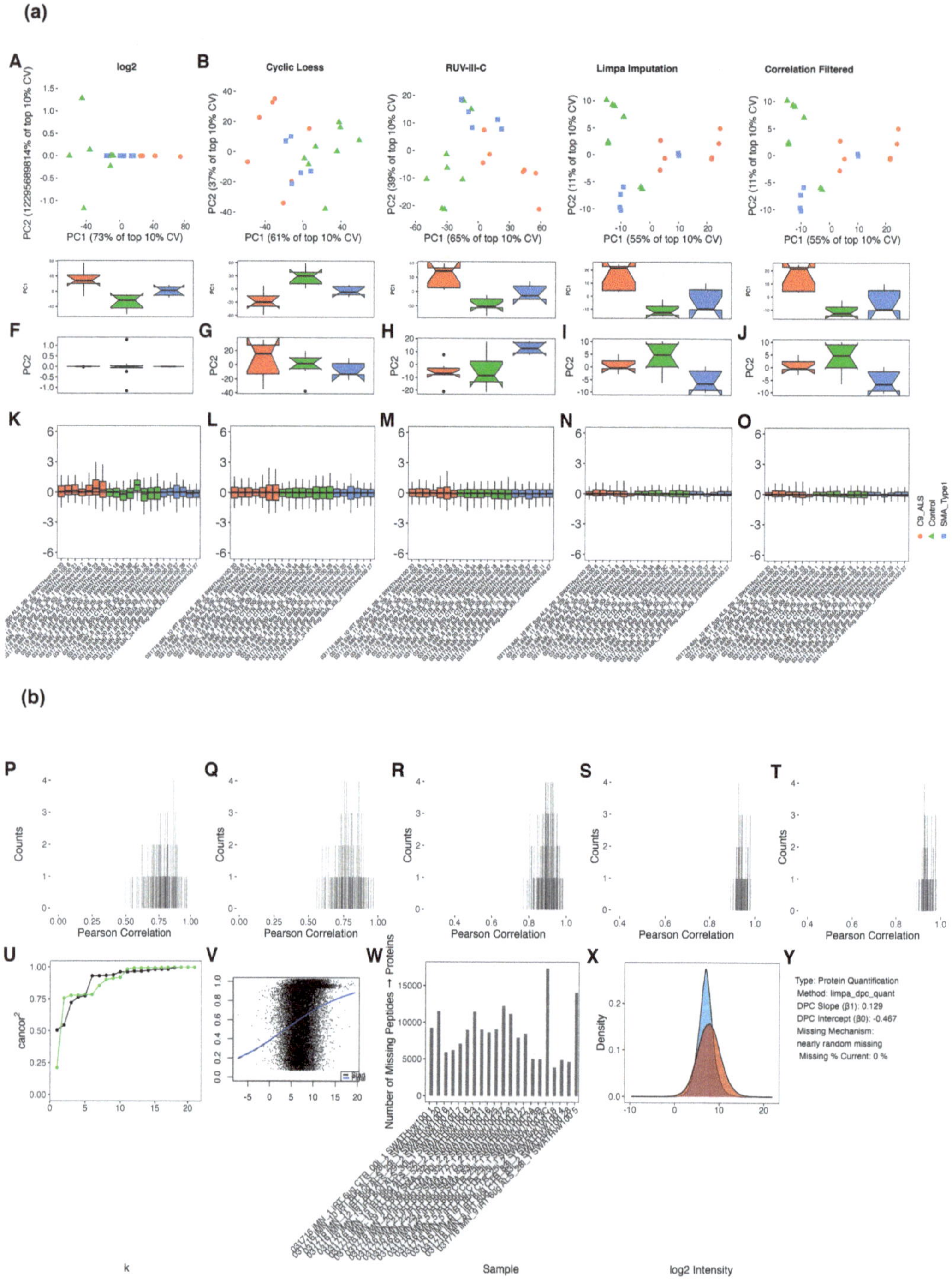

Fig. 9 Composite QC figure ready for publication. Rows are data visualizations across successive rounds of normalization (columns). As can be seen, the normalization strategies used vastly improve the dataset from raw through to curated, indicating bias removal, technical variance reduction, and improved sample correlation

3.10.1 Perform Differential Expression Analysis

The analysis is performed for each biological comparison defined in the design matrix step. These contrasts are specified in the *contrasts_tbl* object that were imported earlier. *In {r 22 DE Analysis Generation}, deAnalysisWrapperFunction()* is called in a loop for each contrast and serves as a wrapper around the *limma* statistical pipeline. Rather than averaging technical replicates, the correlation between replicates of the same biological replicate is explicitly modeled during the linear fitting process. The *block* parameter is utilized within the *dpcDE()* function to identify technical replicate groups. The block parameter is an array of strings in the same order as the rows of the design matrix, where each role represents a sample and replicate samples have the same string value. This approach allows the model to estimate a correlation between technical replicates, which accounts for the non-independence of these samples while retaining the original degrees of freedom and precision weights associated with each individual run. For each contrast, it will fit a linear model to the data for each protein using *limpa::dpcDE() and* compute the specified contrast using *contrasts.fit. It will also* apply empirical Bayes moderation. The *eBayes_trend* and *eBayes_robust* parameters are set to TRUE to apply trended and robust estimation, respectively. Optionally, if a *treat_lfc_cutoff* is provided, it performs a test against a minimum log-fold change threshold using the treat method. It is *not* recommended that this is specified.

3.10.2 Output and Visualize Results

After the statistical analysis is complete for each contrast, the results are saved and visualized in *{r 23 Output DE Analysis Results}*.

1. The *outputDeAnalysisResults()* function is called within the loop.

   This function takes the results from *deAnalysisWrapperFunction*() and merges the statistical results (log-fold change, *p*-value, *q*-value) with the corresponding UniProt annotations (*uniprot_tbl*). It will also write complete, annotated results table to a.tsv file in the *de_output_dir*. Finally, publication quality volcano plots are generated. Proteins exceeding the specified *de_q_val_thresh* are highlighted. These plots are saved in the *publication_graphs_dir* in the format specified by *plots_format*. A typical volcano plot from a comparison can be seen in Fig. 10.

3.11 Functional Enrichment Analysis

Functional enrichment helps to identify biological pathways, molecular functions, and cellular components that are over-represented within the dataset. This provides a systems-level view of the biological response and moves the analysis individual protein quantification to interpretable biological insights. MultiScholaR automates this process by interfacing with tools to offer both rank-based and threshold-based enrichment methods.

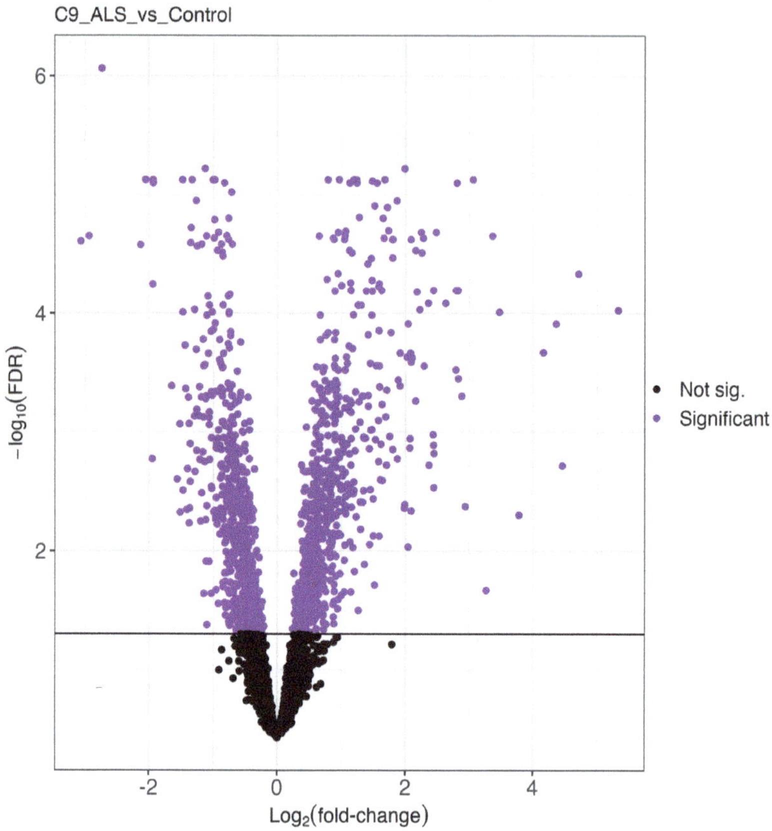

Fig. 10 Volcano plots are generated for user-specified comparisons. These display the relationship between log-fold change and statistical significance for each protein

3.11.1 Submit DE Results for Ranked Enrichment

In *{r 24 STRING DB Functional Enrichment Analysis}*, *runStringDbEnrichmentAllContrasts()* submits the complete list of differentially expressed proteins from each contrast to the STRING database [5] for rank-based functional enrichment analysis. All proteins are ranked using a combined score: sign(log2FC) × -log10(FDR q-value). This scoring preserves both the direction (up vs. down) and statistical confidence of differential expression, providing a more robust representation of the biological signal than discrete thresholding.

3.11.2 Plot StringDB Results

In *{r 25 Plot the StringDB Results}*, stringDB enrichment results are visualized using *plotStringDbEnrichmentResults()*, which generates publication-quality individual and summary plots organized by functional category (as seen in Fig. 11).

3.11.3 Prepare DE Results for Threshold-Based Enrichment

The results from the differential expression analysis across all contrasts must first be collated into a specific format for this step. In *{r 27 Threshold-Based Enrichment Analysis}*, this is achieved by running

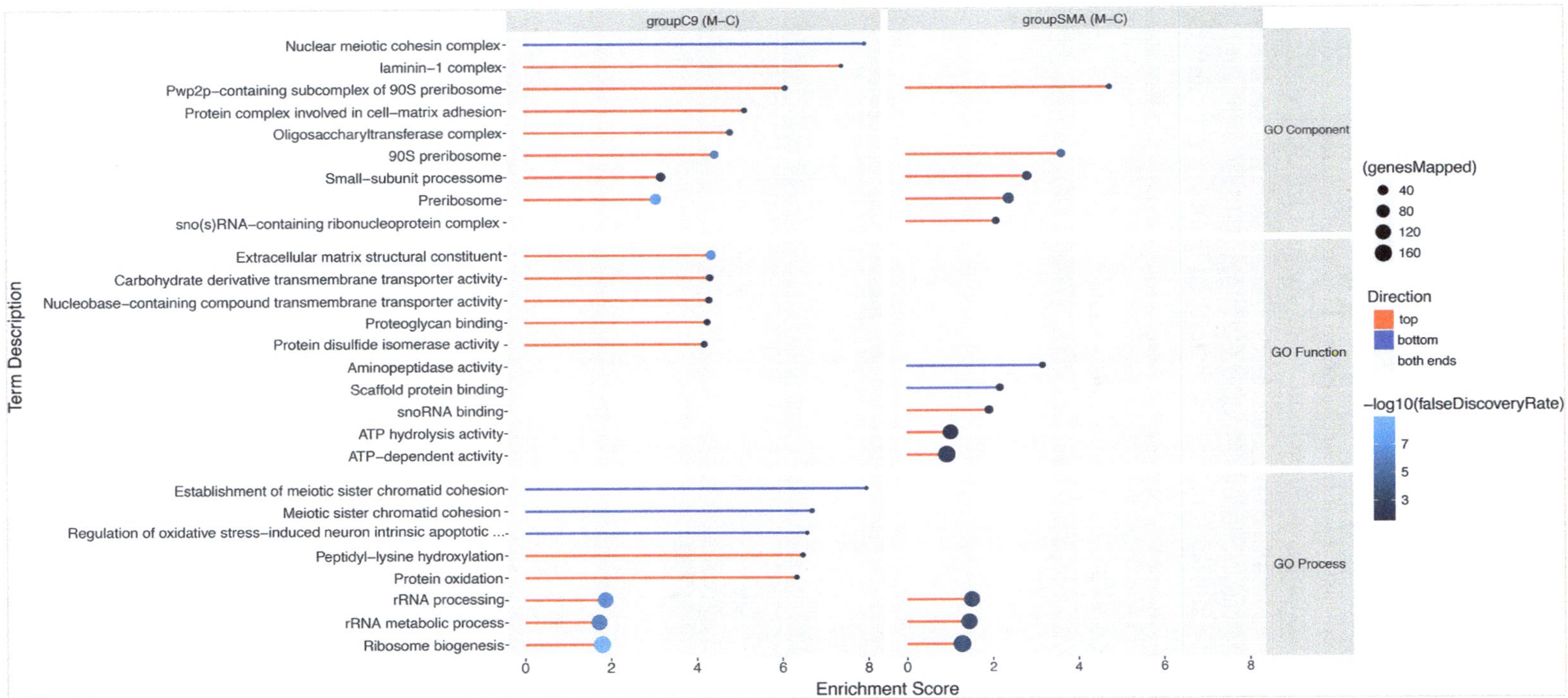

Fig. 11 Faceted bar graph of stringDB enrichment results of Gene Ontology (GO) pathways. Several biological signals are significant in both group comparisons as potential pathological mechanisms of interest. Bar length represents the enrichment score, point color encodes statistical significance as $-\log10(\mathrm{FDR})$, point size reflects the number of proteins from the input list that map to each enriched term

the *createDEResultsForEnrichment()* function, which gathers the necessary DE result files from the output directory and organizes them into a structured S4 object specifically for this step.

3.11.4 Run Enrichment Analysis

This type of enrichment analysis is made simple for the user and is facilitated by *processEnrichments()*, and uses *gprofiler2*. This is the preferred tool for enrichment if the organism's *taxon_id* is in the supported list of organisms. This tool leverages a comprehensive database, which includes reactome and KEGG [6]. If the organism is not supported by *gprofiler2*, the *clusterProfiler* package [7] is used instead, performing a Gene Ontology (GO) enrichment analysis using the annotations previously fetched from UniProt and stored in the

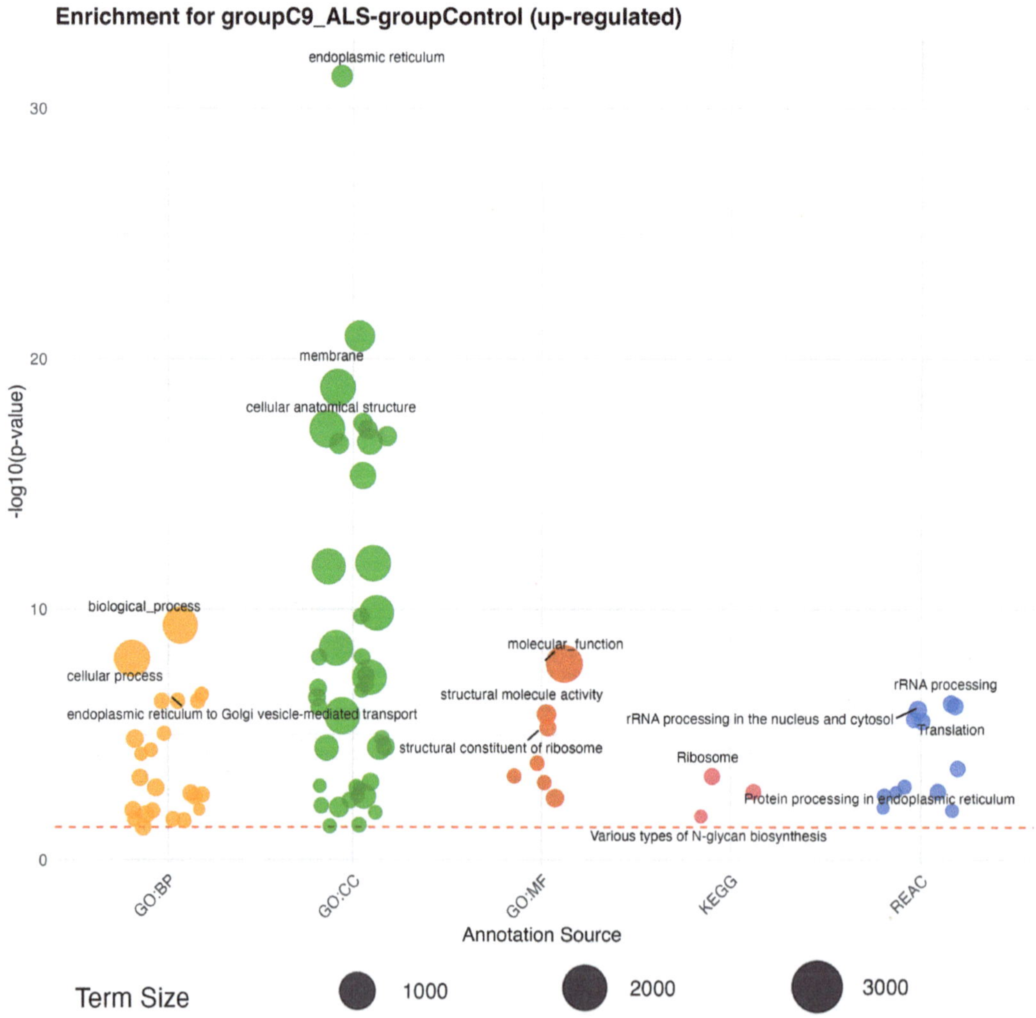

Fig. 12 Manhattan plot displaying the top enriched terms across ontologies. Higher y axis scores denote greater significance of enrichment, whilst the size of term circles represents number of proteins within that term. A plot is displayed for every comparison, and one each for upregulated and downregulated

uniprot_dat_cln object. The analysis is run separately on up-regulated and down-regulated protein lists for each contrast, based on user-defined log-fold change (*up_cutoff*, *down_cutoff*) and significance (*q_cutoff*) thresholds. Users should avoid changing defaults for these unless they have a specific reason to do so as this could introduce bias into the results. Enrichment plots are outputted in two formats: static plots and dynamic html plots for exploring results. A small subset of enriched terms from the Neurolincs data can be seen in Fig. 12.

3.12 Reproducibility and Reporting

The final steps of the *MultiScholaR* workflow focus on systematically capturing all parameters, organizing key results for publication, creating an audit trail, and generating a final summary report. This comprehensive approach ensures that the analysis is transparent, reproducible, and publication ready.

3.12.1 Capture Workflow Parameters

To create a complete record of the analysis, run *createWorkflowArgsFromConfig()*. This function captures all settings and parameters used during the workflow, including filtering thresholds, normalization methods, statistical parameters, and version information for *MultiScholaR*, saving everything into a structured list.

3.12.2 Results Summary

copyToResultsSummary() copies the final QC plots, differential expression result tables, and all pathway enrichment figures/tables into the *publication_graphs* directory. These constitute the most important files pertinent to publication.

3.12.3 Generate Final Report

The workflow concludes by generating a comprehensive Microsoft Word report summarizing the entire analysis. *RenderReport()* will embed all generated figures, tables, and workflow parameters in a publication-ready document. The resultant document contains blocks of text that are ready for insertion into a manuscript.

4 Notes

4.1 Running DIA-NN

Users should always retrieve organism FASTA files from a reliable source like Swiss-Prot, and make sure the headers are in a standard format. Inconsistent or non-standard FASTA headers are a common source of annotation issues downstream. The DIA-NN processing step can be computationally intensive, potentially taking several hours to days for large datasets, and it is recommended that this is performed on hardware that can run the software.

4.2 Installing Dependencies

Installing MultiScholar via *installMultiScholaR()* from a fresh R installation can take some time, as it resolves and installs numerous packages from CRAN and Bioconductor.

4.3 Data Import

A common point of failure is a mismatch between column names in the DIA-NN report file and the names expected by the import functions in the Data Management chunk. Headers of *report.parquet* or *report.tsv* files should always include columns like *Q.Value*, *Protein.Ids*, and *Precursor.Quantity* and are correctly named. If in doubt, users can open these objects as explorable data frames within RStudio.

4.4 Specify Experimental Design

The sample identifiers in your design matrix *must exactly match* the run names in the header of your quantitative matrix (i.e., the column headers of the wide-format data). The design app is engineered to prevent this by populating the sample list directly from the data.

4.5 Normalization

The goal of normalization is to reduce technical variation, which should be visible as samples clustering more tightly by their biological grouping in the PCA plots and as more centered distributions in the RLE plots. If these plots do not show improvement, it may be necessary alter parameters for RUV or skip this step entirely if it does achieve the desired outcome.

4.6 Imputation

The *limpa*-based imputation assumes that missing values are primarily Missing Not At Random (MNAR), where low-abundance peptides fall below the limit of detection. Always review the diagnostic plots from *generateLimpaQCPlots()* to ensure the imputation model fits the data well.

4.7 Two-Peptide Rule

Filtering for proteins identified by at least two peptidoforms is a standard practice, however, is not always "one size fits all." For example, some studies might investigate low abundance proteins where a single-peptide protein is of high biological interest. If so, you may consider relaxing this filter by setting *num_peptidoforms_per_protein_thresh* to 1.

4.8 Differential Expression

The statistical power of the *limma* analysis is highly dependent on the number of biological replicates. While *limma*'s empirical Bayes moderation performs well with small sample sizes, it is highly recommended that users generate data for as many biological replicates as possible in the event of outliers, bad quality samples or other technical issues that might cause a lower group n.

References

1. Demichev V, Messner CB, Vernardis SI, Lilley KS, Ralser M (2020) DIA-NN: neural networks and interference correction enable deep proteome coverage in high throughput. Nat Meth 17(1):41–44. https://doi.org/10.1038/s41592-019-0638-x
2. Li M, Cobbold SA, Smyth GK (2025) Quantification and differential analysis of mass spectrometry proteomics data with probabilistic recovery of information from missing values. bioRxiv 29:651125. https://doi.org/10.1101/2025.04.28.651125
3. Matlock AD, Vaibhav V, Holewinski R, Venkatraman V, Dardov V, Manalo DM, Shelley B, Ornelas L, Banuelos M, Mandefro B, Escalante-Chong R, Li J, Finkbeiner S, Fraenkel

E, Rothstein J, Thompson L, Sareen D, Svendsen CN, NeuroLINCS Consortium NIH, Van Eyk JE (2023) NeuroLINCS Proteomics: defining human-derived iPSC proteomes and protein signatures of pluripotency. Sci Data 10(1):24. https://doi.org/10.1038/s41597-022-01687-7

4. R Core Team (2025). R: A Language and Environment for Statistical Computing. R Foundation for Statistical Computing, Vienna, Austria. https://www.R-project.org/
5. Szklarczyk D, Kirsch R, Koutrouli M, Nastou K, Mehryary F, Hachilif R, Gable AL, Fang T, Doncheva NT, Pyysalo S, Bork P, Jensen LJ, von Mering C (2023) The STRING database in 2023: protein-protein association networks and functional enrichment analyses for any sequenced genome of interest. Nucleic Acids Res 51(D1):D638–D646. https://doi.org/10.1093/nar/gkac1000
6. Kolberg L, Raudvere U, Kuzmin I, Vilo J, Peterson H (2020) gprofiler2-an R package for gene list functional enrichment analysis and namespace conversion toolset g:Profiler. F1000Res 9:709. https://doi.org/10.12688/f1000research.24956.2
7. Yu G, Wang LG, Han Y, He QY (2012) clusterProfiler: an R package for comparing biological themes among gene clusters. Omics: J Integr Biol 16(5):284–287. https://doi.org/10.1089/omi.2011.0118

Check for updates

Chapter 3

Entropy and Fractal Dimensions Application in MALDI-TOF MS Data to Train Classification Models

Lucas Cardoso Lazari, Ghasem Azemi, Livia Rosa-Fernandes, Suely K.N. Marie, Antonio Di Ieva and Giuseppe Palmisano

Abstract

Matrix-Assisted Laser Desorption/Ionization Time-of-Flight (MALDI-TOF) mass spectrometry generates rich biochemical fingerprints ideal for disease screening. However, its spectral output is highly complex, requiring advanced processing before clinical use. Machine learning excels in learning patterns within this high-dimensional data, but it depends on effective feature extraction. This chapter presents preprocessing strategies for MALDI-TOF spectra and demonstrates the use of fractal dimension and entropy analysis as a feature generation method.

Keywords MALDI-TOF, Machine learning, Entropy, Fractal dimension

1 Introduction

Matrix-Assisted Laser Desorption/Ionization Time-of-Flight (MALDI-TOF) mass spectrometry is a technique capable of generating detailed biochemical fingerprints from biological samples. It serves as a tool for the rapid and accurate screening of the protein and peptide content in various diseases, capturing molecular patterns that can be related to diagnosis and prognosis [1, 2]. The raw output of this technique is a spectrum containing thousands of data points representing different molecule intensities. While rich with information, these spectra are too complex for direct clinical interpretation or the immediate identification of specific biomarkers.

The complexity of these data makes machine learning an interesting tool in the analysis process. Machine learning algorithms are capable of identifying subtle patterns in this high-dimensional data, capturing signatures that can be used to differentiate patient groups and develop diagnostic or prognostic tools [3–5]. However, the raw MALDI-TOF data must be processed to extract meaningful features to be used in machine learning applications. This involves cleaning

Antonio Di Ieva and Jennilee M. Davidson (eds.), *Computational Neuroscience*, Methods in Molecular Biology, vol. 3062, https://doi.org/10.1007/978-1-0716-5440-8_3,

the spectra and performing feature extraction, which typically focuses on protein/peptide peaks or the entire binned spectrum [3, 6].

Given that the spectra are composed of intensity measurements over a mass-to-charge ratio range, which are acquired through the time a molecule takes to reach the analyzer, they present a time-series structure and thus can be analyzed using methods developed for time-series data [7]. Therefore, this chapter will demonstrate how to process raw MALDI-TOF MS files and calculate two distinct types of time-series complexity measures: fractal dimensions and entropy. We will illustrate how these metrics can be utilized as features for training machine learning models in classification tasks.

2 Materials

This protocol utilizes raw MALDI-TOF MS files of serum protein profiles from glioblastoma and healthy control patients, which will be available in the PRIDE database (accession: PXD060922) [8] or upon request. The data processing pipeline employs both R (version 4.5.0) and Python (version 3.12.8). Initial spectral processing, which includes normalization, transformation, and baseline correction, is performed in R using the MALDIquant and MALDIQuantForeign libraries [9, 10]. Subsequently, Python is used for all fractal, entropy, and machine learning-related tasks, including feature selection, oversampling, data scaling, and model training. The fractal dimensions and entropy of the data are calculated using the Antropy [11] library in Python. Make sure to install all libraries and their dependencies before running the scripts and read the comments within the code for further explanation about the steps in data processing.

3 Methods

3.1 Initial Processing

The spectrum needs to undergo a few processing steps to reduce inter-sample variability and facilitate the calculations of the complexity measurements. These steps include transformation, base-correction, smoothing, and normalization (*see* **Note 1**). This can be done using the following approach in R:

1. Data is imported in the mzML format using the MALDIquantForeign library (*see* **Note 2**). Then, MALDIquant is used to preprocess the spectra.
2. First, transform the spectra using the square-root function.
3. Then smooth to reduce the effect of noise using the Savitzky-Golay method.

4. Perform base correction and normalization using the total-ion-current. Each processing step is performed within each spectrum, so no information is leaked between the samples.

```
# Importing libraries and data
packages <- c("MALDIquant","MALDIquantForeign")
lapply(packages, require, character.only = TRUE)
gbm <-  import("<path_to_gbm_files>", type="mzML")
controls  <- import("<path_to_controls_files>",
type="mzML")


# Create the function to process the spectra
processData <- function(data_list) {
  samp_transf <- transformIntensity(data_list,
method="sqrt")
  samp_smooth <- smoothIntensity(samp_transf,
method="SavitzkyGolay", halfWindowSize = 10)
  base_corr <- removeBaseline(samp_smooth, method=
"TopHat")
  int_norm <- calibrateIntensity(base_corr, method=
"TIC")
  return(int_norm)}


# Get the final normalized spectra
processed_gbm <- processData(gbm)
processed_ct <- processData(controls)
```

5. Then, save the spectra information to be used in Python for the next steps. The following code is a suggestion to save it as a *txt* file, where each line contains the sample ID, the intensities, and the m/z values separated by a semi-colon.
6. The script iterates over all normalized spectra and separates their IDs, mass-to-charge ratio, and respective normalized intensities measurements.

```
tumor <- "tumor_processed_spectra.txt"
control <- "control_processed_spectra.txt"
for (i in 1:length(processed_gbm)) {
  list_content <- paste(unlist(processed_gbm[[i]]@
intensity), collapse = ",")
  mass_charge <- paste(unlist(processed_gbm[[i]]@
mass), collapse = ",")


# The first element of the 'line' variable is sample
ID. The correct extraction of sample ID depends on
the file path, therefore it is necessary to manually
evaluate if the string splits are being performed
correctly
```

```
 line <- paste(paste(str_split(str_split(processed_
gbm[[i]]@metaData$file, '\\\\')[[1]][8], "_")[[1]]
[1], list_content, sep = ";"), mass_charge, sep =
";")
  write(line, file = tumor, append = TRUE)
}


for (i in 1:length(processed_ct)) {
  list_content <- paste(unlist(processed_ct[[i]]@
intensity), collapse = ",")
  mass_charge <- paste(unlist(processed_ct[[i]]@
mass), collapse = ",")
  line <- paste(paste(str_split(str_
split(processed_ct[[i]]@metaData$file, '\\\\')[[1]]
[8], "_")[[1]][1], list_content, sep = ";"), mass_
charge, sep = ";")


  write(line, file = control, append = TRUE)
}
```

3.2 Fractals and Entropy Calculation

1. With the files saved, read them in Python to calculate the fractal dimensions and entropy of the spectra.
2. For each spectrum, divide them into windows of fixed size based on their mass-to-charge ratio and separate the intensity datapoints (*see* **Note 3**). This means that for a window size of 1000 m/z, the first window for a spectrum that start at 2000 m/z is going to comprise all data points that fall between 2000 and 3000 m/z (*see* **Note 4**)
3. Then, use the corresponding function from Antropy to calculate the fractals and entropy within each window intensity values. This includes the Katz, Higuchi, Petrosian, and Detrended fluctuation for fractal dimensions, and Permutation Entropy, Approximate, Entropy Spectral Entropy, Singular Value Decomposition Entropy, Sample Entropy and Hjorth complexity/mobility for entropy measurements.
4. First, import all necessary libraries for this analysis.

```
import os
import numpy as np
import pandas as pd
from sklearn.model_selection import StratifiedKFold,
StratifiedShuffleSplit, RandomizedSearchCV,
train_test_split
from sklearn.preprocessing import StandardScaler
from sklearn.neural_network import MLPClassifier
from sklearn.metrics import confusion_matrix, roc_
curve, precision_recall_curve, auc
from imblearn.over_sampling import SMOTE
```

```
from boruta import BorutaPy
from scipy.stats import randint, uniform
import matplotlib.pyplot as plt
import antropy as ant
from collections import defaultdict
```

5. Create a function to read the *txt* files and store the data in numpy arrays. This function will iterate over the *txt* files and read them line by line, retrieving all the spectra information during the process.

```
# Create a function to read the generated txt files
and retrieve the spectra in python
def get_ids_and_values(filename):
    with open(filename, 'r') as file:
        ids = []
        values = []
        masses = []
        for line in file:
            parts = line.strip().split(';')
            if len(parts) == 3:
                id, value_str, mzs = parts
                ids.append(id)
                values.append([float(val) for val in
value_str.split(',')])
                masses.append([float(val) for val in
mzs.split(',')])
        ids_array = np.array(ids)
        values_array = np.array(values, dtype=float)
        masses_array = np.array(masses, dtype=float)
    return ids_array, values_array, masses_array
```

6. Apply the functions to retrieve the data separately for each group. Make sure to use the correct path for the MALDI-TOF data when using them in the function.

```
# Then, use this function to load the data, make
sure to use the correct path to your files
tumor_ids, tumor_int, tumor_mz = get_ids_and_
values("<path>tumor_processed_spectra.txt")
ct_ids, ct_int, ct_mz = get_ids_and_values("<path>
control_processed_spectra.txt")
labels_tumor = np.repeat(1, tumor_ids.shape[0])
labels_control = np.repeat(0, ct_ids.shape[0])
```

7. With the data loaded, create the functions to calculate the fractal and entropy measurements. These functions will also be used to trim the spectra, removing regions that do not contain relevant information. In this step, create a class to hold the functions to calculate the measurements.

8. The Hjorth parameters function from Antropy returns a tuple, containing the Hjorth complexity and mobility, therefore create a function to split the Hjorth column into two separate columns.

```
class FeatureExtractor:
    def __init__(self, functions=None, min_mz =
2000, max_mz=15000, interval_size=500):
        self.functions = functions or [
            ant.petrosian_fd, ant.katz_fd, ant.
higuchi_fd, ant.detrended_fluctuation,
            ant.perm_entropy, ant.spectral_entropy,
ant.svd_entropy,
            ant.app_entropy, ant.sample_entropy,
ant.hjorth_params
        ]


        self.max_mz = max_mz
        self.min_mz = min_mz
        self.interval_size = interval_size


        self.bin_edges = np.arange(self.min_mz,
                                   self.max_mz +
self.interval_size,
                                   self.
interval_size)


    def calculate_metrics_interval(self,
                                   code_ids,
                                   spectra_mass_
list,
                                   spectra_
intensity_list,
                                   group_array):
        """
        Non-overlapping bins from min_mz→min_mz+
interval_size, etc.
        Returns:
          - df: DataFrame of features
          - all_index_maps: list of dicts, one per
spectrum,
                            mapping "<lo>_<hi>" →
(start_idx, end_idx)
        """
        records = []
        all_index_maps = []


        # build list of (label, lo, hi)
        windows = []
```

```
        for lo, hi in zip(self.bin_edges[:-1],
self.bin_edges[1:]):
            label = f"mz_{int(lo)}_{int(hi)}"
            windows.append((label, lo, hi))


        for code, mz, intensity, group in
zip(code_ids,

spectra_mass_list,

spectra_intensity_list,

group_array):


            rec = {'ID': code, 'Group': group}
            index_map = {}


            # digitize once
            bin_idx = np.digitize(mz, self.bin_
edges) - 1


            for w, (label, lo, hi) in
enumerate(windows):
                # find contiguous stretch of indices
in this bin
                idxs = np.where(bin_idx == w)[0]
                if idxs.size:
                    start, end = int(idxs[0]),
int(idxs[-1])
                else:
                    start = end = -1


                index_map[label] = (start, end)
                # pull out the intensity slice
                mask_data = intensity[idxs].
astype(float)


                # compute metrics
                for func in self.functions:
                    name = func.__name__
                    col = f"{name}_{label}"
                    rec[col] = (
                        # Spectral entropy needs an
extra argument 'sf'. We are using default here, but
this could be optimized
                        func(mask_data, sf=100)
                        if (name == "spectral_
entropy" and mask_data.size)
```

```
                        else func(mask_data)
                        if mask_data.size
                        else np.nan
                    )

            records.append(rec)
            all_index_maps.append(index_map)

        df = pd.DataFrame(records)
        return df, all_index_maps

    @staticmethod
    def split_hjorth(df):
        def convert(val):
            try:
                if pd.isna(val):
                    return (np.nan, np.nan)
                val_clean = ''.join(c for c in
str(val) if c in '0123456789.,-eE')
                return tuple(float(x.strip()) for x
in val_clean.split(","))
            except Exception:
                return (np.nan, np.nan)

        hjorth_cols = [c for c in df.columns if
'hjorth_params_mz_' in c]
        new_cols_data = {} # 1. Create a dictionary
to hold new columns

        for col in hjorth_cols:
            interval = col.split('_params_mz_')[1]

            # Unzip the converted values into two
lists: mobility and complexity
            mobility, complexity = zip(*df[col].
map(convert))

            # 2. Add the new data to the dictionary
            new_cols_data[f"hjorth_mobility_
{interval}"] = mobility
            new_cols_data[f"hjorth_complexity_
{interval}"] = complexity

        # 3. Create a new DataFrame from all the
new columns at once
        new_df = pd.DataFrame(new_cols_data)
```

```
        # 4. Drop the old columns and concatenate
the new ones in two steps
        df = df.drop(columns=hjorth_cols)
        df = pd.concat([df, new_df], axis=1)


        return df
```

9. Next, set up the class with a desired window size (*interval_size* variable). The optimal window size for each dataset may vary (*see* **Note 5**). For convenience, the fractals and entropy are calculated in the same function. We will refer to them together as complexity measurements.

```
# Now we set up the function to perform the fractal
calculations
feature_processing = FeatureExtractor(min_mz =
2000, max_mz=15000, interval_size=200)
complexity_tumor, index_mapping_tumor = feature_
processing.calculate_metrics_interval(tumor_ids,
tumor_mz, tumor_int, labels_tumor)
complexity_control, index_mapping_ct = feature_
processing.calculate_metrics_interval(ct_ids, ct_
mz, ct_int, labels_control)
complexity_df_tmp = pd.concat([complexity_tumor,
complexity_control])
# hjorth params function returns a tuple, so it is
necessary to split that column
# into mobility and complexity
complexity_df = FeatureExtractor.
split_hjorth(complexity_df_tmp)
```

10. The function returns the dataframe necessary to train and test the machine learning model, and a dictionary mapping the features to the spectral regions. Each column represents a type of fractal dimension or entropy in a section of the spectrum, and each row is one of the samples. The index mapping variable is not used in this chapter, but it is a useful dictionary that can be used to access the exact datapoints of the spectra for each window generated.

3.3 Machine Learning for Classification

With the fractals calculated, the models can be trained for classification. A single model will be used as an example, but the code provided will follow a structure that allows the testing of multiple models in sequence. The training procedure will be carried out using a nested repeated cross-validation strategy. For that, the outer loop will split the data into train and test set; then, the training data will be used for fitting the scaler, oversampling, and feature reduction using BorutaPy [12]; then, it will be split into train and validation for hyperparameter tuning using random search strategy. The best model will be used to predict the unseen test set. The model performance will be averaged across all folds, and the balanced accuracy will be reported (balanced accuracy will be used due to class imbalance in this dataset). This pipeline will be done for fractals and entropy individually for comparison, and then they will be merged for a final classification.

1. First, create a class to hold the functions and models. In this case, only one model will be tested (multi-layer perceptron). More models can be added by expanding the parameters and models' dictionary. The provided functions of this class automatically handle multiple models.
2. Then, create the function for the model training and place it inside the created class. First, split the dataset randomly into train and test with stratification to ensure the classes maintain their proportions in each fold. In this example, five folds will be used in the outer loop and four folds in the inner loop (hyperparameter tunning).
3. The training set is then used by *SMOTE* to oversample by generating synthetic samples, which sometimes allows the models to perform better in the classification of the minority class.
4. Then, perform feature scaling using the *StandardScaler* function from *sklearn* to ensure that all features contribute fairly to the model; this is done in the training set and then passed to the test set.
5. Finally, apply Boruta to filter the most important features to increase model performance.
6. After feature reduction, optimize the hyperparameters by randomly selecting them and evaluating model performance using the *RandomizedSearchCV* function.
7. Then, select the best-scoring model to be evaluated on the test set. The model is evaluated through its balanced accuracy, sensitivity, and specificity, as well as with the ROC and PR curves. All these steps are done within the following function.

```
class ModelPipeline:
    def __init__(self, param_distributions=None,
models=None):
        self.param_distributions = param_
distributions if param_distributions is not None
else {
            'MLP': {
                'hidden_layer_sizes': [(50,),
(100,), (50, 50), (100, 50)],
                'activation': ['relu', 'tanh',
'logistic'],
                'alpha': uniform(0.0001, 0.01),
                'learning_rate': ['constant',
'adaptive'],
```

```
            }
        }
        self.models = models if models is not None
else {
            'MLP': MLPClassifier(random_state=42,
max_iter=1000)
        }


    def run(self, df, n_splits=5):
        # Initialize results list and dictionary to
store selected features per fold
        results, selected_feats = [], {}


        # Separate features (X) and labels (y)
        X, y = df.drop(columns=["ID", "Group"]),
df["Group"]


        # Stratified K-Fold cross-validation to
maintain label proportions
        skf = StratifiedKFold(n_splits=n_splits,
shuffle=True, random_state=42)


        for fold, (train_idx, test_idx) in
enumerate(skf.split(X, y)):
            fold_name = f"Fold_{fold + 1}"
            print(f"\nStarting {fold_name}")


            # Split the data into training and test
sets
            X_train, y_train = X.iloc[train_idx],
y.iloc[train_idx]
            X_test, y_test = X.iloc[test_idx],
y.iloc[test_idx]


            # Apply SMOTE to balance the training
data
            print(f"SMOTE on training data
(original size: {len(y_train)})")
            X_train, y_train = SMOTE(random_state=
42).fit_resample(X_train, y_train)
            print(f"SMOTE done (resampled size:
{len(y_train)})")


            # Standardize the feature values
            scaler = StandardScaler()
            X_train_scaled = scaler.
fit_transform(X_train)
```

```
            X_test_scaled = scaler.
transform(X_test)


            # Run Boruta for feature selection
using a Random Forest base estimator
            print(f"Running Boruta feature
selection")
            selector = BorutaPy(
                RandomForestClassifier(n_jobs=-1,
class_weight='balanced', max_depth=5),
                n_estimators='auto',
                random_state=42,
                alpha=0.001
            )
            selector.fit(X_train_scaled, y_train)


            # Store selected features for this fold
            selected = X.columns[selector.support_]
            selected_feats[fold_name] = selected.
tolist()
            print(f"Boruta selected {len(selected)}
features")


            # Apply the feature selector to the
scaled training and test sets
            X_train_sel = selector.
transform(X_train_scaled)
            X_test_sel = selector.
transform(X_test_scaled)


            # Loop through each model to tune and
evaluate
            for name, model in self.models.items():
                print(f"Tuning model '{name}' with
RandomizedSearchCV")


                # Perform randomized hyperparameter
search with 4-fold CV
                search = RandomizedSearchCV(
                    model, self.param_
distributions[name], n_iter=20, cv=4, random_state=
42
                )
                search.fit(X_train_sel, y_train)


                # Get the best model and parameters
                best_model = search.best_estimator_
```

```
                print(f"Best params for '{name}':
{search.best_params_}")


                # Make predictions and get
prediction probabilities
                y_pred = best_model.
predict(X_test_sel)
                y_prob = best_model.predict_
proba(X_test_sel)[:, 1]


                # Compute performance metrics
                fpr, tpr, _ = roc_curve(y_test,
y_prob)
                prec, rec, _ = precision_recall_
curve(y_test, y_prob)
                cm = confusion_matrix(y_test,
y_pred)


                print(f"Evaluation complete for
model '{name}' on {fold_name}")
                # Store all relevant results
                results.append({
                    "fold": fold_name,
                    "model": name,
                    "best_params": search.
best_params_,
                    "confusion_matrix": cm,
                    "roc_curve": (fpr, tpr,
auc(fpr, tpr)),
                    "pr_curve": (rec, prec,
auc(rec, prec))
                })


        print("\nAll folds completed.")
        return results, selected_feats
```

8. The function returns a dictionary containing the model metrics across all folds. To visualize the training results, create a function to iterate over the folds and calculate the average performance of the model. Then use the average metrics to generate a barplot with the mean balanced accuracy, sensitivity, and specificity, and the ROC and PR curves.

```
class PlotData:
    def __init__(self, results):
        """
        Initialize with a list of result dicts.
Each dict should have keys:
```

```
        'model', 'confusion_matrix', 'roc_curve',
'pr_curve'
        """
        self.results = results
        self.metrics = defaultdict(lambda: {
            'conf_mats': [], 'sens': [], 'spec':
[], 'bal_acc': [],
            'roc': [], 'pr': []
        })
        self.avg_metrics = {}
        self._aggregate()
        self._compute_averages()


    def _aggregate(self):
        """Collect per-fold metrics for each
model."""
        for entry in self.results:
            model = entry['model']
            cm = entry['confusion_matrix']
            tn, fp, fn, tp = cm.ravel()
            sens = tp / (tp + fn)
            spec = tn / (tn + fp)
            bal = (sens + spec) / 2


            self.metrics[model]['conf_mats'].
append(cm)
            self.metrics[model]['sens'].
append(sens)
            self.metrics[model]['spec'].
append(spec)
            self.metrics[model]['bal_acc'].
append(bal)


            fpr, tpr, _ = entry['roc_curve']
            self.metrics[model]['roc'].append((fpr,
tpr))
            rec, prec, _ = entry['pr_curve']
            self.metrics[model]['pr'].append((prec,
rec))


    def _compute_averages(self):
        """Compute average confusion matrix and
metrics per model."""
        for model, vals in self.metrics.items():
            avg_cm = np.mean(vals['conf_mats'],
axis=0)
            avg_sens = np.mean(vals['sens'])
            avg_spec = np.mean(vals['spec'])
            avg_bal = np.mean(vals['bal_acc'])
            self.avg_metrics[model] = {
```

```
                'cm': avg_cm,
                'sens': avg_sens,
                'spec': avg_spec,
                'bal_acc': avg_bal
            }


    def plot_summary(self):
        """Plot barplot of avg sensitivity,
specificity, and balanced accuracy for all
models."""
        labels = list(self.avg_metrics.keys())
        sens_vals = [self.avg_metrics[m]['sens']
for m in labels]
        spec_vals = [self.avg_metrics[m]['spec']
for m in labels]
        bal_vals = [self.avg_metrics[m]['bal_acc']
for m in labels]


        x = np.arange(len(labels))
        width = 0.2
        cmap = plt.get_cmap('OrRd')
        colors = [cmap(0.7), cmap(0.5), cmap(0.3)]


        fig, ax = plt.subplots()
        bars1 = ax.bar(x - width, sens_vals, width,
label='Sensitivity', color=colors[0])
        bars2 = ax.bar(x, spec_vals, width, label=
'Specificity', color=colors[1])
        bars3 = ax.bar(x + width, bal_vals, width,
label='Balanced Acc', color=colors[2])


        # Annotate values
        for bars in [bars1, bars2, bars3]:
            for bar in bars:
                height = bar.get_height()
                ax.text(bar.get_x() + bar.get_
width()/2, height,
                        f'{height:.2f}', ha=
'center', va='bottom')


        ax.set_xticks(x)
        ax.set_xticklabels(labels)
        ax.set_ylabel('Score')
        ax.set_title('Average Metrics by Model')
        ax.legend(loc='upper left', bbox_to_anchor=
(1, 1))
        plt.tight_layout()
        plt.show()
```

```
    def _plot_with_std(self, curves, xlabel,
ylabel, title, auc_text, auc_loc):
        """Helper to plot mean curve with shaded ±1
std deviation and AUC text."""
        grid = np.linspace(0, 1, 100)
        interp_vals = []
        aucs = []
        for x_vals, y_vals in curves:
            interp = np.interp(grid, x_vals,
y_vals)
            interp_vals.append(interp)
            aucs.append(np.trapz(y_vals, x_vals))
        mean = np.mean(interp_vals, axis=0)
        std = np.std(interp_vals, axis=0)
        mean_auc = np.mean(aucs)
        std_auc = np.std(aucs)


        cmap = plt.get_cmap('OrRd')
        line_color = cmap(0.6)


        plt.fill_between(grid, mean - std, mean +
std, alpha=0.2, color=line_color)
        plt.plot(grid, mean, color=line_color)
        plt.xlabel(xlabel)
        plt.ylabel(ylabel)
        plt.title(title)
        # Place AUC text
        plt.text(auc_loc[0], auc_loc[1], f'AUC =
{mean_auc:.2f} ± {std_auc:.2f}',
                ha='right' if auc_loc[0] > 0.5 else
'left',
                va='bottom', transform=plt.gca().
transAxes)


    def plot_roc_pr(self):
        """For each model, plot ROC and PR curves
side by side with AUC annotations."""
        for model, vals in self.metrics.items():
            plt.figure(figsize=(12, 5))
            plt.subplot(1, 2, 1)
            self._plot_with_std(
                vals['roc'], 'False Positive Rate',
'True Positive Rate',
                f'ROC Curve: {model}', auc_text=
None, auc_loc=(0.95, 0.05)
            )
            plt.subplot(1, 2, 2)
            self._plot_with_std(
                vals['pr'], 'Recall', 'Precision',
                f'PR Curve: {model}', auc_text=
None, auc_loc=(0.05, 0.05)
```

```
)
plt.tight_layout()
plt.show()
```

3.3.1 Glioblastoma Classification Using Entropy

1. Separate the entropy and fractal features from the total dataframe.
2. Then use the entropy dataframe to train and test the model.
3. Use the resulting variable to plot the results (Fig. 1).

```
keywords = ['perm_entropy', 'spectral_entropy',
'svd_entropy',
            'app_entropy', 'sample_entropy',
'hjorth_mobility', 'hjorth_complexity']
matching_columns = [col for col in complexity_
df.columns if any(keyword in col for keyword in
keywords)]
etp_df = complexity_df[matching_columns +
[col for col in ['Group', 'ID'] if col not in
matching_columns]]
fd_df = complexity_df.drop(columns=matching_
columns).assign(Group=complexity_df['Group'], ID=
complexity_df['ID'])


model_trainer = ModelPipeline()
result_ml_etp, boruta_features_etp = model_trainer.
run(etp_df)
plot_data_etp = PlotData(result_ml_etp)
plot_data_etp.plot_summary('etp_summary')
plot_data_etp.plot_roc_pr('etp_curves')
```

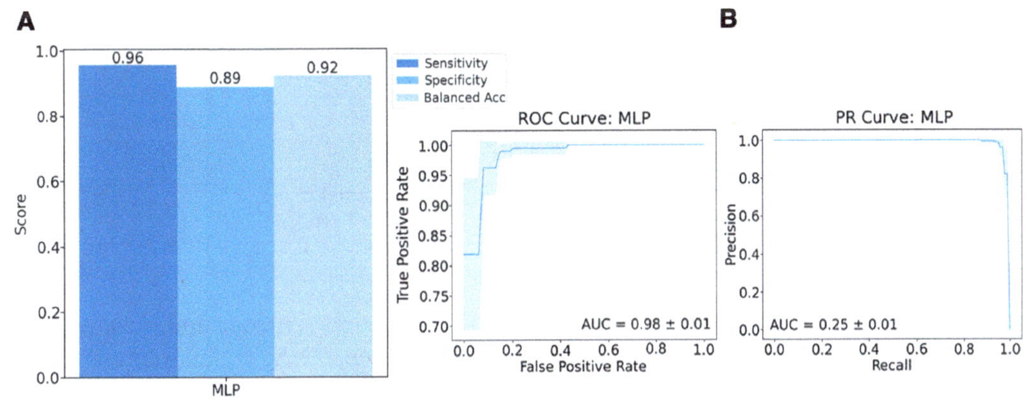

Fig. 1 (**A**) Average performance metrics for training using the entropy dataset. (**B**) Average ROC and PR curves with the standard deviation as a shaded area

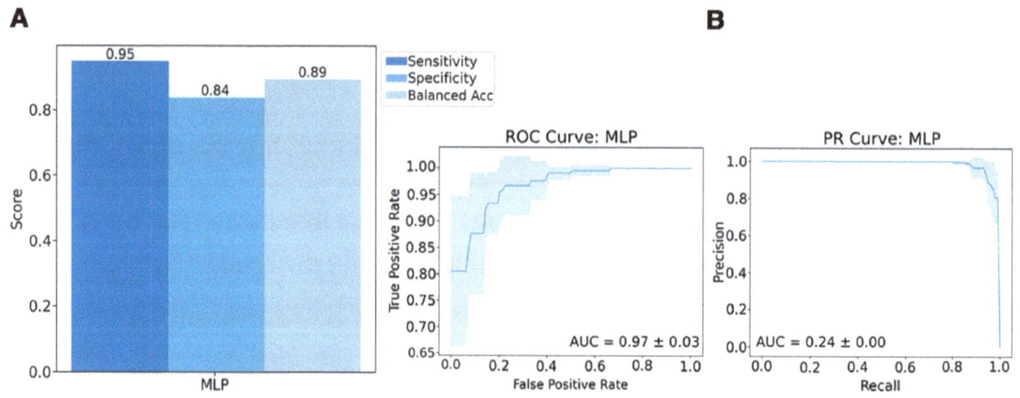

Fig. 2 (**A**) Average performance metrics for the training using the fractals dataset. (**B**) Average ROC and PR curves with the standard deviation as a shaded area

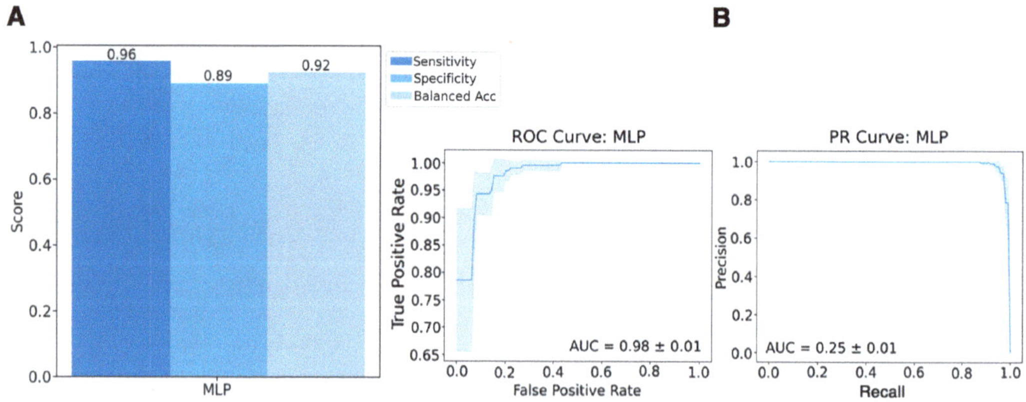

Fig. 3 (**A**) Average performance metrics for the training using the merged dataset. (**B**) Average ROC and PR curves with the standard deviation as a shaded area

3.3.2 Classification Using Fractal Dimensions

1. Similarly, apply the functions on the fractals dataframe and plot the model performance using the dictionary results from the machine learning function (Fig. 2).

```
result_ml_fd, boruta_features_fd = model_trainer.
run(fd_df)
plot_data_fd = PlotData(result_ml_fd)
plot_data_fd.plot_summary('fd_summary')
plot_data_fd.plot_roc_pr('fd_curves')
```

3.3.3 Classification Using the Complete Dataset

1. To train the model with all features (entropy and fractals), simply use the functions in the *complexity_df* dataframe and plot the results afterwards (Fig. 3).

```
result_ml_merge, boruta_features_merge = model_
trainer.run(complexity_df)
```

```
plot_data_merge = PlotData(result_ml_merge)
plot_data_merge.plot_summary('merge_summary')
plot_data_merge.plot_roc_pr('merge_curves')
```

Through this comparison it is possible to observe that fractals and entropy perform differently in this dataset with these defined parameters, with the entropy scoring higher and the merge not impacting the model's performance (*see***Note 6**).

4 Notes

1. Since the scope of this chapter is to focus on fractal dimensions calculation and not in the optimization of preprocessing techniques for MALD-TOF MS data, we will conduct the initial processing using parameters that were already optimized previously [8]. For a comprehensive understanding of the parameters and methods for normalization, transformation and base-correction, refer to the MALDIquant documentation.
2. If the MALDI-TOF MS files are available in their raw format, it is necessary to convert them into mzML. This can be done using the MSConvert application available at the ProteoWizard suite [13].
3. It is also possible to use the whole spectrum to calculate the fractal dimensions. However, it was demonstrated that the accuracy of models trained this way was significantly lower [7].
4. A fixed window size of 1000 datapoints, rather than 1000m/z, can result in misaligned window content across different samples. This discrepancy happens because variations in measurement precision—particularly in samples acquired with different equipment—lead to an inconsistent number of measurements within a given window. For example, in the m/z range of 2000–3000, one sample might contain 2000 datapoints while another has 2500.
5. There is no defined rule of window size to maximize model performance. It is advisable to run several values for these parameters and evaluate which one yields the best scoring models. Consider also testing different types of models. If the dataset is large and computation time is high, consider using a subset of your samples to evaluate these parameters.
6. The analysis of fractals and entropy in MALDI-TOF MS data is still in its early stages. As such, there are no well-established parameters that are considered optimal for this type of analysis. The fact that entropy performed better may be specific to the dataset or due to the chosen parameters—such as window size—favoring entropy over fractals. It is advisable to test a wide range of parameters, ideally evaluating fractals and entropy separately.

Acknowledgments

This work was supported by the São Paulo Research Foundation (FAPESP, grants 2018/15549-1 and 2020/04923-0 to G.P., and 2025/00049-7 to L.C.L.), the Brazilian National Council for Scientific and Technological Development (CNPq, grants 307854/2018-3 and 317353/2021-7 to G.P.), and the Coordination for the Improvement of Higher Education Personnel (CAPES, PIPD 88887.111997/2025-00 to L.C.L.). The funding agencies had no involvement in the study design, data collection and analysis, decision to publish, or manuscript preparation.

References

1. Patel R (2015) MALDI-TOF MS for the diagnosis of infectious diseases. Clin Chem 61:100–111. https://doi.org/10.1373/clinchem.2014.221770
2. Merlos Rodrigo MA, Zitka O, Krizkova S, Moulick A, Adam V, Kizek R (2014) MALDI-TOF MS as evolving cancer diagnostic tool: a review. J Pharm Biomed Anal 95:245–255. https://doi.org/10.1016/j.jpba.2014.03.007
3. Weis C, Cuénod A, Rieck B, Dubuis O, Graf S, Lang C, Oberle M, Brackmann M, Søgaard KK, Osthoff M, Borgwardt K, Egli A (2022) Direct antimicrobial resistance prediction from clinical MALDI-TOF mass spectra using machine learning. Nat Med 28:164–174. https://doi.org/10.1038/s41591-021-01619-9
4. Lazari LC, de Rose Ghilardi F, Rosa-Fernandes L, Assis DM, Nicolau JC, Santiago VF, Dalçóquio TF, Angeli CB, Bertolin AJ, Marinho CRF, Wrenger C, Durigon EL, Siciliano RF, Palmisano G (2021) Prognostic accuracy of MALDI-TOF mass spectrometric analysis of plasma in COVID-19. Life Sci Alliance 4:e202000946. https://doi.org/10.26508/LSA.202000946
5. Lazari LC, Zerbinati RM, Rosa-Fernandes L, Santiago VF, Rosa KF, Angeli CB, Schwab G, Palmieri M, Sarmento DJS, Marinho CRF, Almeida JD, To K, Giannecchini S, Wrenger C, Sabino EC, Martinho H, Lindoso JAL, Durigon EL, Braz-Silva PH, Palmisano G (2022) MALDI-TOF mass spectrometry of saliva samples as a prognostic tool for COVID-19. J Oral Microbiol 14:2043651. https://doi.org/10.1080/20002297.2022.2043651
6. Nachtigall FM, Pereira A, Trofymchuk OS, Santos LS (2020) Detection of SARS-CoV-2 in nasal swabs using MALDI-MS. Nat Biotechnol 38:1168–1173. https://doi.org/10.1038/s41587-020-0644-7
7. Lazari LC, Azemi G, Russo C, Rosa-fernandes L, Suely KN, Di Ieva A, Palmisano G (2025) Fractal and machine learning analyses of MALDI-TOF mass spectrometry data in glioblastoma. bioRxiv at https://doi.org/10.1101/2025.03.03.641183
8. Lazari LC, Silva JM, Donado PRS, Shinjo SMO, Rosa-Fernandes L, Di Ieva, A, Marie SKN, Palmisano G (2025) Identification of astrocytomas through serum protein fingerprint using MALDI-TOF MS and machine learning. bioRxiv at https://doi.org/10.1101/2025.02.21.639567
9. Gibb S (2015) MALDIquantForeign: import/export routines for MALDIquant
10. Gibb S, Strimmer K (2012) Maldiquant: a versatile R package for the analysis of mass spectrometry data. Bioinformatics 28:2270–2271. https://doi.org/10.1093/bioinformatics/bts447
11. Vallat R (2023) AntroPy: entropy and complexity of (EEG) time-series in Python. https://github.com/raphaelvallat/antropy
12. Kursa MB, Rudnicki WR (2010) Feature selection with the boruta package. J Stat Softw 36:1–13. https://doi.org/10.18637/jss.v036.i11
13. Matthew CC, Brendan M, Robert B, Daria A, Daniel LR, Steffen N, Laurent G, Bernd F, Brian P, Jarrett E, Katherine H, Darren K, Natalie T, Nicholas S, Barbara F, Tahmina AB, Mi-youn B, Christopher P, David C, Lisa F, Parag M (2012) A cross-platform toolkit for mass spectrometry and proteomics. Nat Biotechnol 30:1–8. https://doi.org/10.1038/nbt.2377.A

Chapter 4

An Introduction to Structural Phylogenetics for Biomedical Sciences

Caroline Puente-Lelievre, Nicholas Matzke and Jordan Douglas

Abstract

Structural phylogenetics offers a powerful framework for understanding the molecular basis of neural function and disease by combining evolutionary analysis with three-dimensional protein structure. While traditional protein-based phylogenetics relies mostly on amino acid sequences, recent advances in computational structure prediction have accelerated the integration of protein tertiary structures into phylogenetic studies. Since structure is more conserved than sequence, this approach may provide increased resolution in cases of high sequence divergence or mutational saturation, common issues in the study of complex, ancient, or fast-evolving protein families. In biomedical research, phylogenetics is a powerful approach for tracing the evolutionary origins and diversification of disease-related proteins. It provides a comparative framework to identify conserved structural motifs, function shifts, and predict potential targets for drug discovery and innovation. In neuroscience, these insights are especially valuable for uncovering the evolutionary basis of neural signaling, plasticity, and dysfunction. This chapter serves as a user-friendly practical tutorial for conducting structural phylogenetic analyses using accessible and computationally efficient tools, using synaptotagmins—calcium-sensing proteins essential for synaptic vesicle fusion—as a running example. This resource is designed for biomedical researchers seeking to integrate phylogenetics as a tool to build a comparative basis to strengthen their experimental design and improve their understanding of protein diversity and function.

Keywords Phylogenetics, Structural phylogenetics, Neuroscience, Biomedicine, Tutorial, Synthetic Biology, Synaptotagmins, 3Di, Protein, Homology search

1 Introduction

Protein phylogenetics has an important role to play in improving our understanding of biomedicine, including neuroscience. This approach has the potential to improve the selection and interpretation of model organisms [1], deepen insights into underlying molecular mechanisms [2], function prediction [3], and may even potentially reveal drug targets.

Historically, protein phylogenetics has relied exclusively on amino acid sequence alignments. However, the recent breakthrough driven

Antonio Di Ieva and Jennilee M. Davidson (eds.), *Computational Neuroscience*, Methods in Molecular Biology, vol. 3062, https://doi.org/10.1007/978-1-0716-5440-8_4,

by artificial intelligence (AI), high-accuracy structure prediction, including AlphaFold [4] and ESMFold [5], has led to the adoption and further development of structural phylogenetics [6, 7]. Because protein tertiary structures are often far more conserved than their sequences [8–10], structural comparisons can rescue evolutionary signal in deeply divergent or highly variable gene families, which can hinder sequence-based analyses in biomedical research.

To successfully implement phylogenetics as a tool in biomedical research, evolutionary processes ought to be considered for data sampling strategies and phylogeny interpretation. Even when the evolutionary relationships between species are well-understood, individual gene trees can still come with surprises, revealing their own evolutionary histories with direct relevance to human health and disease. **Orthologous** genes, those that diverged by speciation, tend to preserve ancestral biochemical functions, whereas **paralogs**, genes that arise from duplication, often acquire novel or tissue-specific roles [11]. Moreover, multidomain proteins often evolve by shuffling autonomously-folding modules (i.e., **domains**) that can recombine across genes and create new functional assemblies [12–15]. These processes and more [16, 17] contribute to a complex relationship between the histories of genes and species.

This is an introductory pipeline for performing phylogenetic analyses incorporating three-dimensional structure information with simple, fast, and computationally inexpensive bioinformatic tools. This approach utilizes **3Di characters**, a 20-state alphabet that represents protein structures in a linear format [18], but there are other ways of capturing protein structure into phylogenetic analyses [19]. In this tutorial, we will use a dataset of C_2 domains found in the synaptotagmins; a family of calcium (Ca^{2+}) sensor proteins that regulate membrane fusion events, especially during neurotransmitter release in the nervous system [20], and Rabphilin-3, a cytosolic neuronal protein involved in the neurotransmitter release cycle [21]. These two proteins contain two or more copies of this domain, alluding to the complex evolutionary history of this fold [22]. This pipeline has been conceived for beginner users with a background in biomedical research and an interest in incorporating phylogenetic tools in their research (Fig. 1).

2 Materials

> Data repository
>
> To follow the protocol with our running example—the synapotagmin C_2 domain— download the data files and scripts from the Zenodo repository [23]

at: https://doi.org/10.5281/zenodo.16415858

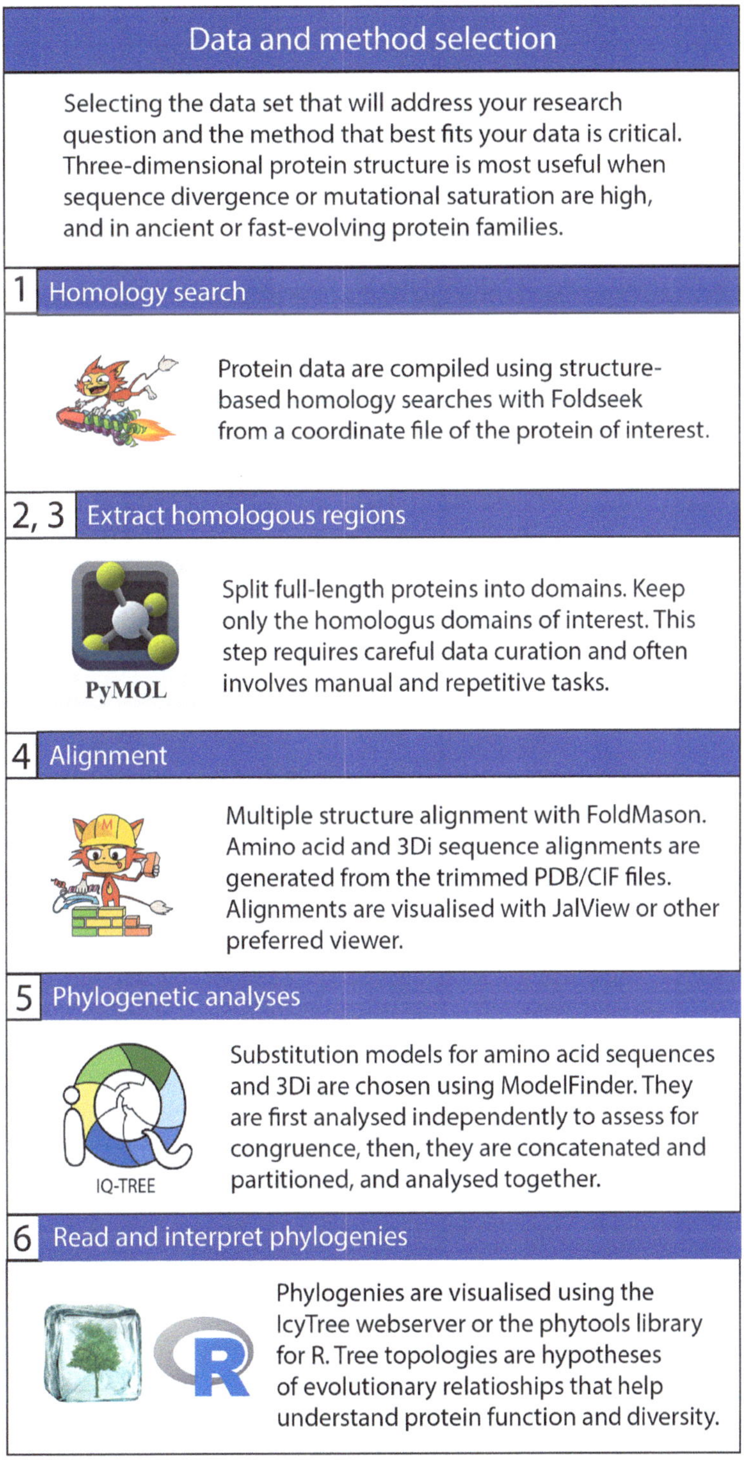

Fig. 1 Workflow for 3Di-based structural phylogenetics

Software

Download and install the following programs. This tutorial is intended for Mac and Linux operating systems. If you are using Windows, it is recommended to perform the steps below using the Ubuntu virtual environment application freely available on the Mircosoft store.

2.1 To View and Edit Protein Structures

- PyMOL: one of the most widely used programs in structural biology for exploring and editing protein coordinate files [24]. PyMOL can be installed from https://www.pymol.org/. PyMOL v2.3.0 was used in this tutorial.

2.2 To Build Alignments

- FoldMason: a scalable tool for building protein 3D structure alignments [25]. It first converts PDB/mmCIF files into a structural database, then aligns them by representing local backbone geometry with a 3Di alphabet and running optimised sequence-search algorithms to produce both amino acid and structure-based alignments. This simple and fast command-line program can be downloaded from https://github.com/steineggerlab/foldmason. Fold-Mason v3-954d202 was used.
- jq: a useful command for extracting variables from JSON files within the terminal environment. The library can be installed following the instructions at https://jqlang.org/download/.
- R: we will use a simple R script to trim the alignment down to the region that overlaps the query structure.

2.3 Phylogenetic Inference

- IQ-TREE: This command-line program takes a multiple sequence alignment as input, selects the best-fitting statistical model of sequence evolution (unless provided by the user), and then uses a fast algorithm to infer a maximum likelihood phylogenetic tree (https://IQ-TREE.github.io/doc/Quickstart). In a single run, you get a ready-to-view tree file, evolutionary model details, and support values, making IQ-TREE a quick and user-friendly way to infer reliable evolutionary relationships. IQ-TREE v3.0.1 was used in this tutorial.

2.4 Visualization Tools

- JalView: to visualize alignments using the 3Di color palette https://www.jalview.org/.
- IcyTree: an open-source web-based application for visualizing and interacting with phylogenetic trees (https://icytree.org/).
- R: we will also R to generate tree figures.

3 Methods

3.1 Obtain Structural Data

Input files

query/Homo sapiens 5H4Y.pdb

Output files
40 coordinate files in the structures/directory, detailed further in accessions/accessions.tsv

The first procedure involves compiling a collection of protein structure coordinate files (CIF or PDB format). There are many ways to go about this initial step, depending on the research question of interest, as well as the availability of existing structural models on both the Protein Databank[a] (PDB) and the AlphaFold structural prediction database[b]. Therefore, for the purposes of this phylogenetic tutorial, we will provide just one simple and straightforward approach for obtaining this dataset; and we refer the reader to previous studies that have employed various approaches for compiling structural data for phylogenetics [26–28].

Our running example will be a survey of homologs of the C_2 domains, found in synaptotagmins [22], protein kinase C [29], and various other proteins [30]. We will begin our search using the C_2A domain of Synaptotagmin V (PDB 5H4Y [31]) as a query structure; confining our search to model organisms (*see* **Notes 1 and 2**).

1. Launch Foldseek [18] in the web browser https://search.foldseek.com/search (*see* **Note 3**).
2. Upload the query structure file Homo sapiens 5H4Y.pdb.
3. Expand the Databases & search settings section, and deselect all databases except for AlphaFold/UniProt50 v4.
4. Enter *Homo sapiens* under Taxonomic filter (Fig. 2).
5. Press Search to find homologs.
6. Observe the similar structures identified (Fig. 3). These are sorted in decreasing order of E-value; which describes the expected number of hits with a score greater than or equal to the one observed from a database of this size by chance alone.
7. Individual structural models can be inspected and their coordinate files can be downloaded by following the links to the databases under Target, and all of the hits can be downloaded as an m8 file from the download button in the top left corner.
8. We do not expect the reader to reproduce the following three steps, but for the purposes of this tutorial, we arbitrarily selected seven hits with an E-value less than 10–7, as well as the hit with the smallest E-value.
9. We repeated Steps 4–8 for all of our model organisms—*Mus musculus*, *Danio rerio*, *Drosophila melanogaster*, and

[a] https://www.rcsb.org/contains under 250,000 structures as of July 2025.
[b] https://alphafold.ebi.ac.uk/contains over 200 million structures as of July 2025

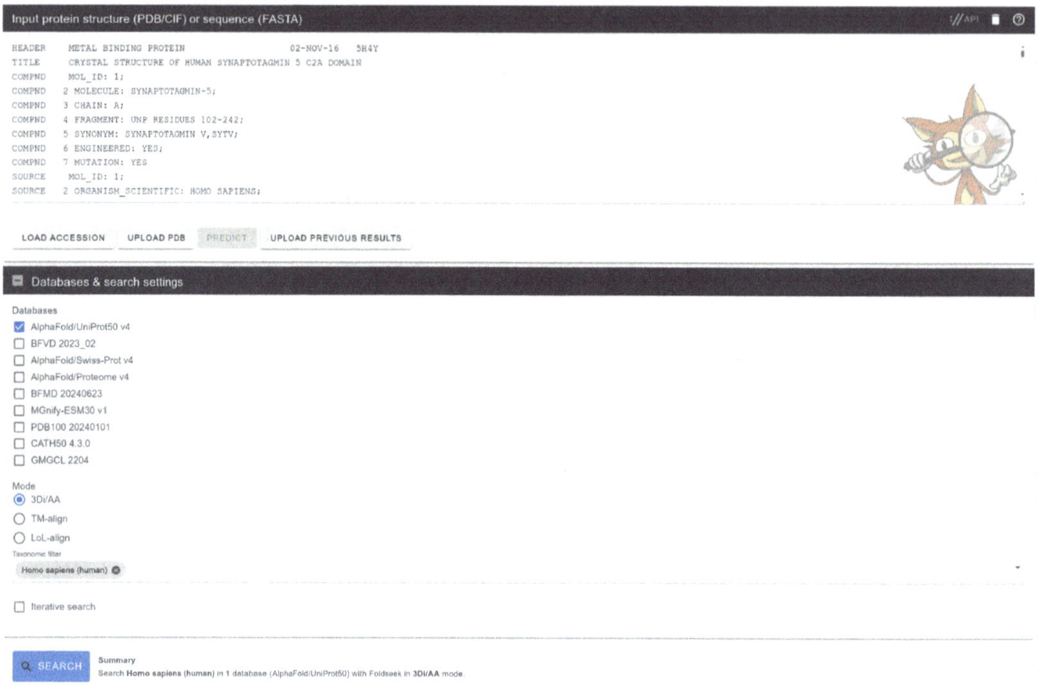

Fig. 2 Searching for structural homologs using the Foldseek web-server

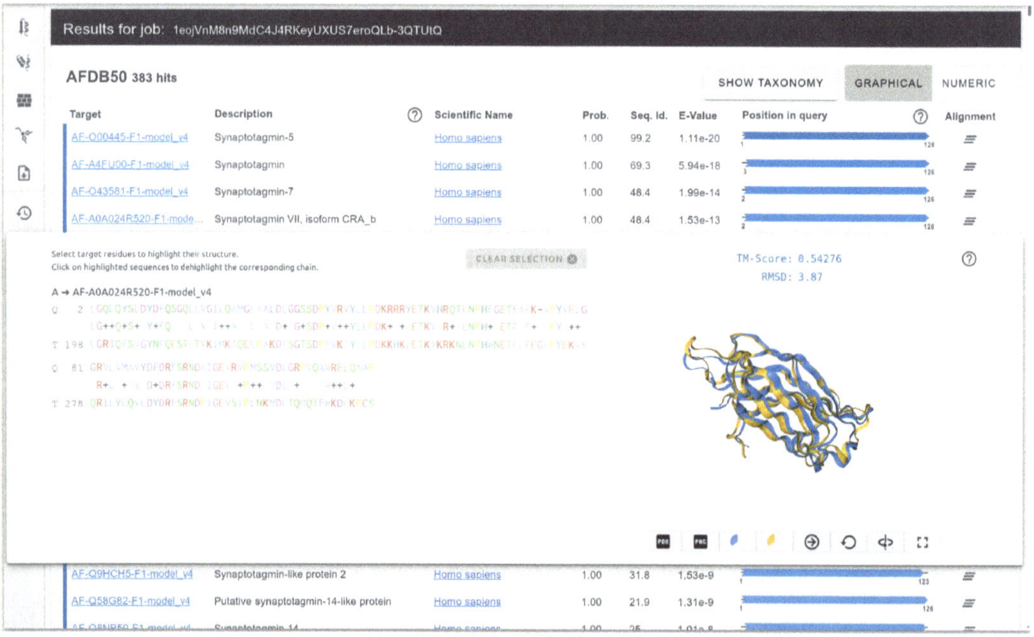

Fig. 3 List of Foldseek hits identified from our query structure, sorted in decreasing order of E-value. Each hit was identified in *Homo sapiens* protein models in the AlphaFold/UniProt50 v4 database (AFDB50). Structural pairwise alignments can be identified by pressing the icon in the Alignment column

Caenorhabditis elegans; downloading eight structural models in each search (*see* **Note 4**).

10. The 40 protein coordinate files (CIF format) can be found in the structures/ directory, while their accession numbers and functional annotations are available at accessions/accessions.tsv.

3.2 Truncate the Proteins into Domains

Input files
40 coordinate files in the structures/ directory

Output files
70 coordinate files in the split/ directory

The synaptotagmins, like many other proteins, evolve at the levels of their domains. Many of our hits in fact are proteins with at least two C_2 domain homologs, including various synaptotagmins, which have a C_2A domain at the N-terminus and C_2B at the C-terminus [22]. Therefore, we will split our full-length protein coordinate files into smaller files that contain atmost just one C_2 domain. Data cleaning and processing requires care, and sometimes involves manual and repetitive tasks that are not easy to automate. As the Foldseek webserver currently does not enable more than one hit to be identified in any given protein structure, we will identify the domains manually.

1. Open *Caenorhabditis elegans* AF-G4RXP2-F1.cif in PyMOL [24].
2. We will recolor the structure using the rainbow spectrum, such that the N-terminus is blue and the C-terminus is red. To do this with the mouse, first select the C button under All on the Names Panel on the top right side of PyMOL. Then, select spectrum and rainbow (Fig. 4). Alternatively, you can type the following into the PyMOL command line:

```
spectrum count, rainbow, all;
bg_color white; # Recolour background to white
(optional)
```

3. Observe that our β-strand-rich C_2 domain homolog appears two times in this protein structure. Hold down and drag the left mouse button to rotate, and the right button to zoom.
4. Ensure the sequence is visible. To do this, open the Display menu and select Sequence.
5. Highlight one of the two domains by dragging across the amino acid sequence (Fig. 5). This is somewhat of a judgement

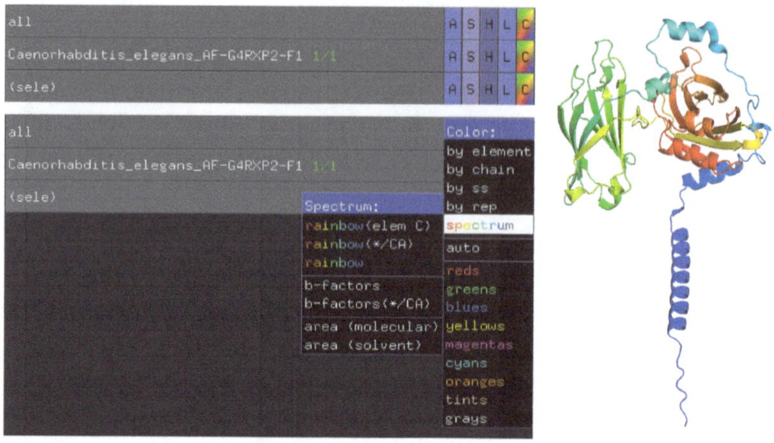

Fig. 4 Using PyMOL, select the C option (Color) under All, and then select rainbow under spectrum to recolor the protein by position. Shown on the right is the ribbon representation of AlphaFold model AF-G4RXP2-F1, from *Caenorhabditis elegans*. Visual inspection suggests that this structure has two domains that are homologous to the query structure 5H4Y; one on the left hand side in green/yellow and the other on the right in red/orange

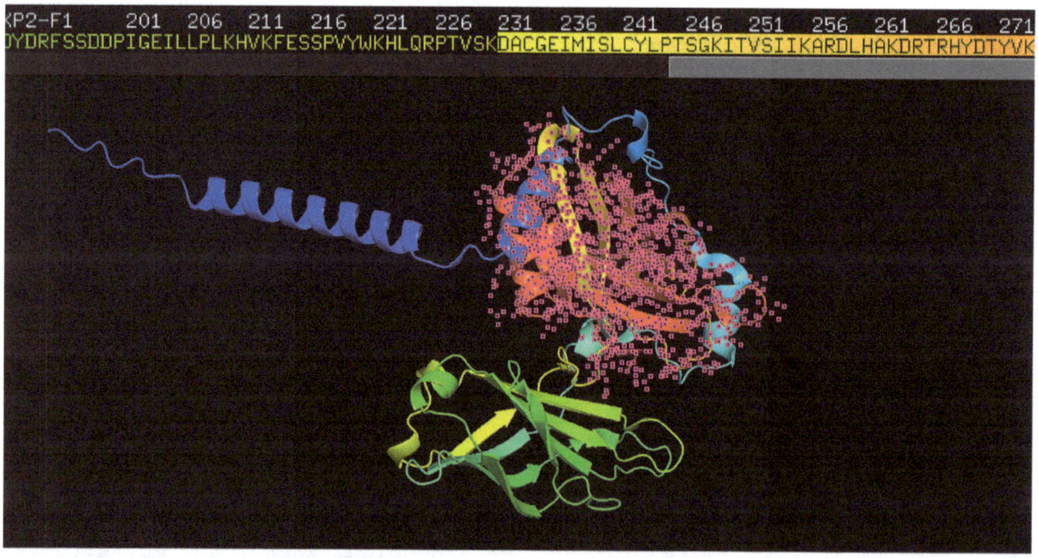

Fig. 5 Highlight one of the two domains using PyMOL

call as to where the domain boundary ends. In this case, we will set the boundary in the flexible linker between the two domains at around position 231.

6. Extract this selected region into a new object by clicking on A under the (sele) section in the Name Panel on the right, and then click on extract object. Alternatively, use the following

command to extract to a new object called p2. `extract p2, sele`

7. Observe that our protein has been split into two different objects. Color the two chains independently by clicking on the C button in the Names Panel, under the *Caenorhabditis elegans* AF-G4RXP2-F1 and p2 rows, respectively, and then picking a color (Fig. 6).
8. Extract these two truncated structures into respective files by clicking File, Export molecule, Multi-file, then one file per object, and finally Save. Save each file into the split folder with a suitable name (Fig. 6). Here, we will use append a suffix of -P1 and -P2 onto the base file name to denote there are multiple truncations of this protein (i.e., *Caenorhabditis elegans* AF-G4RXP2-F1-P1.cif and *Caenorhabditis elegans* AF-G4RXP2-F1-P2.cif).
9. Optional: superimpose these two structures onto one another using the following command. Observe how similar their three dimensional structures are (Fig. 7).super p1, Caenorhabditis_elegans_AF-G4RXP2-F1
10. Repeat Steps 1–9 on the remaining protein structures. When doing so, look for any domains that appear to be structurally similar to the query structure. It is okay if there are some false positives, as they will be tested more robustly in the next procedure. If there does not appear to be more than one homolog, then copy the full protein structure into the split

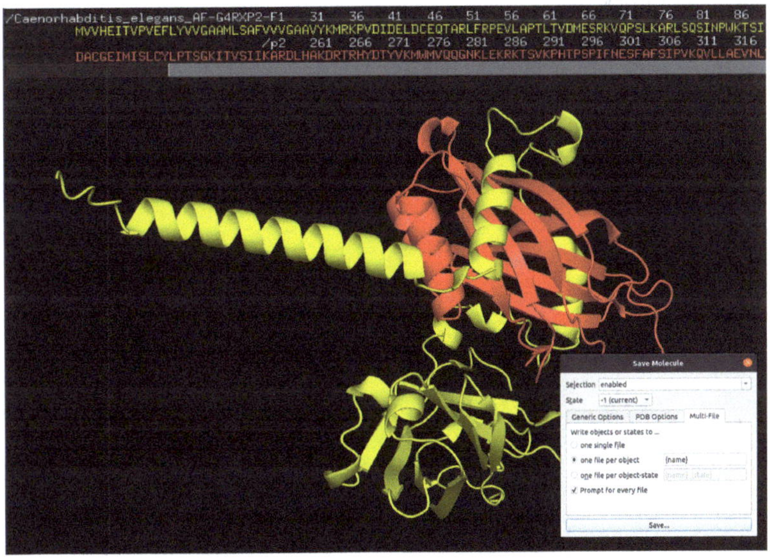

Fig. 6 The protein has been split into two domains, each of which appear to be homologous with our query structure. They can be exported as separate coordinate files by clicking File and then Export molecule

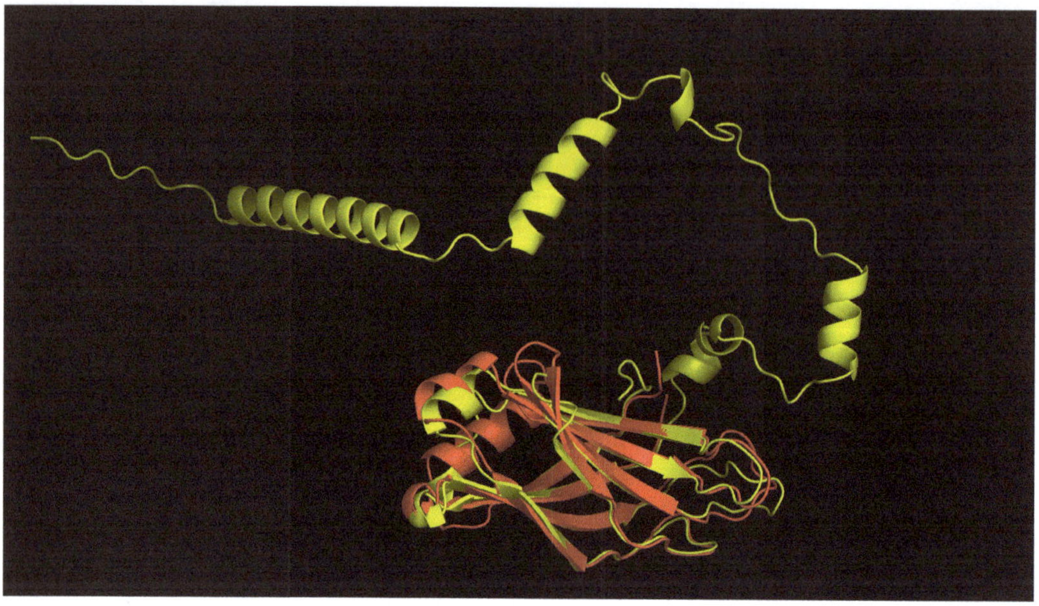

Fig. 7 The two domain truncations have been superimposed onto each other. Observe how similar their structures are

folder. We do not expect the reader to repeat this on every protein structure, noting that the precooked coordinate files are already available in this folder.

3.3 Filter Out Non-Homologous Domains

Input files
70 coordinate files in the split/ directory

Output files
51 coordinate files in the splitfilter/ directory

Next, we will determine robustly which of the truncated segments in the split directory are likely to be homologous with our query structure. We will do this by aligning each of these coordinate files with the query one at a time, using FoldMason [25]. Each pairwise alignment will be evaluated using the lDDT score [32], which is a number between 0 and 1 where more similar structures have higher scores. The low scoring target proteins will be filtered out for the rest of the tutorial.

In the following steps, we will walk through this process looking at just one of the protein structures. Skip to the end to do the entire filtering step with just one command.

1. Open the command line terminal and navigate to the Zenodo repository.
2. We will align our query structure with the newly generated truncated structure *Caenorhabditis elegans* AF-G4RXP2-F1-P1.cif, and save the results in two temporary folders tmp1 and tmp2. To initialize this, enter the following commands into the terminal:

```
QUERY="query/Homo_sapiens_5H4Y.pdb" TARGET="split/
Caenorhabditis_elegans_AF-G4RXP2-F1-P1.cif" mkdir
-p tmp1
mkdir -p tmp2
cp $QUERY tmp1/. cp $TARGET tmp1/.
```

3. Align these two structures using FoldMason. We will set --report-mode to 2 to ensure that the alignment score is saved to a JSON file, and we will set –pair-threshold to 1.0 so that the score will only be calculated using positions in the two structures that are aligned with one another.

```
foldmason easy-msa --report-mode 2 --pair-threshold
1.0 tmp1 tmp2/results tmp2/tmp
```

4. The pairwise alignments have been saved as tmp2/results aa.fa (amino acid sequence) and tmp2/results 3di.fa (3Di structure sequence). These can be viewed using a multiple sequence alignment visualization program, such as JalView [33] or AliView [34].
5. Extract the alignment score from tmp2/results.json using the command below. The lDDT score is 0.932626, which is above our threshold of 0.85. This provides confidence that our target structure is most likely homologous to the query.

```
MSASCORE=$(jq.statistics.msaLDDT tmp2/results.json)
echo $MSASCORE
```

6. Clean up the two temporary directories. We won't be using them anymore.

```
rm -r
tmp1 rm
-r tmp2
```

7. Run the script below to align every target structure with the query, one at a time. Any structures that have an lDDT score greater than 0.85 will be copied to the folder called splitfilter. If this threshold is lowered, there may be some false positives; and if raised, there will be fewer structures included later downstream. Advanced readers might want to tweak this

threshold and explore its effects on the subsequent steps of this workflow.

```
bash scripts/filterdomains.sh 0.85
```

3.4 Align the Structures

Input files
coordinate files in the splitfilter/ and query/ directories

Output files:
alignment files in the alignments/ directory

Next, we will produce a multiple structure alignment using FoldMason (*see* **Note 5**). We will describe this using the command line, noting that Steps 1–4 below can also be performed using the FoldMason webserver at https://search.foldseek.com/foldmason. The following will walk through this process step by step. Skip to Step 6 to do everything in one command.

1. Open the command line terminal and navigate to the Zenodo repository.
2. Copy all of our structures into the same temporary folder, and create a new temporary folder to save the results into.

```
mkdir -p foldmasonout
mkdir -p structuresin
cp query/* structuresin/.
cp splitfilter/* structuresin/.
```

3. Align these structures using:

```
foldmason easy-msa structuresin foldmasonout/
alignment foldmasonout/
```

4. Copy the two alignments (sequence and structure) into the alignments folder, and clean up the temporary directories:

```
mv foldmasonout/alignment_3di.fa alignments/. mv
foldmasonout/alignment_aa.fa alignments/. rm -r
foldmasonout
rm -r structuresin
```

5. Trim the two alignments to the region that is aligned with our query structure, using the R script:

```
cd alignments
```

```
Rscript../scripts/trim.R alignment_aa.fa
Homo_sapiens_5H4Y Rscript../scripts/trim.R
alignment_3di.fa Homo_sapiens_5H4Y
```

6. The steps above can all be executed at once using the provided bash script:

```
bash scripts/align.sh
```

7. Open JalView. Load the amino acid alignment (alignment aa.trimmed.fa). Set the colour scheme by opening the Colour menu and then select Zappo. Observe how the sites appear to be poorly conserved, reflecting significant sequence divergence (Fig. 8). Also observe that this alignment contains many gaps ('–' symbol, denoting insertions/deletions); however, these are not an issue because most phylogenetic algorithms, including the ones we will be using, treat gaps as missing data and effectively ignore them from the calculation [35, 36].
8. Now load the 3Di alignment into JalView (alignment 3di. trimmed.fa). We need to change the color scheme because the default schemes are describing similarities between amino acids, not 3Di characters. To do this, open the Colour menu and select User defined, followed by Load scheme, open the scripts/3Di-gecos-palette.jc file available in this repository, and then press OK. Observe the strong degree of structure similarity in this alignment (Fig. 9), compared with the amino acid sequences (Fig. 8).

3.5 Build the Phylogenetic Trees

Input files
trimmed alignment files in the alignments/ directory

Output files
tree files in the trees/ directory

Now that the amino acid and 3Di alignments have been produced in fasta format, they will be used as input for phylogenetic analysis in IQ-TREE [37] or another program for phylogenetic reconstruction (*see* **Notes**). IQ-TREE has extensive documentation available if a more tailored analysis is needed (tutorials at https://IQ-TREE.github.io/doc/Tutorial and an AI assistant at https://chatgpt.com/g/g-aZvnPPUW1-iq-gpt).

Initially, amino acid sequence (AA) and protein structure (3Di) will be analyzed separately for comparison. Then, they will concatenated and partitioned (AA+3Di).

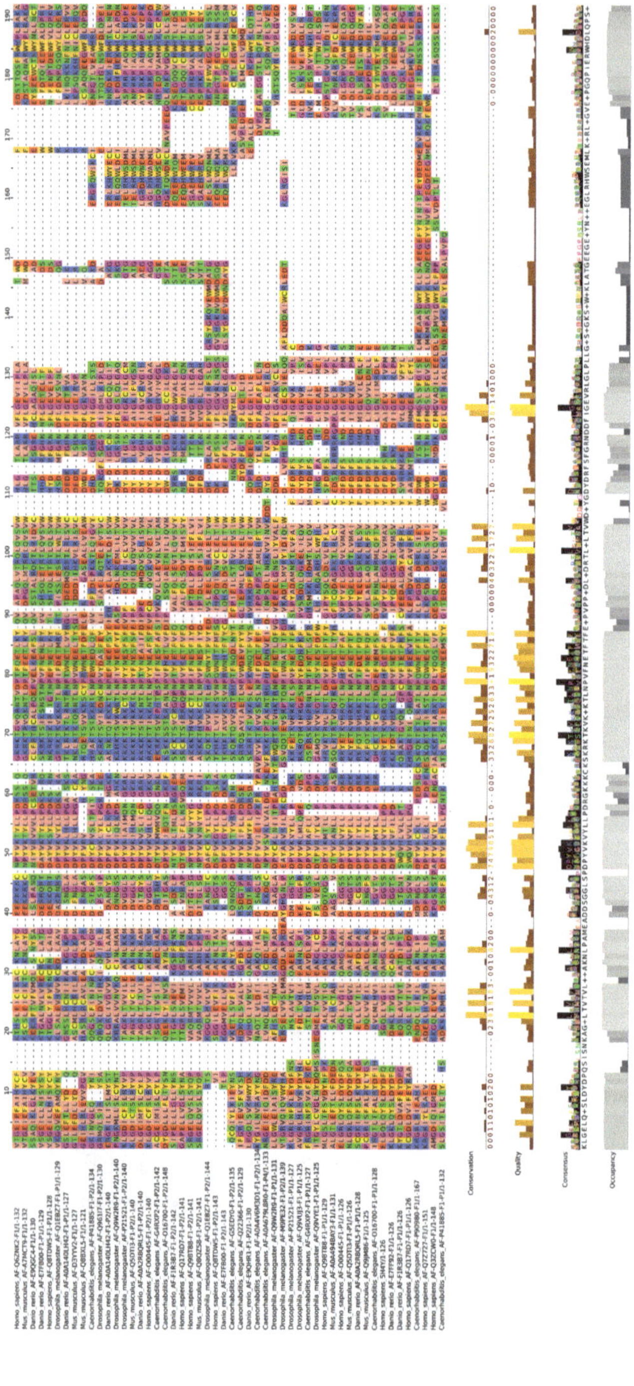

Fig. 8 Amino acid alignment displayed in JalView with the Zappo color scheme

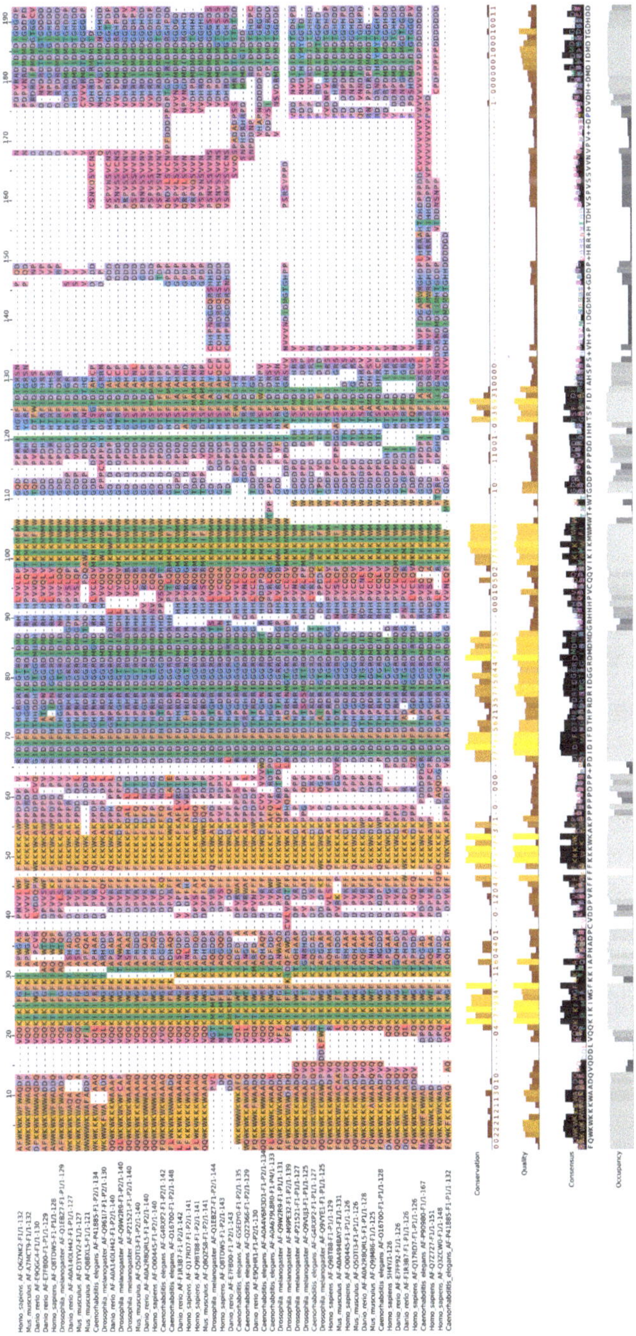

Fig. 9 3Di alignment displayed in JalView with the Foldseek 3Di color scheme

1. Navigate to the trees directory in the command line terminal. Delete all the existing files in this directory so we can make everything from scratch.

```
cd trees
rm *
```

2. Build a phylogeny from the trimmed amino acid alignment, with automated model selection and 1000 ultrafast random bootstrap replicates. First, we will copy the alignment into this folder, so that the outputs will be generated here too:

```
cp../alignments/alignment_aa.trimmed.fa.
iqtree -s alignment_aa.trimmed.fa -m MFP -bb 1000
```

-s alignment aa.trimmed.fa specifies your input alignment file (in this case, a FASTA of trimmed amino-acid sequences).
-m MFP invokes automated model-selection with ModelFinder.
-bb 1000 requests 1000 ultrafast bootstrap replicates (UFBoot).

3. To determine the substitution model that best fit your data set, open the alignment aa.trimmed.fa.iqtree file in a text editor (e.g., BBEdit for Mac OS, Notepad for Windows) for a summary of the IQ-TREE run. Observe that the LG [38] substitution model (with four rate categories) was selected for the synaptotagmin amino acid alignment:

```
ModelFinder
Best-fit model according to BIC: LG+R4
```

4. For 3Di, model selection and bootstrap analysis have to be run separately to be able to select a specialized substitution model for 3Di. To date, 3Di substitution matrices are not available directly from IQ-TREE. To test the currently available substitution models—the classic Foldseek 3Di alignment matrix (FOLDSEEK 3DI) and two 3Di matrices by Garg and Hochberg [39] (GH AF 3DI and GH LLM 3DI); *see* **Note 6**. Use the matrices.nex file available in the matrices folder of the repository and run ModelFinder in IQ-TREE with these models, using the following command:

```
cp../alignments/alignment_3di.trimmed.
fa.iqtree -s alignment_3di.trimmed.fa -mset
GH_LLM_3DI,GH_AF_3DI,FOLDSEEK_3DI
-m MF -mdef ../matrices/matrices.nex
```

5. Open the newly generated alignment 3di.trimmed.fa.iqtree file in a text editor. Observe that the GH AF 3DI substitution

model was selected (with frequencies estimated, invariant sites, and 4 gamma rate heterogeneity categories) for the synaptotagmin 3Di alignment:
ModelFinder

```
<Best-fit model according to BIC: GH_AF_3DI+F+I+G4
```

6. Once the substitution model has been selected (GH AF 3DI in this case), run the following command to produce a phylogenetic tree (*see* **Notes 7 and 8**):

```
iqtree -s alignment_3di.trimmed.fa -mdef ../
matrices/GH_AF_3DI.nexus -bb 1000 -m GH3DIAF+F+I+
G4 -redo
```

   - s alignment edi.trimmed.fa specifies your input alignment file (in this case, a FASTA of trimmed 3Di sequences).
   - p SYTs sets.nex tells IQ-TREE to read a NEXUS-format partition file (here, SYTs sets.nex) that defines your CHARSET blocks (e.g., which columns are AA vs. 3Di).
   - mdef GH AF 3DI.nexus points IQ-TREE to a Nexus file that defines your custom 3Di substitution matrix under the model name (e.g., GH AF 3DI). Make sure the name in this nexus file is exactly the same as in the command line.
   - bb 1000 runs 1000 ultrafast bootstrap replicates (UFBoot) to assess node support on the phylogenetic tree.
   - redo tells IQ-TREE to overwrite any existing results and start fresh. It ensures that you don't have to manually delete old.iqtree,.treefile, or.log files before re-running.

7. Finally, we will build a phylogeny from both trimmed alignments (amino acids and 3Di). To do this, first we will concatenate the two alignments into one using the following script, which will generate a new file called concatenated alignment.fa:

```
bash../scripts/concat.sh alignment_aa.trimmed.fa
alignment_3di.trimmed.fa
```

8. Generate a nexus file that defines these two partitions, and what site model, we will use for each partition. For our working example, this file is scripts/SYTs sets.nex and contains the following content:

```
#nexus begin sets;
charset aa = 1-192; charset str3di = 193-384;
charpartition mine = LG+I+G4:aa, GH3DiAF+F+I+
G4:str3di; end;
```

9. Build a phylogeny from the concatenated alignments with 1000 bootstrap replicates:

```
cp../scripts/SYTs_sets.nex.
iqtree -s concatenated_alignment.fa -mdef../
matrices/GH_AF_3DI.nexus -bb 1000 -spp SYTs_sets.
nex
```

10. Our three consensus trees, with bootstrap support indicated, are in the trees directory with the file extension.contree.

3.6 Reading and Interpreting Phylogenies

Input files
consensus tree files in the trees/ directory

Output files
pictures in the figures/ directory

The final step is to visualize and interpret our phylogenies. There are many tools available to assist with this task, including FigTree [40], DensiTree [41], and the APE library for R [42]. Here, we will use IcyTree [43] and the phytools package for R [44]. One of the most challenging aspects in phylogenetics is interpreting the tree and making sense of its story (if it has one). It is often insufficient to plot a picture of a tree without further input; this task frequently involves illustrating and annotating the tree in a manner that delivers a key message without getting lost in unwanted detail. Please note that these trees are unrooted, meaning there is no ancestral root node in the tree. Also note that your IQ-TREE consensus tree files will have some random variation from our own displayed here, due to the stochastic nature of bootstrapping.

1. Open https://icytree.org in the web-browser.
2. Open the amino acid consensus tree file (alignment aa.trimmed.fa.contree) in IcyTree.
3. Display the bootstrap support by opening the File menu, and then pressing Internal node text and selecting Label. Display the x-axis by opening File, and then selecting Axis and then Age (or alternatively press a on the keyboard as a hotkey). Observe the moderate bootstrap support values (close to 50%), and observe the long branches (often more than 1 substitution per site; leading to mutational saturation), as shown in Fig. 10.
4. Open the 3Di consensus tree file (alignment 3di.trimmed.fa.contree) in IcyTree, and show bootstraps and x-axis again. Observe how the branches are much shorter now reflecting

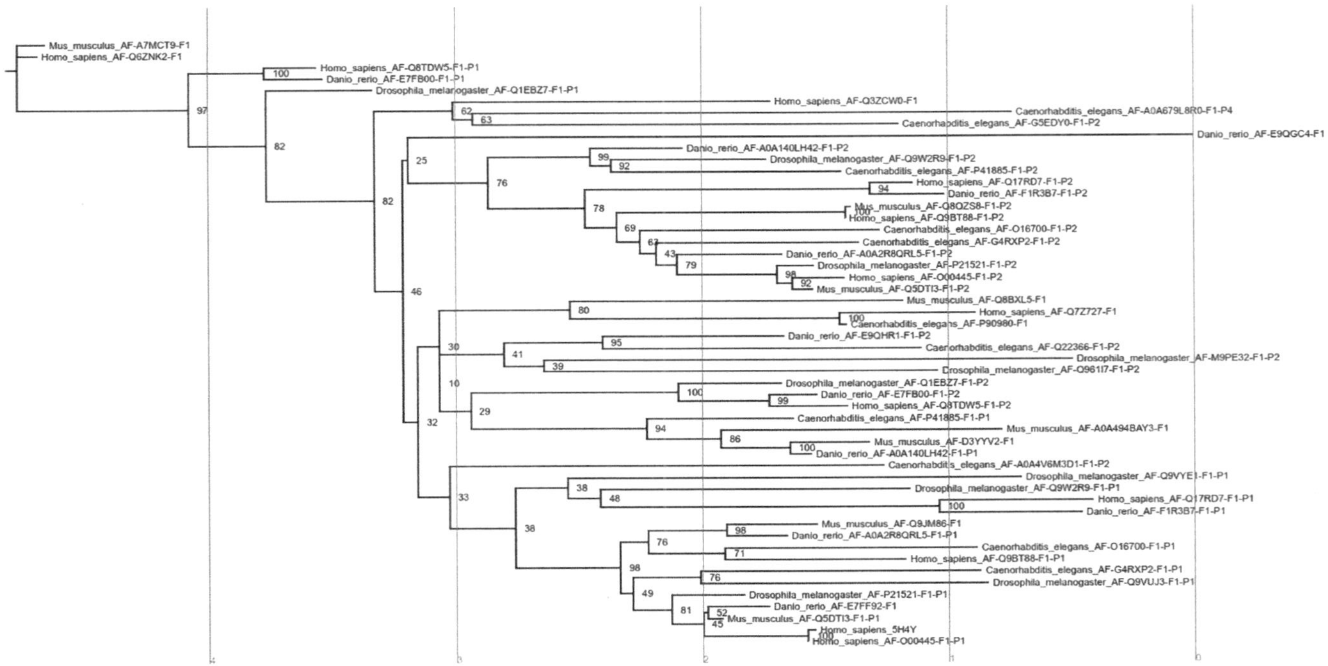

Fig. 10 C_2 domain phylogeny based on amino acids, displayed in IcyTree. Internal node labels show bootstrap support (out of 100), and the x-axis is in units of substitutions per site

less structural divergence compared with sequence, while the boot-strap values are at a comparable magnitude to before.

5. Now, open the command line terminal and navigate to the figures directory.
6. Generate a tanglegram of the two trees side by side, so we can compare them directly. This can be done using our R script, which is using the phytools library [44]. This script will require some modification if being applied on a different dataset. Executing the command below will generate a file called tanglegraml.png, also shown in Fig. 11.

```
Rscript../scripts/plotTanglegram1.R
```

7. Observe that there are very few clades with high support coming from both trees. It is difficult to detect any large clades that are well-supported (over 90% support) by either method, let alone both. Resolving the phylogeny of this family appears to be challenging, while sequence and structure are telling slightly different stories.
8. Now we will look at the AA+3Di tree. Open the concatenated tree in IcyTree (SYTs sets.nex.contree). Show the bootstrap supports and display the x-axis, as before. Observe that the bootstrap supports are generally much higher in this tree, reflecting the additional confidence that came from using both datasets together.
9. Generate another tanglegram, this time comparing the AA and AA+3Di trees. This time, some of the well-supported clades with at least 90% support in the AA+3Di tree will be assigned their own color. Output shown in Fig. 12.

```
Rscript../scripts/plotTanglegram2.R
```

10. Observe that th=e AA+3Di phylogeny has several features that make it appealing over either the AA or 3Di trees. Notably, it has much stronger support, particularly on older branches. It also is grouping the synaptotagmin C_2A (pink) and B (blue) domains into two respective clades, in contrast with the AA alignment, which has split either clade into two or more groups each with low support. Caution is advised when interpreting these functional annotations, as they are often generated automatically and not curated. This evolutionary model, therefore, serves as a basis for making functional predictions about uncharacterized proteins (*see* **Note 9**).

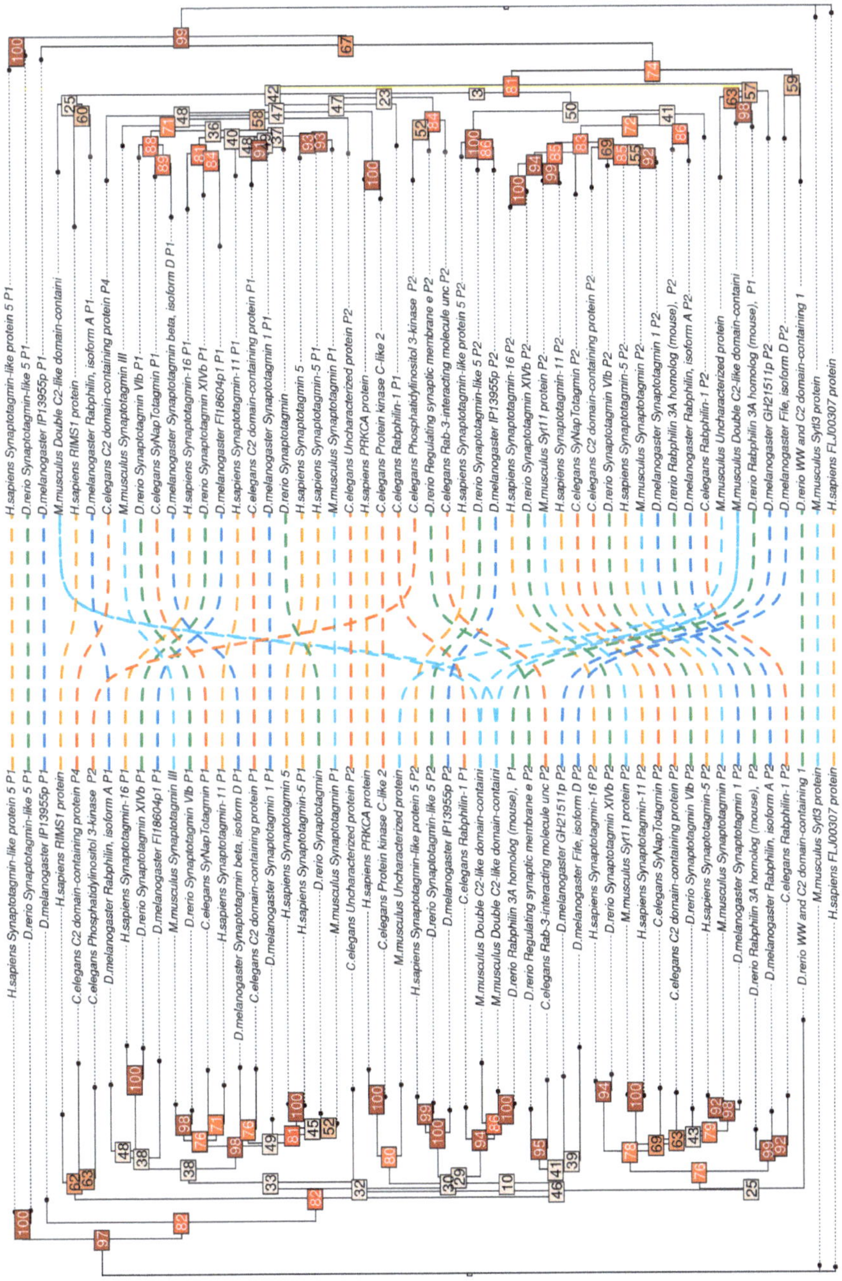

Fig. 11 Tanglegram showing the amino acid phylogeny (left) against the 3Di phylogeny (right). Bootstrap supports are indicated on internal nodes (darker red = higher support). The links in the middle are colored by organism. Branch lengths are proportional to substitutions per site

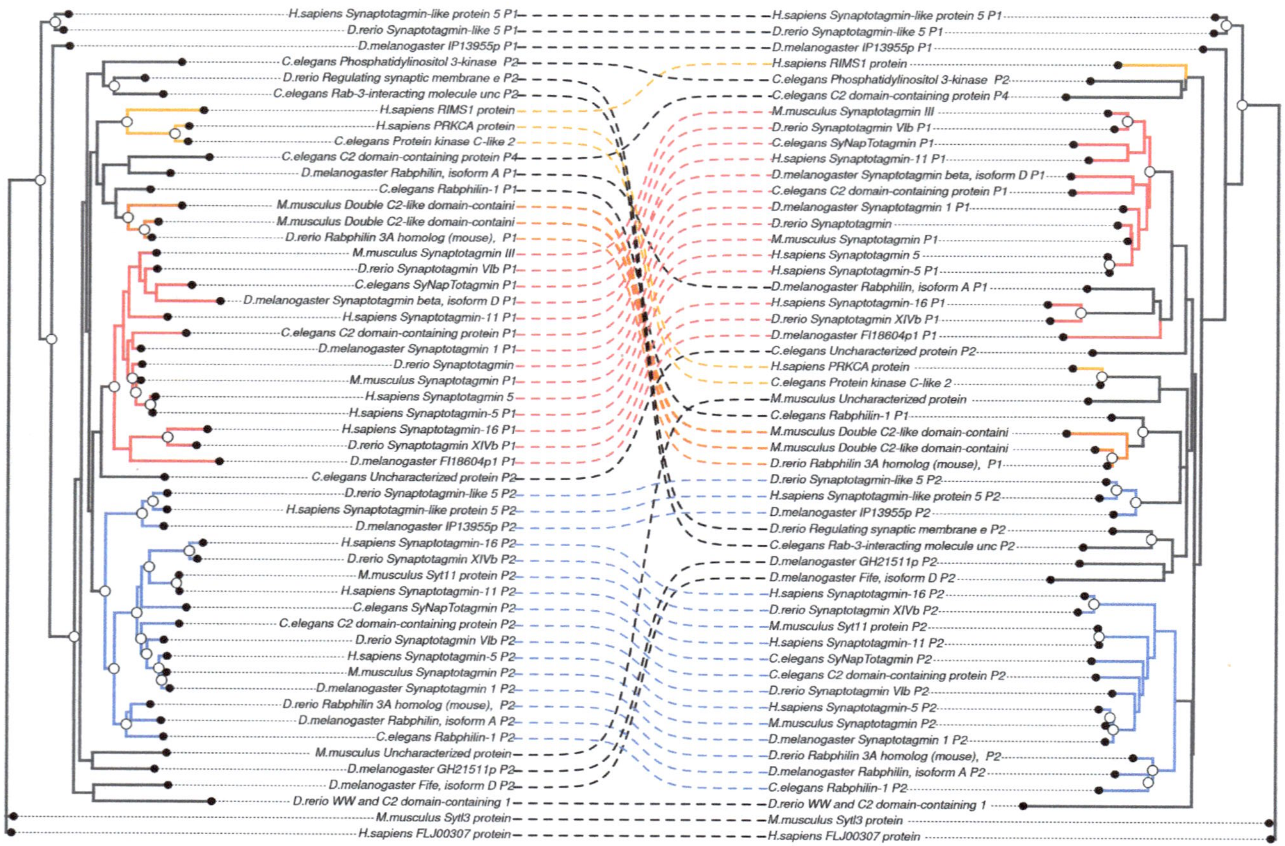

Fig. 12 Tanglegram between the AA+3Di (left) and AA (right) phylogenies. Nodes with at least 90% bootstrap support are indicated with a white circle. The links in the middle are colored by four major well-supported clades described by AA+3Di—pink is associated with the synaptotagmin C_2A) domain, blue with the synaptotagmin C_2B domain, yellow is found in protein kinase C, orange is found in Rabphilin 3A; while black is not associated to any of these groups

4 Notes

1. Three-dimensional protein structure is most useful in cases of high sequence divergence or mutational saturation, including ancient or fast-evolving protein families. If your protein of interest does not show any of these characteristics, amino acid sequence by itself might be sufficient.
2. If available, favor experimental structures over AI predicted ones, and favor structural models that have reasonable confidence scores (e.g., a pLDDT score over 70%).
3. In this protocol, we used the web version of Foldseek, but it can also be downloaded and run on a local database of structures, available at https://github.com/steineggerlab/foldseek.
4. We restricted our search to model organisms, which are enriched in the AlphaFold databases. Using proteins that are not available in databases can be achieved by generating new structural predictions, for example, by using AlphaFold Server at https://alphafoldserver.com among other resources.
5. Phylogenetics assumes that every site in the alignment describes homology (i.e., the sequences and structures share common ancestry at that position). If the alignments are looking poor, they can be corrected through manual adjustment. In some cases, misaligned structures can result from having too many non-homologous regions in the protein coordinate files. This can be fixed by truncating further non-homologous regions from the protein structures using PyMOL. For example, in Fig. 5, the blue N-terminal helix could have been removed as it is not part of the C_2 domain homologous region.
6. As of July 2025, there are just three published 3Di substitution models (one from Foldseek [18] and two from Garg and Hochberg 2025 [39]). As further models become available in the future, they should be added to the matrices folder so they may be used during model selection.
7. If your amino acid data come from very divergent proteins with high heterogeneity in their frequencies across sites, using mixture models that account for site-heterogeneous substitution processes might be necessary to identify models that better fit your data [45]. However, these models are computationally more expensive so we have not used them here. For more information on how to implement these models, go to https://iqtree.github.io/doc/Complex-Models#mixture-models.
8. Be aware of the IQ-TREE version you are using. Some commands have changed between versions. We used IQ-TREE 3 [46] for this tutorial because it has the most advanced workflows

for model testing, mixture models, and scaling. It is generally recommended to use the latest version for all software.

9. The 3Di approach in phylogenetics, and whether it makes sense to concatenate amino acid and 3Di alignments together, is a rapidly evolving area that still requires considerable investigation, as extensively discussed elsewhere [19, 39].

Acknowledgments

CPL and NJM were supported by the HFSP grant RGP0060/2021 and the University of Auckland, Faculty of Science Research Development Fund, FoS RDF 3732317. NJM was additionally supported by New Zealand Royal Society RDF 21-UOA-040, Marsden Grant 18-UOA-034, and University of Auckland DRDF Research Fund 3727963. JD was supported by the Australian Research Council.

References

1. Enard W, Gehre S, Hammerschmidt K et al (2009) A humanized version of Foxp2 affects cortico-basal ganglia circuits in mice. Cell 137:961–971
2. Bridgham JT, Ortlund EA, Thornton JW (2009) An epistatic ratchet constrains the direction of glucocorticoid receptor evolution. Nature 461:515–519
3. Eisen JA (1998) Phylogenomics: improving functional predictions for uncharacterized genes by evolutionary analysis. Genome Res 8:163–167
4. Jumper J, Evans R, Pritzel A et al (2021) Highly accurate protein structure prediction with AlphaFold. Nature 596:583–589
5. Lin Z, Akin H, Rao R et al (2023) Evolutionary-scale prediction of atomic-level protein structure with a language model. Science. 379:1123–1130
6. Moi D, Bernard C, Steinegger M et al (2025) Structural phylogenetics unravels the evolutionary diversification of communication systems in gram-positive bacteria and their viruses. Nature Structural & Molecular Biology, 32(12), 2492-2502.
7. Puente-Lelievre C, Malik AJ, Douglas J et al (2023). Tertiary-interaction characters enable fast, model-based structural phylogenetics beyond the twilight zone. BioRxiv pp 2023–12
8. Chothia C, Lesk AM (1986) The relation between the divergence of sequence and structure in proteins. EMBO J 5:823–826
9. Flores T, Orengo C, Moss D et al (1993) Comparison of conformational characteristics in structurally similar protein pairs. Protein Sci 2:1811–1826
10. Illerg°ard K, Ardell DH, Elofsson A (2009) Structure is three to ten times more conserved than sequence—a study of structural response in protein cores. Proteins 77:499–508
11. Koonin EV (2005) Orthologs, paralogs, and evolutionary genomics. Annu Rev Genet 39:309–338
12. Björklund AK, Ekman D, Light S et al (2005) Domain rearrangements in protein evolution. J Mol Biol 353:911–923
13. Moore AD, Björklund AK, Ekman D et al (2008) Arrangements in the modular evolution of proteins. Trends Biochem Sci 33:444–451
14. Bornberg-Bauer E, Alb'a MM (2013) Dynamics and adaptive benefits of modular protein evolution. Curr Opin Struct Biol 23:459–466
15. Dohmen E, Klasberg S, Bornberg-Bauer E et al (2020) The modular nature of protein evolution: domain rearrangement rates across eukaryotic life. BMC Evol Biol 20:1–13
16. Scornavacca C, Galtier N (2017) Incomplete lineage sorting in mammalian phylogenomics. Syst Biol 66:112–120
17. Hotopp JCD (2011) Horizontal gene transfer between bacteria and animals. Trends Genet 27:157–163
18. Van Kempen M, Kim SS, Tumescheit C et al (2024) Fast and accurate protein

structure search with Foldseek. Nat Biotechnol 42:243–246
19. Puente-Lelievre C, Malik A, Douglas J (2025) Protein structural phylogenetics. Genome Biology and Evolution, 17(8)evaf139
20. Wolfes AC, Dean C (2020) The diversity of synaptotagmin isoforms. Curr Opin Neurobiol 63:198–209
21. Montaville P, Coudevylle N, Radhakrishnan A et al (2008) The PIP2 binding mode of the C2 domains of rabphilin-3A. Protein Sci 17:1025–1034
22. Südhof TC, Rizo J (1996) Synaptotagmins: C2-domain proteins that regulate membrane traffic. Neuron 17:379–388
23. Douglas J, Puente-Lelievre C (2025) Dataset for: an Introduction to Structural Phylogenetics for Biomedical Sciences. Version 1.0.0. Zenodo
24. DeLano WL et al (2002) Pymol: an open-source molecular graphics tool. CCP4 Newsl Protein Crystallogr 40:82–92
25. Gilchrist CL, Mirdita M, Steinegger M (2026) Multiple protein structure alignment at scale with FoldMason. Science, 391(6784):485-488.
26. O'Donoghue P, Luthey-Schulten Z (2003) On the evolution of structure in aminoacyl tRNA synthetases. Microbiol Mol Biol Rev 67:550–573
27. Lundin D, Poole AM, Sjöberg B-M et al (2012) Use of structural phylogenetic networks for classification of the ferritin-like superfamily. J Biol Chem 287:20565–20575
28. Mifsud JC, Lytras S, Oliver MR, Toon K, Costa VA, Holmes EC, Grove, J (2024) Mapping glycoprotein structure reveals Flaviviridae evolutionary history. Nature, 633(8030):695-703
29. Kikkawa U, Kishimoto A, Nishizuka Y (1989) The protein kinase C family: heterogeneity and its implications. Annu Rev Biochem 58:31–44
30. Brose N, Hofmann K, Hata Y et al (1995) Mammalian homologues of *Caenorhabditis elegans* unc-13 gene define novel family of C2-domain proteins. J Biol Chem 270:25273–25280
31. Qiu X, Ge J, Gao Y et al (2017) Structural analysis of Ca2+-binding pocket of synaptotagmin 5 C2A domain. Int J Biol Macromol 95:946–953
32. Mariani V, Biasini M, Barbato A et al (2013) lDDT: a local superposition-free score for comparing protein structures and models using distance difference tests. Bioinformatics 29:2722–2728
33. Waterhouse AM, Procter JB, Martin DM et al (2009) Jalview Version 2—a multiple sequence alignment editor and analysis workbench. Bioinformatics 25:1189–1191
34. Larsson A (2014) AliView: a fast and lightweight alignment viewer and editor for large datasets. Bioinformatics 30:3276–3278
35. Truszkowski J, Goldman N (2016) Maximum likelihood phylogenetic inference is consistent on multiple sequence alignments, with or without gaps. Syst Biol 65:328–333
36. Machado DJ, Castroviejo-Fisher S, Grant T (2021) Evidence of absence treated as absence of evidence: the effects of variation in the number and distribution of gaps treated as missing data on the results of standard maximum likelihood analysis. Mol Phylogenet Evol 154:106966
37. Minh BQ, Schmidt HA, Chernomor O et al (2020) IQ-TREE 2: new models and efficient methods for phylogenetic inference in the genomic era. Mol Biol Evol 37:1530–1534
38. Le SQ, Gascuel O (2008) An improved general amino acid replacement matrix. Mol Biol Evol 25:1307–1320
39. Garg SG, Hochberg GK (2025) A general substitution matrix for structural phylogenetics. Mol Biol Evol 42:msaf124
40. Rambaut A (2009) FigTree. Tree figure drawing tool. http://tree.bio.ed.ac.uk/software/figtree/
41. Heled J, Drummond AJ (2012) Calibrated tree priors for relaxed phylogenetics and divergence time estimation. Syst Biol 61:138–149
42. Paradis E, Claude J, Strimmer K (2004) APE: analyses of phylogenetics and evolution in R language. Bioinformatics 20:289–290
43. Vaughan TG (2017) IcyTree: rapid browser-based visualization for phylogenetic trees and networks. Bioinformatics 33:2392–2394
44. Revell LJ (2024) phytools 2.0: an updated R ecosystem for phylogenetic comparative methods (and other things). PeerJ 12:e16505
45. Lartillot N, Philippe H (2004) A Bayesian mixture model for across-site heterogeneities in the amino-acid replacement process. Mol Biol Evol 21:1095–1109
46. Wong TK, Ly-Trong N, Ren H et al (2026) IQ-TREE 3: phylogenomic inference software using complex evolutionary models. Molecular Biology and Evolution, 43(5):msag117

Chapter 5

Segmentation-Free Multi-Fractal and Radiomics Response Profiling of Live Unstained Confluent Microglia In Vitro Using Phase-Contrast Microscopy Images

Millijoy D. Villanueva, Ghasem Azemi, Ion Andronache, Benjamin Heng, Anna Guller and Antonio Di Ieva

Abstract

This chapter presents a segmentation-free, cell ensemble-level workflow for detecting treatment-induced responses in confluent microglial cultures in vitro, using standard phase-contrast microscopy and Python-based imaging data analysis. Entire fields of view are analyzed after minimal preprocessing, avoiding fixation, staining, and single-cell segmentation. We implement multifractal, lacunarity, and texture analysis (MFTA) to characterize the microglial monolayers. In parallel, we introduce cell radiomics as the adaptation of radiomics principles to label-free microscopy, extracting a high-dimensional panel of intensity- and texture-based features directly from grayscale images of live cells. Together, MFTA and cell radiomics provide complementary, quantitative readouts that are sensitive to subtle changes in ensemble organization. Statistical comparisons employ false discovery rate control to identify robust discriminative features between conditions, with optional validation by conventional morphometry on a subset of segmented cells. The workflow is demonstrated on an experimental dataset by the successful reveal of the effect of a moderate static magnetic field on human C20 microglia cell line.

Keywords Multi-fractal analysis, Radiomics analysis, Phase contrast microscopy images, Cell culture, Microglia

1 Introduction

Microglia are central nervous system (CNS) resident immune cells whose morphology and spatial organization reflect the state of the neural microenvironment [1]. Conventional, length-based cell morphometry, as well as fractal, lacunarity, and texture analyses have been employed to capture the geometric complexity and heterogeneity of microglia and to detect disease- or treatment-associated changes. These approaches typically require prior segmentation of cells or lesions of interest [2–6]. To obtain images suitable for robust segmentation, cell or tissue samples are usually chemically fixed, stained, and

Antonio Di Ieva and Jennilee M. Davidson (eds.), *Computational Neuroscience*, Methods in Molecular Biology, vol. 3062,
https://doi.org/10.1007/978-1-0716-5440-8_5,

imaged using bright-field, epifluorescence, or confocal fluorescence microscopy. This workflow (fixation → staining → high-contrast imaging → segmentation) is labor-intensive and not inherently compatible with phase-contrast transmitted-light microscopy (PCM)—the main workhorse for label-free, nondestructive monitoring of live cells cultured in vitro. Due to the mechanism of image generation, in standard PCM, the contrast and signal-to-noise ratios are relatively low, and the images are prone to certain artifacts [7, 8]. These challenges were addressed in several studies to allow cell tracking and quantitative evaluation of adherent cells' confluency—the fraction of the cell culture vehicle surface area covered by cells [7, 9–12]. A few studies have applied fractal-type analyses directly to PCM or quantitative phase images of cultured cells, still requiring the delineation of individual cell contours [11]. However, these strategies have not been adapted to analyze PCM images of high-density in vitro cultures, where cells form fully confluent monolayers, making detection of individual cell borders technically challenging, error-prone, and highly sensitive to preprocessing choices.

In many neuropathological settings, including brain tumors, neurodegeneration, stroke, trauma, and foreign body or implant reactions, microglia do not remain evenly distributed as single surveyors. They frequently organize into fronts, cuffs, and dense clusters around lesions, abnormal vessels, plaques, or materials [1, 13–17]. Such collective arrangements reflect coordinated sensing and response and may be better captured by population-level complexity and texture metrics than by cataloguing a limited number of individually segmented cells. A robust, label-free, segmentation-independent approach to quantify microglial ensembles in vitro would therefore offer a practical route to monitor functionally relevant responses in near real time without perturbing the system under study. Therefore, confluent cell cultures in vitro represent an accessible instrument to explore pathology-associated microglia.

Luckily, in parallel, the quantitative image analysis is increasingly shifting from a single object-centric to ensemble-level descriptors. Fractal and multi-fractal metrics are particularly well-suited for these purposes as a way to summarize scale-dependent complexity. Lacunarity characterizes the degree of "gappiness" and spatial inhomogeneity [18, 19]. The gray-level co-occurrence, run-length, and related texture features quantify intensity patterns [2, 4, 5]. In parallel, a set of analytical solutions called radiomics has taken a major role in clinical imaging. Radiomics frameworks allow the high-throughput extraction of image texture features [20–24]. To the best of our knowledge, the radiomics-style feature extraction has not been systematically validated as a segmentation-free strategy for PCM images of confluent live cell monolayers, representing an opportunity for methodological development.

Despite recent progress in fractal and radiomics-based microscopy analyses, there remains a methodological gap: a robust, segmentation-free strategy for quantifying microglial responses in confluent monolayers using standard PCM is still lacking. To address this gap, we developed and applied a segmentation-free, *cell ensemble-level* Python code-based workflow to detect static magnetic field (SMF)-induced responses in human microglia using conventional phase-contrast microscopy (Fig. 1), also introducing an ensemble-level image analysis workflow that combines multiple fractal-based features and radiomic descriptors. The aim is to sensitively detect treatment-induced microglial changes without cell fixation, staining, or segmentation, enabling label-free, minimally disruptive monitoring of live cultures.

Static magnetic fields (SMF) are considered a promising therapeutic factor in neuropathology [25, 26]. An effect of a similar modality, an extremely low-frequency magnetic field on embryonic yolk sac (which is the source of brain microglia [27, 28]) was previously revealed using fractal metrics and lacunarity analysis [29].

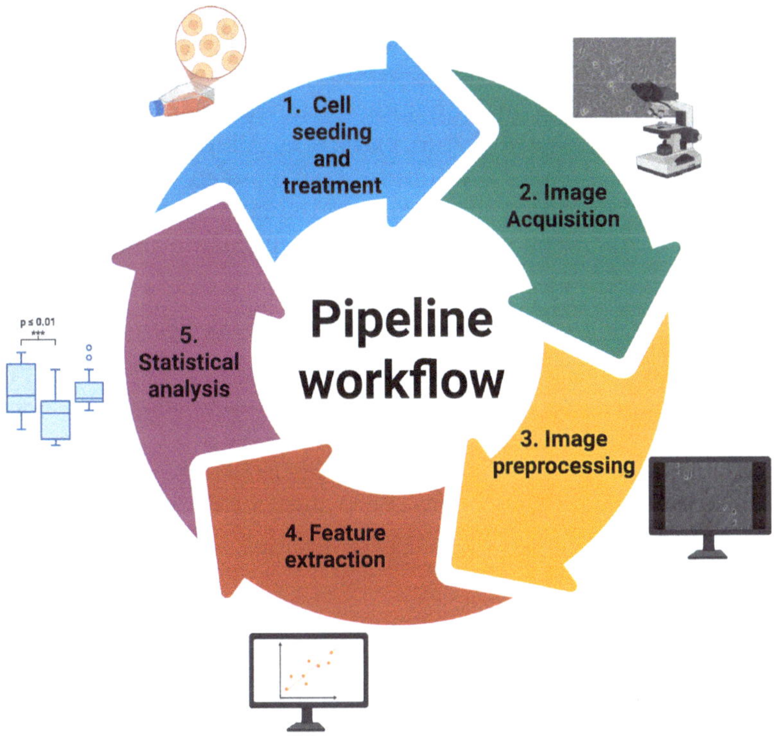

Fig. 1 Visualization summary of the provided pipeline workflow. The steps include: (1) Cell seeding of the relevant microglial cell line. (2) Image acquisition using phase contrast light microscopy. (3) Image preprocessing. (4) Multifractal-lacunarity-texture and radiomics feature extraction. (5) Statistical analysis. Created in https://BioRender.com

In the current protocol, C20 human microglia were seeded at high density and exposed, from the time of plating, either to the ambient geomagnetic field or to a moderate-strength SMF of 142 ± 9 mT for 21 h, at which point both conditions formed confluent monolayers. PCM images were acquired and subjected to minimal preprocessing, and the entire field of view was treated as the region of interest. We combined multifractal and lacunarity metrics derived from a FracLac-based feature set with lacunarity and texture analysis [30, 31] and termed it *"Multi-fractal and Texture analysis" (MFTA)* and adapted the PyRadiomics framework for microscopy images (termed below as *"Radiomics"*), thereby extracting a broad panel of complexity and texture descriptors directly from the grayscale images, without segmenting individual cells. False discovery rate (FDR) control was used to identify features that significantly discriminated SMF-exposed from control cultures. In parallel, we validated that these image-level signatures correspond to genuine cellular changes by performing conventional morphometry on manually segmented cells (*see* **Note 1**).

2 Materials

This section outlines the required materials for culturing human microglial cells, preparing them for experimental treatments, performing imaging under phase-contrast microscopy, and imaging quantitative feature extraction. All materials should be handled in accordance with institutional biosafety and cell culture guidelines. Ensure that all materials and equipment are prepared and organized in advance to ensure an efficient experimental workflow.

2.1 Cell Line and Culture Reagents

1. Prepare an immortalized human microglial cell line. For detailed information on the specific microglial cell line used in the development of this protocol (*see* **Note 2**).
2. Prepare a complete cell culture medium. Use Dulbecco's Modified Eagle Medium/Nutrient Mixture F-12/Nutrient Mixture F-12 (DMEM/F-12) supplemented with 10% fetal bovine serum (FBS) and 1% penicillin–streptomycin (P/S).
3. Use Trypsin–EDTA solution for detaching adherent cells for cell culturing.
4. Use phosphate-buffered saline (PBS) for washing and resuspension steps.
5. You will need cell culturing vessels during this stage. The following materials are required:
   a. 75 cm^2 U-shaped canted neck cell culture flask with vent cap to culture cells.
   b. 12-well clear TC-treated multiple well plates, individually wrapped, and sterile for cell seeding, treatment exposure, and subsequent imaging.

6. When pelleting and changing medium, use 15 mL centrifuge tubes.
7. A 37 °C humidified tissue culture incubator with 5% CO_2 is required for optimum cell culture.
8. For warming reagents prior to usage, use a water bath, reaching 37 °C.
9. A cell culture hood (a biosafety cabinet) will be required for all cell culture handling.
10. Cell culture consumables: 5 mL Eppendorf tubes, sterile pipette tips, and serological pipettes.
11. When counting cells to reach the right cell seeding density, use a Trypan blue solution: 0.4% for cell counting. Use a disposable Countess cell counting chamber slide for cell counting. For cell counting, use and check of viability, use an automated cell counter.

2.2 Materials for Phase Contrast Microscopy Protocol and Image Analysis

1. For imaging, use an Inverted phase-contrast microscope (e.g., Nikon Eclipse TS100 or equivalent) equipped with a 20×/0.40 NA phase-contrast objective and phase-contrast Ph1 condenser annulus.
2. A color digital camera (e.g., DS-Vi1 or equivalent) and attached working computer workstation must be attached to the microscope for image capture and digital storage.
3. After image acquisition, ensure access to a computer workstation with adequate storage space and a reliable internet connection to perform the following feature extraction and statistical analysis.

3 Methods

3.1 Cell Culture and Maintenance

This section outlines the procedures for microglia cell maintenance and subculturing. This protocol is best suited for microglial cell cultures maintained in 75 cm^2 flasks.

1. Turn on the water bath, let the temperature rise, and stay at 37 °C. When the temperature has been reached, place the media solution bottle in the water bath until needed.
2. Retrieve a 75 cm^2 flask containing a confluent microglia monolayer and place it in a cell culture hood. Remove and discard culture medium.
3. Rinse the cell layer with 5 mL of PBS. After adding PBS, swirl the flask around gently for PBS to cover all internal surfaces of the flask. Rest for 1 min.

4. Discard PBS. Add 3 mL of trypsin–EDTA. Gently swirl the flask around for the trypsin–EDTA to cover all internal surfaces of the flask.
5. Place the flask inside a 37 °C incubator for 1–3 min or until all cells have detached from the cell culture vessel bottom. This can be observed under a bright-field/phase-contrast light microscope.
6. Once cells are detached and dispersed, place 5 mL of media solution in the flask. Transfer the media-trypsin solution to a 15 mL centrifuge tube.
7. Centrifuge the cells for 5 min at 1000 RPM. Once finished, gently return the 15 mL tube to the cell culture hood so as not to disturb the cell pellet.
8. Gently discard the supernatant. Once removed, resuspend the cells in the same volume of media solution. Resuspend the cells by gently pipetting 3–5 times.
9. Add appropriate aliquots of the new cell suspension to a new 75 cm^2 flask. Add media solution until the volume inside the flask reaches 15 mL. Place in a 37 °C incubator and repeat the subculturing process when cells reach 70%–80% confluency.

3.2 Cell Preparation for Treatment

The following steps outline the preparation of three technical replicates from a single passage of microglial cells, seeded into 12-well plates. Twelve-well plates were utilized as they provided adequate space for cell growth and an appropriate number of wells for replicates. For all experimental studies, we recommend using a minimum of three independent passages to ensure biological reproducibility. For details regarding the specific experimental treatment conditions used in developing this protocol (*see* **Note 3**).

1. Retrieve a 75 cm^2 flask containing a confluent cell layer of microglia cells. Place in a cell culture hood. Discard media in the flask.
2. Repeat steps 4–9 from the subculturing section (Sect. 3.1).
3. Once the cell suspension is created, place 10 μL of the cell suspension in an Eppendorf tube. Place 10 μL of trypan blue solution in the same Eppendorf tube and resuspend the solution by gently pipetting 3–5 times.
4. Retrieve 10 μL of trypan-cell suspension and pipette gently into both chambers of the cell counting chamber slides.
5. Count the cells in the suspension using an automated cell counter. Adjust the suspension to a final cell density of 1.6×10^5 cells/mL. For further information on the reasoning for this selected cell density (*see* **Note 4**). Once the final cell suspension has been made, dispense 2 mL of cell suspension

into each well. For each passage, a minimum of three wells is recommended for experimentation. For all experiments, at least three independent cell passages should be included.

3.3 Phase Contrast Microscope Imaging Protocol

This section outlines the steps for how the images (Fig. 2) are obtained, which were used to develop the protocol for this study.

1. At 21 h post-seeding and treatment, retrieve the 12-well plate from the incubator and transfer it to the microscope stage.
2. Turn on the inverted microscope, with the color digital camera attached.
3. Rotate the nosepiece to select the LWD 20×/0.40 NA Ph1 ADL phase-contrast objective. Ensure that the phase-contrast annulus (Ph1) in the condenser is aligned with the objective.
4. Focus on a representative field of view containing evenly distributed monolayer of cells. Set the image acquisition to automatic exposure and white balance to minimize operator bias.
5. Acquire the images of the cultured cells, ensuring consistent microscope and camera settings are applied across all wells and passages.
6. Save the images as a Tagged Image File Format (.TIFF) file format for downstream analysis.

3.4 Image Preprocessing and Normalization

The preprocessing and feature extraction steps described in this section were performed using Python. Although various programming languages are capable of image analysis, Python was selected for its accessibility, flexibility, and extensive library packages commonly used in image processing and analysis. The following instructions describe the installation of Visual Studio Code (VS Code), which is a free source code editor that easily integrates with Python environments. While several open-access code editors are compatible with Python, VS Code is used in this protocol due to its flexibility

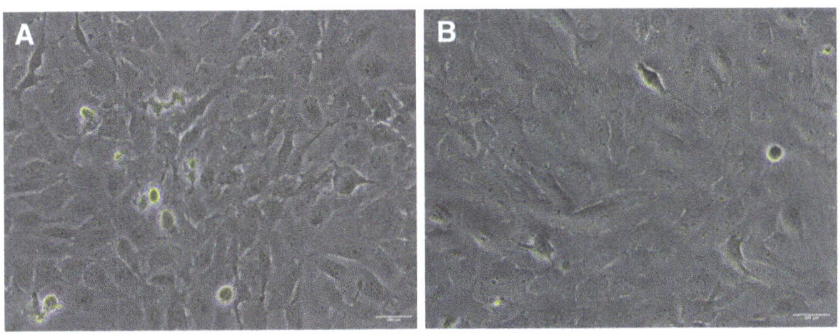

Fig. 2 Example of brightfield phase light contrast microscope images of monolayer segmentation and label-free C20 cells in vitro. (**A**) Control group. (**B**) SMF-treated group. Scale bar 20 μm

and user-friendly interface. However, other code editors may also be used, provided they are fully compatible with Python and the required library dependencies.

3.4.1 Installing Python and Code Editor, Visual Studio Code (VS Code)

1. Go to the Python.org website to download Python. Navigate to the downloads section, select the installer for your operating system (Windows, macOS, Linux, etc.) Once the download is complete, run the Python installer. Select the option to "Add Python to PATH" to use Python from your command line.
2. To verify that you have installed Python successfully on your device, run one of the following commands depending on your operating system.
3. For Linux or macOS devices, open a terminal window and type the command "`python3-version.`" For Windows devices, open a command prompt and run the command "`py-3-version-`." If the installation was successful, the output window will show the version of Python that you installed.
4. To download Visual Studio (VS) Code, visit the official website at code.visualstudio.com/download. On the download page, click the version that matches your current operating system and download the installer file. Once the download is complete, open the file to start the installation process.
5. Open VS Code. On the left-hand tab, click on "Extensions." Search for "Python" and install "Python" (Microsoft) and "Jupyter" (Microsoft).
6. Once installation is complete, VS Studio Code is all ready to go.

3.4.2 Preparing Your Files and Python Notebook for Analysis

The following section outlines the setup of a practical "Input–Output" system for data organization and analysis, requiring a clear directory structure for managing raw data and processed results. In this structure, files to be processed (image files or.CSV datasets, depending on the analysis stage) are stored in a designated "Input" folder. The resulting processed files are directed to an allocated "Output" folder, which acts as a repository for all generated results

1. After collecting all brightfield phase-contrast microscope images in.TIF format, organize them into two separate folders. Label the control and treatment group images as "Treatment_1" and "Treatment_2", respectively. Ensure that accurate records are maintained regarding which treatments were applied to each group.
2. In a preferred directory on your computer, create two folders, titled "Input" and "Output" (Fig. 3A). During feature extraction and statistical analysis, all files to be processed will be transferred into the "Input" folder. The Output folder will contain all resulting files, including .CSV and .PNG outputs from the

feature extraction and subsequent analysis. Depending on the stage of the workflow, files may need to be transferred between these folders to ensure proper organization and processing.

3. Transfer the folders containing your images, "Treatment_1" and "Treatment_2," to the Input folder (Fig. 3B).
4. Open VS Code. Press "File" in the top menu bar and click "New File" (Fig. 4A). In the drop-down window, choose the option "Jupyter Notebook" (see Fig. 4A). This will open a new window showing a notebook file ("Untitled.ipynb"). Save the file as "MFTA_analyses.ipynb." This will save it as a Python Notebook.
5. Once the file is saved, click "Code" to add your first line of code as described in the following steps (see Fig. 3B).

3.4.3 Renaming Image File Names and Application of Min–Max Normalization on Images

For all subsequent steps, Python scripts are accessed from the GitHub repository: https://github.com/MillijoyVillanueva/Multi-fractal_and_radiomics_profiling_in_microglia_in_vitro/tree/main. This section provides instructions for generating randomized identifiers or file names for images within each treatment group, which is recommended for any analysis to minimize bias and ensure blinded analysis. Following file organization, it is also advisable to perform image normalization to enable consistent comparisons across datasets.

1. Open the following link: https://github.com/MillijoyVillanueva/Multi-fractal_and_radiomics_profiling_in_microglia_in_vitro/blob/main/1_Renaming_files_and_Min_Max_Normalisation.ipynb. In the "Preview" window, this will display the cells corresponding to the steps described in this section.
2. Open your Python Notebook file. Copy and execute the provided Python code from the cell "`# Cell 1—Install required packages`" in the provided GitHub link. This will install the required packages.

Fig. 3 Organization of folders in a macOS computer system. **(A)** In a main folder, create two folders titled "Input," which will contain your images, and "Output," where all output files will appear. **(B)** Inside the "Input" folder, place the images of your two treatment groups. Label them as "Treatment_1" and "Treatment_2"

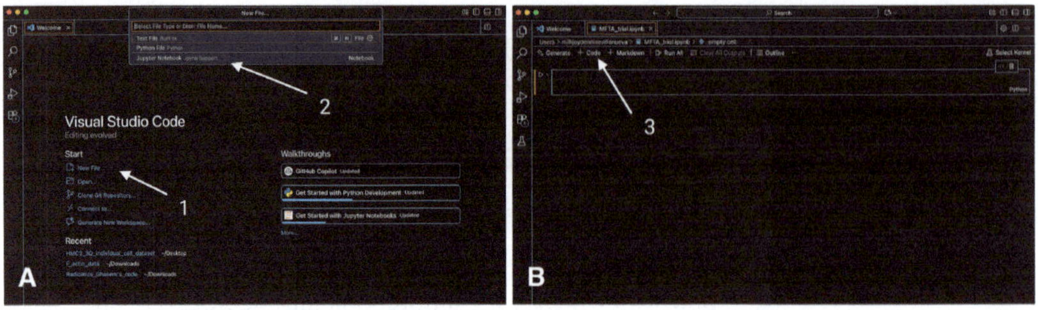

Fig. 4 Example of VS Code Window in a macOS computer system. **(A)** New window in VS Studio Code. Click on 1 to open a new file. A drop-down window will open, and click on 2 To create a Jupyter Notebook. **(B)** The new window of the newly created notebook. Save it as "MFTA.ipynb." Press on 3 To create a cell code

3. After all the packages have been installed successfully, open a new code cell. Copy and execute the provided Python code from the cell "`# Cell 2—Import packages`" to import the required Python libraries.
4. After completing the previous step, open a new code cell in the programming environment. Copy and execute the Python script from the cell "`# Cell 3—Renaming files`". Update the input directory argument to specify the folder path corresponding to the `Treatment_1` image set. Run the script to automatically assign randomized identifiers to each image file. Repeat the process for the `Treatment_2` dataset.
5. Open a new code cell in the programming environment. Copy and execute the Python code from the cell "`# Cell 4—Min-Max Normalization of images`". Update the `User Input` section with the directory paths of the `Treatment_1` and `Treatment_2` folders and specify the output directory under `Output` base folder. Execute the script to process all.TIF images, rescaling their pixel intensity values to an 8-bit range (0–255). Normalized images will be automatically saved into two new subdirectories, "`Treatment_1_norm`" and "`Treatment_2_norm`", located within the specified output directory.
6. Transfer the "`Treatment_1_norm`" and "`Treatment_2_norm`" folder into the main input folder for further analysis.

3.5 Multifractal and Radiomics Feature Extraction

3.5.1 Multifractal Feature Extraction

This section provides a comprehensive Python-based code that includes the implementation of computing and image-processing libraries to perform multifractal feature extraction. For more information on the multifractal features extracted (*see* **Note 5**).

1. Open the following GitHub link: https://github.com/MillijoyVillanueva/Multi-fractal_and_radiomics_profiling_in_microglia_in_vitro/blob/main/2_MFTA_feature_extraction.ipynb. In the "Preview" window, this will display the cells corresponding to the steps described in this section.
2. In the same Python notebook open in VS Code, create a new code cell. Copy and execute the Python code from the cell "`# Cell 1—Installing required libraries for feature extraction`" from the provided GitHub link to install the libraries required for MFTA feature extraction.
3. Once complete, create a new code cell in the same Python notebook. Copy and execute the Python code from the cell "`# Cell 2—Importing packages for feature extraction`" to import all required packages for this step.
4. In a new code cell, paste the Python code from the cell "`# Cell 3—Feature extraction code.`" In the script, specify the file pathnames for the "`Treatment_1_norm`" and "`Treatment_2_norm`" folders. In the section labeled "OUTPUT_CSV," provide the pathname of the output directory, followed by the filename for the resulting dataset (e.g., /User/yourname/Output.MFTA_results.csv). Execute the cell to initiate MFTA feature extraction from the input images.
5. The resulting output CSV file (Fig. 5) will contain the extracted 131 MFTA features for all input images, organized according to their respective groups. In this example, the two groups are labeled "`Treatment_1_norm`" and "`Treatment_2_norm.`" It is recommended to save or relocate the output CSV file to a separate directory to prevent it from being overwritten in subsequent analyses.

3.5.2 Radiomics Feature Extraction

This section outlines the radiomic features extracted using the open-source Pyradiomics library, which is an open-source Python package for the extraction of Radiomics features from 2D and 3D images and binary masks. For more information on the Pyradiomics package (*see* **Note 6**).

1. Before performing radiomics feature extraction, a PyRadiomics environment must be created, and the PyRadiomics package installed. In VS Code, select *File→ New Text File* to open an empty text editor window. Save this file as "Radiomics_analysis.ipynb" to create a new notebook for the analysis.
2. Once the Python notebook has been created, open the Command Palette in VS Code by pressing Ctrl + *Shift* + *P* (Windows) or *Cmd* + *Shift* + *P* (macOS). Type "Python: Create Environment" and press *Enter*. Choose either "Venv" or "Conda", depending on your preference and select the

File	Group	ParDim	CubeCount	TriangDim
736958.tif	Treatment_1_norm	0.4307637257388740	2.031196400902160	0.2853374351539300
631549.tif	Treatment_1_norm	0.41556198694687400	2.066532741744000	0.518948340562921
363638.tif	Treatment_1_norm	0.38777410530879200	1.9762023442914700	0.17444061417151900
844906.tif	Treatment_1_norm	0.3148636529320040	2.0116195417766400	0.13420361818234000
607918.tif	Treatment_1_norm	0.3929263086607320	2.0295153949997800	0.30737261191631100
448326.tif	Treatment_1_norm	0.36308129569276000	1.967757358030300	0.15303846637640500
144660.tif	Treatment_1_norm	0.3782884344574610	1.992126399886470	0.15712450233873000
856755.tif	Treatment_1_norm	0.3287335279238550	1.9907210753149200	0.17015002333070700
720422.tif	Treatment_1_norm	0.31598537999338300	2.0182610958561800	0.14232320930818300
891397.tif	Treatment_1_norm	0.3429778556423120	2.0183962301772800	0.1410235929681500
703594.tif	Treatment_1_norm	0.45674603111860100	1.9804257532611400	0.17796036663971000
899359.tif	Treatment_1_norm	0.3740709122110090	2.0105020652875900	0.14695015376166600
786949.tif	Treatment_2_norm	0.5084873734311800	1.9427200958000400	0.1948248915741770
853821.tif	Treatment_2_norm	0.2908098868062310	2.008275993448450	0.10380062425279900
718633.tif	Treatment_2_norm	0.33593535401292500	2.0402842991192600	0.2233805989982370
793746.tif	Treatment_2_norm	0.4145738306688710	1.9826556981221800	0.2726260743765810
790703.tif	Treatment_2_norm	0.41825826993932200	1.9430816969220300	0.3780498455857380
937599.tif	Treatment_2_norm	0.4194091958986480	2.028574685171320	0.19997195344569700
231894.tif	Treatment_2_norm	0.4519934986808140	2.005340950624810	0.12245433208803500
435249.tif	Treatment_2_norm	0.445243614712887	1.999490331715840	0.16832946786418800
229162.tif	Treatment_2_norm	0.36327101754315600	2.0010512100036700	0.10977745789307800
835206.tif	Treatment_2_norm	0.46539646583016900	2.0151983397993000	0.2053290623820090
797181.tif	Treatment_2_norm	0.34656481413198200	2.0033137107823100	0.18277087188602900
319766.tif	Treatment_2_norm	0.45066724941914200	1.9731241206182100	0.1640154679982780

Fig. 5 Example output.CSV file generated from the MFTA feature extraction code opened in the "Numbers" software on a macOS computer system. The first column lists the file names, while the second column ("Group") classifies the files according to their treatment groups ("`Treatment_1_norm`" and "`Treatment_2_norm`"). Subsequent columns contain the extracted MFTA features along with their corresponding numerical values

desired Python interpreter. The environment creation process is complete once the selected interpreter appears in the bottom-right corner of the VS Code window.

3. Open a new terminal in VS Code by selecting *Terminal→ New Terminal.* A terminal window will appear at the bottom of the VS Code interface, allowing you to install required packages and execute the code within the newly created environment.
4. Open a new cell. To install the PyRadiomics package, type "`!pip install pyradiomics`" (Venv environment)

or "`conda install -c radiomics pyradiomics`" (Conda environment) and execute the code. This will install the required package using the "`pip`" function. To verify if the package is complete, open a Python interactive session by putting "`python`" into your terminal and press "*Enter*". If no error occurs, PyRadiomics is successfully installed and accessible within your VS Code environment.

5. Open the following GitHub link: https://github.com/MillijoyVillanueva/Multi-fractal_and_radiomics_profiling_in_microglia_in_vitro/blob/main/3_Radiomics_feature_extraction.ipynb. In the "Preview" window, this will display the cells corresponding to the steps described in this section.
6. Within the Python notebook in your PyRadiomics environment, create a new code cell, copy and execute the Python code from the cell "`# Cell 1—Installing required libraries for feature extraction`" from the provided GitHub link and execute it. This will install the libraries required for the radiomics feature extraction.
7. Once complete, open a new cell and execute the Python code from the cell "`# Cell 2—Importing packages for feature extraction`" to import the required packages for feature extraction.
8. Open a new cell code. Copy and paste the provided Python code from the cell "`# Cell 3—feature extraction code.`" In the "FOLDER_1" and "FOLDER_2" sections, add the file pathnames for your normalized images in the "`Treatment_1_norm`" and "`Treatment_2_norm`" folders. In the "OUTPUT_BASE" section, add your output file pathname. Once edited, run the code. This should start extracting radiomics features from your images and will save an output. CSV file titled "Radiomics_results.csv" into your output folder.
9. Once complete, a.CSV file (Fig. 6) will appear in your output folder containing 102 radiomics features extracted from your images.

3.6 Statistical Analysis

The section below provides a Python code to statistically analyze the multi-fractal and radiomics feature textures extracted from the previous sections (see Sect. 3.5). The code entails the removal of outliers, a two-sided Mann–Whitney *U* test, and a false discovery rate (FDR) correction. For more information on the statistical analysis methods (*see* **Note 7**).

1. You should now have two output.CSV files, one for MFTA features (saved as "MFTA_results.csv") and another for radiomics

Filename	Group	original_shape2D_Elongation	original_shape2D_MajorAxisLength
736958.tif	Treatment_1_norm	0.7500013320814120	1847.518583163150
631549.tif	Treatment_1_norm	0.750000024898505	1847.5203397864000
363638.tif	Treatment_1_norm	0.7499999583599880	1847.520832277950
844906.tif	Treatment_1_norm	0.7499995932962230	1847.5208191003900
607918.tif	Treatment_1_norm	0.7499998386989500	1847.520320391460
448326.tif	Treatment_1_norm	0.7499998562026050	1847.5203730340000
144660.tif	Treatment_1_norm	0.74999989976821	1847.5204609896
856755.tif	Treatment_1_norm	0.7499996736773400	1847.520563717670
720422.tif	Treatment_1_norm	0.7499995525223930	1847.520908210780
891397.tif	Treatment_1_norm	0.7499989243530150	1847.521441338050
703594.tif	Treatment_1_norm	0.7500000103556640	1847.5198399953500
899359.tif	Treatment_1_norm	0.7500002356448690	1847.5200067178600
786949.tif	Treatment_2_norm	0.7499993860432800	1847.5202222719500
853821.tif	Treatment_2_norm	0.7499999170511690	1847.5208490638100
718633.tif	Treatment_2_norm	0.7500002843257570	1847.5200989975
793746.tif	Treatment_2_norm	0.7499996887408000	1847.5205854685400
790703.tif	Treatment_2_norm	0.7499995071552290	1847.5209778040200
937599.tif	Treatment_2_norm	0.750000042340511	1847.5202857544900
231894.tif	Treatment_2_norm	0.749999839952626	1847.5206412943600
435249.tif	Treatment_2_norm	0.7499996961063100	1847.520979306040
229162.tif	Treatment_2_norm	0.7499994777946330	1847.5208928253300
835206.tif	Treatment_2_norm	0.7500001264179100	1847.5203870116400
797181.tif	Treatment_2_norm	0.7499997839966140	1847.5212523683200
319766.tif	Treatment_2_norm	0.7499998217520160	1847.520980546440

Fig. 6 Example output.CSV file generated from the Radiomics feature extraction code opened in the "Numbers" software on a macOS computer system. The first column lists the file names, while the second column ("Group") classifies the files according to their treatment groups ("Treatment_1_norm" and "Treatment_2_norm"). Subsequent columns contain the extracted Radiomics features along with their corresponding numerical values

features (saved as "Radiomics_results.csv"). Transfer your CSV files to your output folder.

2. Open the following GitHub link: https://github.com/MillijoyVillanueva/Multi-fractal_and_radiomics_profiling_in_microglia_in_vitro/blob/main/4_Statistical_analysis.ipynb. In the "Preview" window, this will display the cells corresponding to the steps described in this section.

3. In the Python Notebook file where you conducted MFTA feature extraction, or Radiomics feature extraction, open in VS Code, open a new cell. Copy and paste the Python code from the cell "`# Cell 1– Installing required libraries for statistical analysis`" from the provided GitHub link. This will install the libraries required for statistical analysis. You may have already installed these libraries, depending on whether you are using your MFTA Python Notebook or your Radiomics Python Notebook. Using the code below will allow the installation of libraries required for downstream analysis in either notebook files.
4. When complete, copy and paste the Python code from the cell "`# Cell 2–Importing packages required for statistical analysis`" below and execute it. These will import the required packages.
5. Once complete, copy and paste the Python code from the cell "`# Cell 3–Statistical analysis code`" below. The code conducts a Mann–Whitney *U* test, followed by multiple testing correction using the Benjamini–Hochberg false discovery rate (FDR) method. Before statistical analysis, outliers were removed using a Robust regression and outlier removal method. The bottom of "`input_csv`", add the pathname to your MFTA_results.csv file or your Radiomics_results.csv file, depending on what you want to analyze. If you want to analyze both, add the pathname of both files to the "`input_csv`" found at the bottom of the code. As you can see, the "#" remains for one of them. For the first analysis, run the code. Once the analysis is done for the first.CSV file, add a "#" to the first "`input_csv`" line, and remove it from the second one. Python ignores lines that start with "`input_csv`", so the code will analyze whichever.CSV does not have it. Before the analysis, add the pathname to your output folder, where you want all the output files to go. When done, run the cell.
6. If the code is successful, you will find two new.CSV files in your output folder: (1.) "`Mann-Whitney_raw.csv`" (Fig. 7) and (2) "`Mann-Whitney_FDR.csv`" (Fig. 8). The "`Mann-whitney_raw.csv`" will contain the direct results of the Mann–Whitney *U* test. You will also find a new folder called "`compiled_significance_plots.`" If the code fails to run, it is likely due to an incorrect directory path to the extracted features CSV file. Ensure that the correct path to the extracted features.CSV generated from the previous step is specified before executing the code.
7. Repeat this step for your other.CSV file. Make sure to save your "`Mann-Whitney_raw.csv`" and "`Mann-Whitney_FDR.csv`" a separate folder, or rename it to

Feature	U_stat	raw_p_value	Mean_Treatment_1_norm	Mean_Treatment_2_norm	Median_Treatment_1_norm	Median_Treatment_2_norm	Direction	Significance_raw
Mink_Blan_Sq	16.0	0.0013539414430926300	0.612098459035096	0.6327384962262880	0.6149913094492930	0.6334191450955870	↑ Treatment_2_norm	**
Mink_VarDil_Sq	16.0	0.0013539414430926300	0.612098459035096	0.6327384962262880	0.6149913094492930	0.6334191450955870	↑ Treatment_2_norm	**
Hig1D_Meander	108.0	0.0017358602386388000	-0.5977255823967790	-0.6393910541015450	-0.605174110743428	-0.6437915101279860	↓ Treatment_2_norm	**
Mink_Blan_Diamond	20.0	0.0029456461505329800	0.6611872302282620	0.6827224740707690	0.6638710976587780	0.6817734844038380	↑ Treatment_2_norm	**
Mink_VarDil_Disk	20.0	0.0029456461505329800	0.698968279776643	0.721185398049829	0.7020330371999520	0.7207114492459200	↑ Treatment_2_norm	**

Fig. 7 Example output.CSV file "mann_whitney_raw.csv" generated from the Statistical Analyses code opened in the "Numbers" software in a macOS computer system. The first column lists the feature names, the second column states the test statistic from the Mann–Whitney *U* test ("U_stat"), the third column lists the raw *p*-value, while the fourth and fifth columns list the mean value of the two treatment groups. Column six indicates whether there is an increase in the second group relative to the first, and the seventh column denotes the statistical significance using an asterisk system (*$p<0.05$, **$p<0.01$, ***$p<0.001$, **** $p<0.0001$, *ns* not significant)

Feature	U_stat	raw_p_value	Mean_Treatment_1_norm	Mean_Treatment_2_norm	Median_Treatment_1_norm	Median_Treatment_2_norm	Direction	Significance_raw	adj_p_value	Significance_FDR
Mink_Blan_Sq	16.0	0.0013539414430926300	0.612098459035096	0.6327384962262880	0.6149913094492930	0.6334191450955870	↑ Treatment_2_norm	**	0.03711572833745910	*
PyrDim_GPGM_Bicub2	22.0	0.0074231456674918300	-0.035271425695075200	0.0034308631004848500	-0.0362909960483456	0.0017496446422603	↑ Treatment_2_norm	**	0.03711572833745910	*
PyrDim_Kolm_Aver	22.0	0.0074231456674918300	-1.954385427933140	-1.9400964469397100	-1.9547608860208000	-1.9357194286934600	↑ Treatment_2_norm	**	0.03711572833745910	*
PyrDim_Kolm_Bilin	22.0	0.0074231456674918300	-1.954385427933140	-1.9400964469397100	-1.9547608860208000	-1.9357194286934600	↑ Treatment_2_norm	**	0.03711572833745910	*
PyrDim_GPGM_Bicub	22.0	0.0074231456674918300	-0.035271425695075200	0.0034308631004848500	-0.0362909960483456	0.0017496446422603	↑ Treatment_2_norm	**	0.03711572833745910	*

Fig. 8 Example output.CSV file "mann_whitney_FDR.csv" generated from the Statistical Analyses code opened in the "Numbers" software in a macOS computer system. The first seven columns are similar to the "mann_whitney_raw.csv" file, followed by the eighth column, which shows the adjusted *p*-value after FDR correction, and the ninth column denotes statistical significance using an asterisk system (*$p<0.05$, **$p<0.01$, ***$p<0.001$, ****$p<0.0001$, *ns* not significant)

"`Mann-Whitney_raw_MFTA.csv`" or "`Mann-Whitney_raw_Radiomics,`" depending on which analysis you did, to avoid any override of files.

3.7 Expected Results and Interpretation

At the end of the feature extraction section and the statistical analysis, you should now have two.CSV files for each type of feature extraction process: The "`Mann_Whitney_raw.csv`" (*see* Fig. 7), which contains the unadjusted Mann–Whitney *U* test, and the "`Mann_Whitney_FDR.csv`" (see Fig. 8), which contains the same results with additional columns for false discovery rate (FDR)-corrected *p*-values using the Benjamini–Hochberg method. For the columns, from left to right, they will contain:

A. Column 1—The name of the feature extracted from the image dataset.

B. Column 2—The Mann–Whitney *U* statistic ("`U_stat`"), which indicated the degree of rank-based difference between the two distributions.

C. Column 3—The unadjusted *p*-value for the test ("`raw_p_value`"). Smaller values indicate stronger evidence of a statistically significant difference between groups before multiple testing correction.

D. Columns 4–7—The mean and median value of the features of both treatment groups. This helps identify which groups show higher average results.

E. Column 8—The "*Direction*" which shows which group shows an overall higher feature value based on median comparison. An upward arrow (↑) denotes an increase in the second group ("`Treatment_2`") relative to the first, while a downward arrow (↓) denotes a decrease.

F. Column 9—The "Significance_raw" column, which is the significance level based on the raw p-value, represented as asterisks.

G. Column 10 (FDR results only)—The "`adj_p_value`" column-, which is the p-value after correction for multiple comparisons using the Benjamini–Hochberg FDR method. This value accounts for the false discovery rate when multiple features are tested simultaneously.

H. Column 11 (FDR results only)—The "`Significance_FDR`" column which is the adjusted significance level based on the FDR corrected *p*-value, using the same asterisk system as above.

Additionally, you will also find a folder called "`compiled_significance_plots`," which will contain a PNG file that visually summarizes the significant features as horizontal bar plots. You will have two.PNG files, one for the Mann–Whitney *U* results ("`MannWhitney_directional.png`") (Fig. 9) and one for the FDR analysis ("`FDR_directional.png`") (Fig. 10).

When comparing the two approaches, in our experiment, we observed that the MFTA feature set yielded a larger number of statistically significant features (after FDR correction) than the Radiomics set. However, the two methods provided complementary information: MFTA features tended to capture intensity-structure relationships rooted in spatial scaling (e.g., Higuchi, box-counting, lacunarity), while Radiomics features reflected intensity distributions and textural co-occurrence. Only a minority of significant features overlapped across the two sets, suggesting that combining fractal-based and conventional radiomics features increases the sensitivity to detect subtle morphological responses in confluent microglial monolayers.

The following is the description of the visual summary:

A. The *x*-axis represents the *p*-value (If it is the Mann–Whitney analysis, it is the raw *p*-value; if it is the FDR analysis, it is the adjusted *p*-value) displayed in a logarithmic scale.

B. The *y*-axis lists the statistically significant feature names.

C. Each horizontal bar corresponds to a statistically significant feature, with the length of the bar indicating the strength of

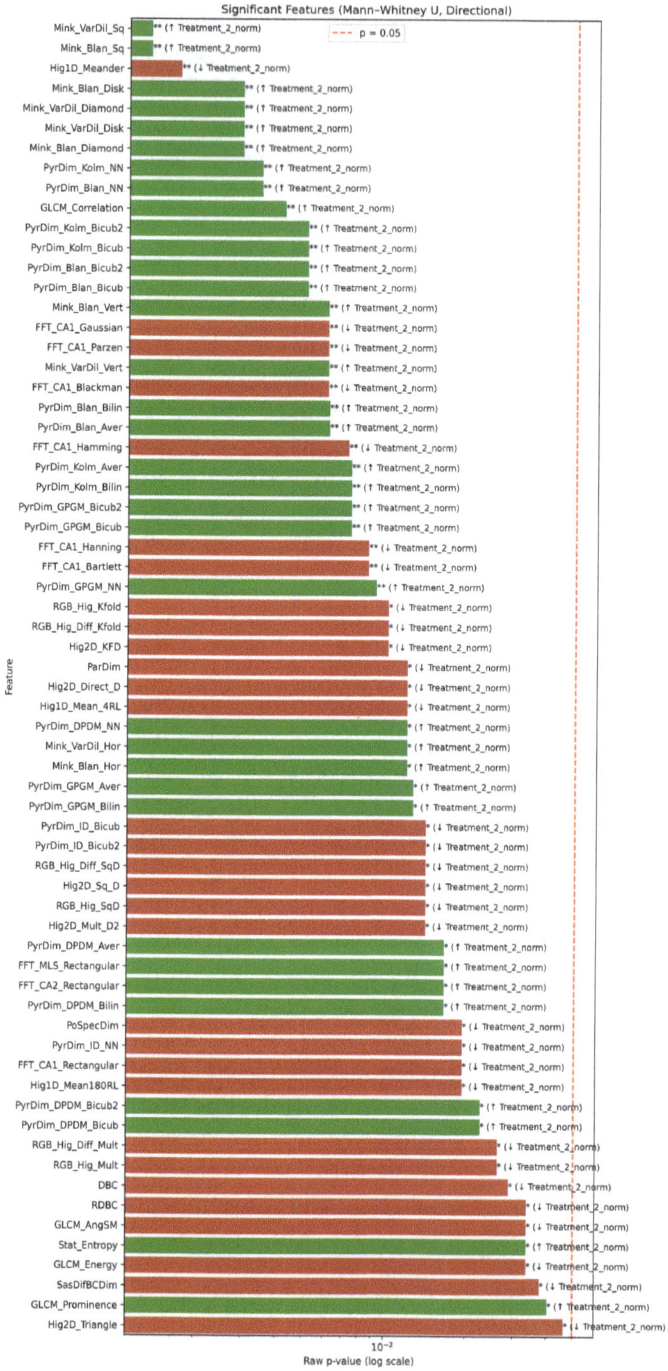

Fig. 9 Example output.PNG file "`MannWhitney_directional.png`" which displays the ranked list of features that differ significantly between the two treatment groups based on their raw Mann–Whitney *Up*-values (from most significant to least significant). For each feature, the *x*-axis shows the *p*-value on a log scale, the direction of change in Treatment 2 relative to Treatment_1 (increase or decrease), and their corresponding significance level denoted using the asterisk system ($^{*}p<=0.05$, $^{**}p<0.01$, $^{***}p<0.001$, $^{****}p<0.0001$)

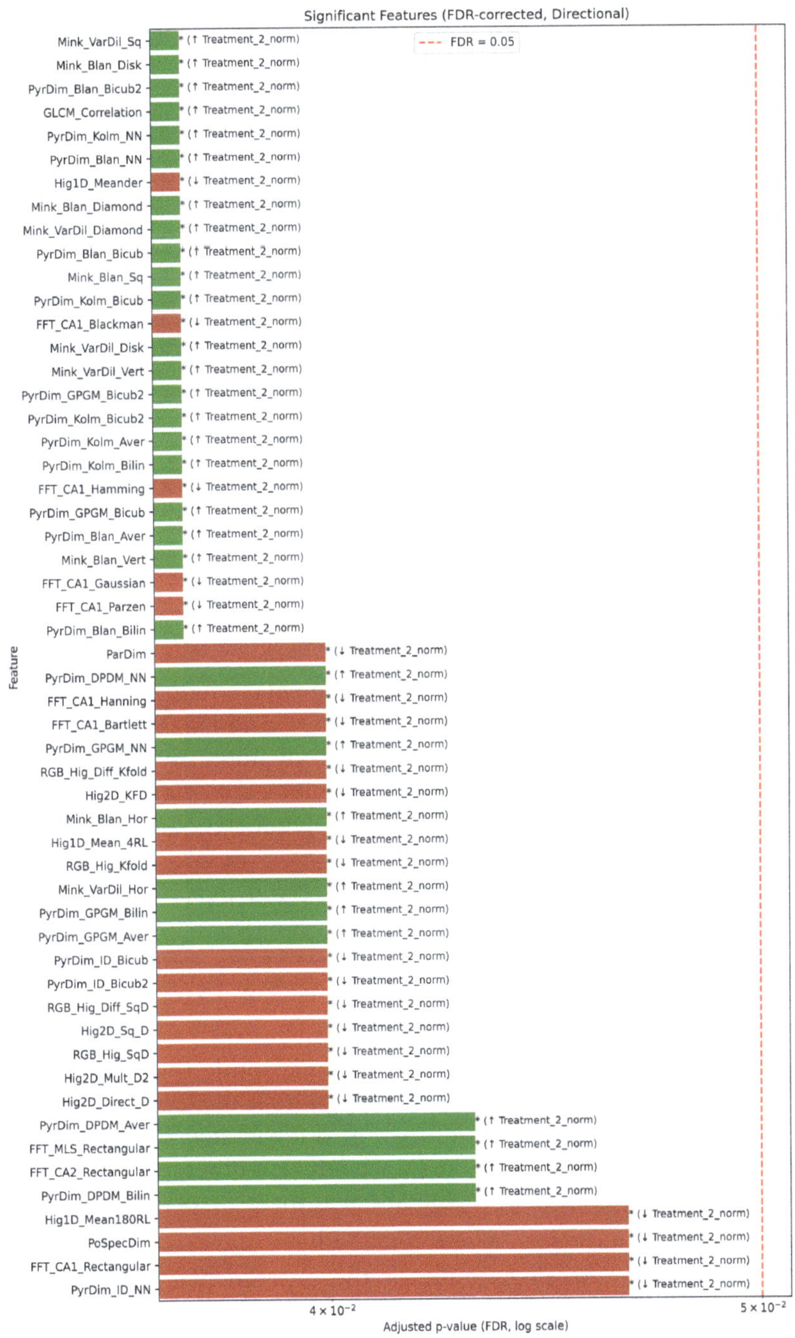

Fig. 10 Example output.PNG file "`FDR_directional.png`" which displays the ranked list of features that differ significantly between the two treatment groups based on their FDR-adjusted p-values (from most significant to least significant). For each feature, the x-axis shows the p-value on a log scale, the direction of change in Treatment 2 relative to Treatment_1 (increase or decrease), and their corresponding significance level denoted using the asterisk system ($^*p < 0.05$, $^{**}p < 0.01$, $^{***}p < 0.001$, $^{****}p < 0.0001$)

statistical significance, where shorter bars are considered more significant.

D. A red dashed vertical line marks the significance threshold (0.05 for both visuals). Features to the left of this line are considered statistically significant.

E. Green bars indicate features that are increased in your "`Treatment_2`" group relative to the first. Red bars indicate features that are decreased.

F. Each bar is annotated with a significance level, which is similar to the asterisk system used, and the direction of the change.

Overall, the plot provides a compact visual summary of all significant features and their directional trends to allow interpretation of complex datasets.

4 Notes

1. To validate the findings from the multifractal-lacunarity-texture analysis, a complementary morphological assessment of C20 cells was performed. Using Fiji (v2.16.0), the boundaries of individual cells within the monolayer were manually traced to evaluate basic morphological parameters. The measured features included cell area, perimeter, circularity, and cell roughness (calculated as $\frac{\text{Perimeter}}{\text{Area}}$). Results show that exposure to SMF significantly reduced cell area, perimeter, and roughness of C20 microglial cells compared to the control group (all $p<0.0001$, ****). However, no significant differences were observed in cell circularity (Fig. 11).
2. The human microglial cell line used when developing this protocol is C20 [32]. When adapting this protocol for other cell lines, it is advisable to optimize the seeding density to achieve a uniform monolayer. It is recommended that preliminary trials be conducted to account for differences in cell size, growth rate, and proliferation, as these factors affect image quality and feature extraction accuracy.
3. As an example of the experimental treatment, we exposed cells to a moderate-strength static magnetic field (SMF). The ambient magnetic field was weak, and equaled $45 \pm 20\,\mu T$ in the laboratory and $35 \pm 10\,\mu T$ in the incubator. As a source of experimental treatment, SMF, a 20×26 "Mega Magnetic Plate" by Oz Biosciences (MF14000, #NC2382291) was used. According to the manufacturer, the plate comprises multiple rare-earth square magnets, each with a remanence of 1080–1150 mT and a maximum energy of 263–279 kJ/m^3. Adjacent magnets have alternating polarity to produce a

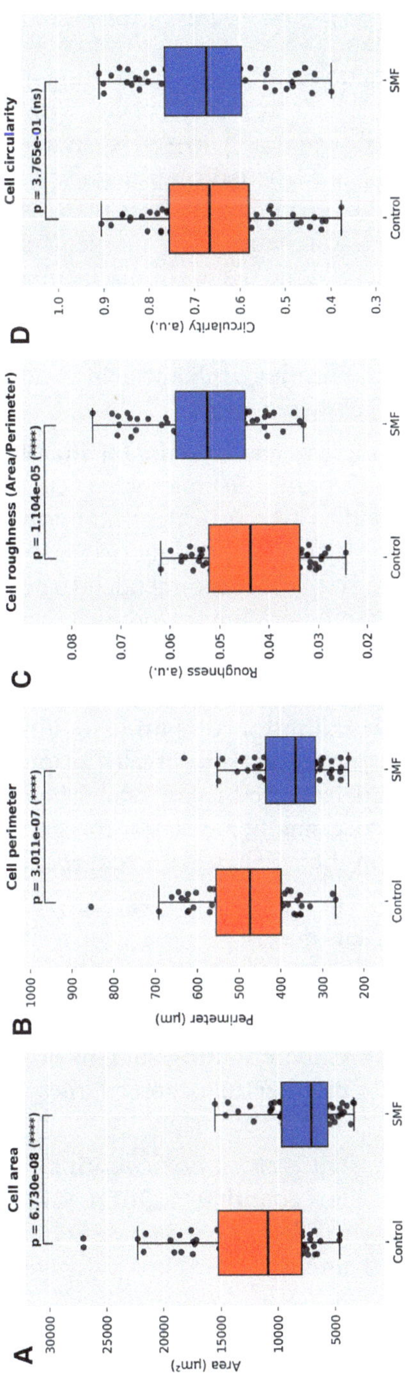

Fig. 11 Basic morphological assessment of C20 cells after exposure to SMF. (**A**) Cell area. (**B**) Cell perimeter. (**C**) Cell roughness (calculated as $\frac{\text{Perimeter}}{\text{Area}}$). (**D**) Cell circularity

more homogeneous field. Our measurements indicated that, at the "Mega Magnetic Plate" surface, the magnetic flux density (MFD) is nonuniform across and between magnets, with a mean of $|235 \pm 40|$ mT. The cells cultured in the 12-well plate (Costar, #3513) are positioned 3.43 mm above the "Mega Magnetic Plate" surface, resulting in exposure to an MFD of $|142 \pm 9|$ mT at the cell plane, with the vertical gradient, ∇Bz (at cell plane) ≈ $|36 \pm 3|$ T/m and the lateral gradient, ∇Bx (within-well, across a magnet) ≈ $|1.2–14.1|$ T/m [33]. For the SMF group, cell culture plates were placed centrally on the "Mega Magnetic Plate" immediately after seeding C20 cells and kept in the tissue culture incubator for 21 h. Negative control (SMF-untreated) cell culture plates were kept in the same incubator, at a distance of ≥15 cm from the "Mega Magnetic Plate," ensuring ambient field conditions. After 21 h of exposure, cells were imaged.

4. Imaging was conducted 21 h after cell seeding and treatment, at which point C20 microglial cells formed a stable monolayer under the SMF treatment and control conditions. This time point may vary depending on the microglial cell line and treatment used. It is recommended to perform a preliminary experiment to determine the optimal incubation period required to achieve consistent monolayer formation and accurate feature extraction.

5. The workflow for computing multifractal, texture, and lacunarity features was conceptually adapted from prior studies [30] and implemented using established computational and imaging processing libraries. For the summary and further information on the features extracted, access the cited references.

   A. Fractal geometry features (NumPy, SciPy, OpenCV, scikit-image):

      - Partitioning dimension (ParDim)—scaling of local variance across subdivided tiles within the image.
      - Cube counting dimension (CubeCount) [34, 35]—number of occupied tiles across multiple grid sizes within the image
      - Differential box counting (DBC), relative differential box counting (RDBC), Sasaki differential box counting (SDBC)—Converts 2D image into 3D and estimates intensity-based box coverage at multiple scales [36–38].
      - Raster FracLac box counting (FLBC), Sliding FracLac sliding box (FLSB), FracLac local box counting and sliding box (FLLBC, FLSBL)—divides image non overlapping boxes or windows to measure fractal occupancy [39].

- Triangulation dimension (TriangDim)—Constructs Delaunay triangles over the intensity surface and scales the total area versus the sampling step.
- Correlation dimension (CorDim).
- Minkowski dimensions (balance and variance of dilation using square, disk, diamond, horizontal, and vertical structuring elements)—uses dilation and erosion with structuring elements to measure morphological balance and boundary roughness [40].
- Mass-radius dimension (Mass_Rad_Dim)—plots cumulative pixel intensity versus radius from the image's center of mass to measure radial intensity scaling [39, 41].

B. Spectral and frequency-domain features (SciPy FFT, SciPy Signal):

- Fast Fourier Transform (FFT) based descriptors—Circular average slope (CA1), CIrcular Average Integral (CA2), and mean level of spectrum (MLS)—analyses spatial frequency distributions via FFT [42].
- Power spectrum dimension (PoSpecDim)—fractal slope of spectral power decay.

C. Fractal curve and surface roughness features (NumPy, SciPy):

- Higuchi 1D Fractal Dimension—Single, Meander, Mean 4 Row and Column (4RC), 4 Radial Lines (4RL), and 180 Radial Lines profiles - Measures roughness of 1D intensity profiles derived from the image [43].
- Higuchi 2D Fractal Dimension—K-Fold Difference (KFD), Multiplicative Difference 1 and 2 (MultD_1, MultD_2), Squared Difference (Sq_D), Direct Difference (Direct_D), Triangle—Extends Higuchi analysis to 2D surfaces.
- Fractal Anisotropy Index (FAI)—Absolute difference between Higuchi 1D values.

D. Morphological and Multi-Resolution Features (scikit-image, OpenCV):

- Pyramidal-based Fractal Descriptors—Difference of Pixel Difference Mean (PyrDim_DPDM), Gradient Pattern Gradient Mean (PyrDim_Gradient Pattern Gradient Mean), Balance (PyrDim_Blan), Allometry (PyrDim_Allo), Information Dimension (PyrDim_ID), Kolmogorov (PyrDim_Kolm)—Assess scale-dependent texture and morphological transformations via image pyramids [44].

E. Texture Features (scikit-image):

- Gray-level co-occurrence matrix (GLCM) features—Contrast, Dissimilarity, Homogeneity, Inverse Difference Moment (IDM), Energy, Entropy, Variance, Correlation, Shade, Prominence, and Maximum Probability—Quantify spatial relationships between pixel intensities.

F. Lacunarity Features (NumPy,scikit-image):

- Raster Pattern Lacunarity (RP_Lacunarity)—uses non-overlapping grids to quantify global gap variation.
- Sliding Window Lacunarity (SV_Lacunarity)—employs overlapping windows for local heterogeneity.
- Lacunarity Dimension—Slope of log–log lacunarity versus inverse box size

G. Complexity and Compression-Based Features (zlib, gzip, cv2. Imencode):

- Normalized Kolmogorov Complexity using ZIP (NKC_ZIP) and GZIP (NKS_GZIP) compression algorithms—ratio of compressed to uncompressed byte size using ZIP and GZIP.

6. Radiomic features were extracted using the open-source PyRadiomics library, implemented in Python. In the Python code, the "radiomics.featureextractor" module was implemented to provide a standardized implementation of radiomic feature classes that quantify image intensity distributions, geometric properties, and higher order textural relationships, also present at the earlier MFTA feature extraction analysis. Information on the mathematical definitions and description of the extracted radiomics features can be accessed through the PyRadiomics website: https://pyradiomics.readthedocs.io/en/latest/.
7. The provided Python code statistically analyzes the MFTA and Radiomics feature files extracted from the previous sections. Firstly, outliers were removed using a method inspired by the ROUT method, which is a robust statistical procedure implemented in GraphPad Prism, a well-known scientific graphing and statistical analysis software. ROUT stands for "Robust Regression and Outlier Removal" [45]. After outlier removal after this method, a two-sided Mann-Whitney *U* test is conducted for each extracted feature in both MFTA and radiomics analyses to assess differences between treatment groups, which is a non-parametric statistical test between groups. To further

enhance the reliability of statistical inference, false discovery rate (FDR) correction was applied using the Benjamini–Hochberg method, which adjusts for multiple hypothesis testing and reduces the likelihood of false positives. Adjusted p-values are then returned, ensuring the features are identified as significant. To enhance interpretability, statistical significance in the output.CSV files was annotated with an asterisk system; $^{*}p<0.05$, $^{**}p<0.01$, $^{***}p<0.001$, $^{****}p<0.0001$, ns not significant.

Acknowledgments

The development of this protocol has been supported by a Macquarie University Research Fellowship for Anna Guller and a Macquarie University PhD scholarship for Millijoy D. Villanueva.

References

1. Hanisch UK, Kettenmann H (2007) Microglia: active sensor and versatile effector cells in the normal and pathologic brain. Nat Neurosci 10(11):1387–1394. https://doi.org/10.1038/nn1997
2. Smith Jr TG, Lange GD, Marks WB (1996) Fractal methods and results in cellular morphology—dimensions, lacunarity and multifractals. J Neurosci Methods 69(2):123–136. https://doi.org/10.1016/S0165-0270(96)00080-5
3. Soltys Z, Orzylowska-Sliwinska O, Zaremba M, Orlowski D, Piechota M, Fiedorowicz A, Janeczko K, Oderfeld-Nowak B (2005) Quantitative morphological study of microglial cells in the ischemic rat brain using principal component analysis. J Neurosci Methods 146(1):50–60. https://doi.org/10.1016/j.jneumeth.2005.01.009
4. Karperien AL, Jelinek HF (2015) Fractal, multifractal, and lacunarity analysis of microglia in tissue engineering. Front Bioeng Biotechnol 3:51. https://doi.org/10.3389/fbioe.2015.00051
5. Karperien A, Ahammer H, Jelinek HF (2013) Quantitating the subtleties of microglial morphology with fractal analysis. Front Cell Neurosci 7:3. https://doi.org/10.3389/fncel.2013.00003
6. Gunasegaran B, Krishnamurthy S, Chow SS, Villanueva MD, Guller A, Ahn SB, Heng B (2025) Comparative analysis of HMC3 and C20 microglial cell lines reveals differential myeloid characteristics and responses to immune stimuli. Immunology 175(1):84–102. https://doi.org/10.1111/imm.13900
7. Jaccard N, Griffin LD, Keser A, Macown RJ, Super A, Veraitch FS, Szita N (2013) Automated method for the rapid and precise estimation of adherent cell culture characteristics from phase contrast microscopy images. Biotechnol Bioeng 111(3):504–517. https://doi.org/10.1002/bit.25115
8. Yin Z, Kanade T, Chen M (2012) Understanding the phase contrast optics to restore artifact-free microscopy images for segmentation. Med Image Anal 16(5):1047–1062. https://doi.org/10.1016/j.media.2011.12.006
9. Pirone D, Bianco V, Miccio L, Memmolo P, Psaltis D, Ferraro P (2024) Beyond fluorescence: advances in computational label-free full specificity in 3D quantitative phase microscopy. Curr Opin Biotechnol 85:103054. https://doi.org/10.1016/j.copbio.2023.103054
10. Vicar T, Balvan J, Jaros J, Jug F, Kolar R, Masarik M, Gumulec J (2019) Cell segmentation methods for label-free contrast microscopy: review and comprehensive comparison. BMC Bioinform 20(1):360. https://doi.org/10.1186/s12859-019-2880-8
11. Bayandin Y, Nikitiuk A, Voronina A, Beloglazova Y, Gagarskikh O, Grishko V, Naimark O (2021) Analysis of morphometric and fractal properties of phase-contrast images of native and apoptotically changed MCF-7 cells. Russ J Biomech 24(3):268–276. https://doi.org/10.15593/RJBiomech/2020.3.04

12. Debeir O, Van Ham P, Kiss R, Decaestecker C (2005) Tracking of migrating cells under phase-contrast video microscopy with combined mean-shift processes. IEEE Trans Med Imaging 24(6):697–711. https://doi.org/10.1109/tmi.2005.846851
13. Lier J, Streit WJ, Bechmann I (2021) Beyond activation: characterizing microglial functional phenotypes. Cells 10(9):2236. https://doi.org/10.3390/cells10092236
14. Condello C, Yuan P, Schain A, Grutzendler J (2015) Microglia constitute a barrier that prevents neurotoxic protofibrillar Aβ42 hotspots around plaques. Nat Commun 6(1):6176. https://doi.org/10.1038/ncomms7176
15. Wang G, Zhong K, Wang Z, Zhang Z, Tang X, Tong A, Zhou L (2022) Tumor-associated microglia and macrophages in glioblastoma: from basic insights to therapeutic opportunities. Front Immunol 13:964898. https://doi.org/10.3389/fimmu.2022.964898
16. Noorani I, Sidlauskas K, Pellow S, Savage R, Norman JL, Chatelet DS, Fabian M, Grundy P, Ching J, Nicoll JAR, Boche D (2023) Clinical impact of anti-inflammatory microglia and macrophage phenotypes at glioblastoma margins. Brain Commun 5(3):fcad176. https://doi.org/10.1093/braincomms/fcad176
17. Eles JR, Vazquez AL, Snyder NR, Lagenaur C, Murphy MC, Kozai TDY, Cui XT (2017) Neuroadhesive L1 coating attenuates acute microglial attachment to neural electrodes as revealed by live two-photon microscopy. Biomaterials 113:279–292. https://doi.org/10.1016/j.biomaterials.2016.10.054
18. Karperien A, Jelinek H, Milosevic N (2011) Lacunarity analysis and classification of microglia in neuroscience. In Proceedings of the 8th European Conference on Mathematical and Theoretical Biology Krakow, Poland, p. MS#88. doi:https://doi.org/10.13140/2.1.3576.6082
19. Roy A, Perfect E (2014) Lacunarity analyses of multifractalmulti-fractal and natural grayscale patterns. Fractals-Complex Geometry Patterns and Scaling in Nature and Society 22(3):1440003. https://doi.org/10.1142/S0218348x14400039
20. Aerts HJ, Velazquez ER, Leijenaar RT, Parmar C, Grossmann P, Carvalho S, Bussink J, Monshouwer R, Haibe-Kains B, Rietveld D, Hoebers F, Rietbergen MM, Leemans CR, Dekker A, Quackenbush J, Gillies RJ, Lambin P (2014) Decoding tumour phenotype by noninvasive imaging using a quantitative radiomics approach. Nat Commun 5: 4006. https://doi.org/10.1038/ncomms5006
21. van Griethuysen JJM, Fedorov A, Parmar C, Hosny A, Aucoin N, Narayan V, Beets-Tan RGH, Fillion-Robin J-C, Pieper S, Aerts HJWL (2017) Computational radiomics system to decode the radiographic phenotype. Cancer Res 77(21):e104–e107. https://doi.org/10.1158/0008-5472.Can-17-0339
22. Jang K, Russo C, Di Ieva A (2020) Radiomics in gliomas: clinical implications of computational modeling and fractal-based analysis. Neuroradiology https://doi.org/10.1007/s00234-020-02403-1
23. Khalili N, Kazerooni AF, Familiar A, Haldar D, Kraya A, Foster J, Koptyra M, Storm PB, Resnick AC, Nabavizadeh A (2023) Radiomics for characterization of the glioma immune microenvironment. NPJ Precis Oncol 7(1):59. https://doi.org/10.1038/s41698-023-00413-9
24. Inglese M, Conti A, Toschi N (2025) Radiomics across modalities: a comprehensive review of neurodegenerative diseases. Clin Radiol 85:106921. https://doi.org/10.1016/j.crad.2025.106921
25. Di Lazzaro V, Ranieri F, Doretti A, Boscarino M, Maderna L, Colombo E, Soranna D, Zambon A, Ticozzi N, Musumeci G, Capone F, Silani V (2024) Transcranial static magnetic stimulation for amyotrophic lateral sclerosis: a bicentric, randomised, double-blind placebo-controlled phase 2 trial. Lancet Reg Health Eur 45:101019. https://doi.org/10.1016/j.lanepe.2024.101019
26. Di Pino G, Pellegrino G, Assenza G, Capone F, Ferreri F, Formica D, Ranieri F, Tombini M, Ziemann U, Rothwell JC, Di Lazzaro V (2014) Modulation of brain plasticity in stroke: a novel model for neurorehabilitation. Nat Rev Neurol 10(10):597–608. https://doi.org/10.1038/nrneurol.2014.162
27. Barry-Carroll L, Gomez-Nicola D (2024) The molecular determinants of microglial developmental dynamics. Nat Rev Neurosci 25(6):414–427. https://doi.org/10.1038/s41583-024-00813-1
28. Paolicelli RC, Sierra A, Stevens B, Tremblay ME, Aguzzi A, Ajami B, Amit I, Audinat E, Bechmann I, Bennett M, Bennett F, Bessis A, Biber K, Bilbo S, Blurton-Jones M, Boddeke E, Brites D, Brône B, Brown GC, Butovsky O, Carson MJ, Castellano B, Colonna M, Cowley SA, Cunningham C, Davalos D, De Jager PL, de Strooper B, Denes A, Eggen BJL, Eyo U, Galea E, Garel S, Ginhoux F, Glass CK, Gokce O, Gomez-Nicola D, González B, Gordon S, Graeber MB, Greenhalgh AD, Gressens P, Greter M, Gutmann DH, Haass C, Heneka MT, Heppner FL, Hong S, Hume DA, Jung S, Kettenmann H, Kipnis J, Koyama R, Lemke G, Lynch M, Majewska A, Malcangio M, Malm T, Mancuso R, Masuda T, Matteoli M, McColl BW, Miron VE, Molofsky AV, Monje M, Mracsko E,

Nadjar A, Neher JJ, Neniskyte U, Neumann H, Noda M, Peng B, Peri F, Perry VH, Popovich PG, Pridans C, Priller J, Prinz M, Ragozzino D, Ransohoff RM, Salter MW, Schaefer A, Schafer DP, Schwartz M, Simons M, Smith CJ, Streit WJ, Tay TL, Tsai LH, Verkhratsky A, von Bernhardi R, Wake H, Wittamer V, Wolf SA, Wu LJ, Wyss-Coray T (2022) Microglia states and nomenclature: a field at its crossroads. Neuron 110(21):3458–3483. https://doi.org/10.1016/j.neuron.2022.10.020

29. Costa EV, Nogueira Rde A (2015) Multifractal dimension and lacunarity of yolk sac vasculature after exposure to magnetic field. Microvasc Res 99:1–7. https://doi.org/10.1016/j.mvr.2015.02.002
30. Andronache I (2025) Application of fractal analysis in interpreting 2D and 3D grayscale images: methodologies and case studies. J Bulg Geogr Soc 52:157–182
31. Andronache I, Peptenatu D, Ahammer H, Radulovic M, Djuricic GJ, Jelinek HF, Russo C, Di Ieva A (2024) Fractals in the neurosciences: a translational geographical approach. Adv Neurobiol 36:953–981. https://doi.org/10.1007/978-3-031-47606-8_47
32. Garcia-Mesa Y, Jay TR, Checkley MA, Luttge B, Dobrowolski C, Valadkhan S, Landreth GE, Karn J, Alvarez-Carbonell D (2017) Immortalization of primary microglia: a new platform to study HIV regulation in the central nervous system. J Neurovirol 23(1):47–66
33. Villanueva MDM (2023) Exploring the impact of static magnetic fields on human glioblastoma cells in vitro [Internet]. Macquarie University; [cited 2026Jul10]. Available from: https://figshare.mq.edu.au/articles/thesis/Exploring_the_impact_of_static_magnetic_fields_on_human_glioblastoma_cells_in_vitro/25040072/1
34. Douketis C, Wang Z, Haslett TL, Moskovits M (1995) Fractal character of cold-deposited silver films determined by low-temperature scanning tunneling microscopy. Phys Rev B 51(16):11022–11031. https://doi.org/10.1103/PhysRevB.51.11022
35. Zahn W, Zösch A (1999) The dependence of fractal dimension on measuring conditions of scanning probe microscopy. Fresenius J Anal Chem 365(1):168–172. https://doi.org/10.1007/s002160051466
36. Sarkar N, Chaudhuri BB (1992) An efficient approach to estimate fractal dimension of textural images. Pattern Recogn 25(9):1035–1041. https://doi.org/10.1016/0031-3203(92)90066-R
37. Jin XC, Ong SH, Jayasooriah (1995) A practical method for estimating fractal dimension. Pattern Recognit Lett 16(5):457–464. https://doi.org/10.1016/0167-8655(94)00119-N
38. Sasaki H, Shibata S, Hatanaka T (1994) Method for evaluation of Japanese lawn grass (Zoysia japonica Steud.) ecotypes for different purposes. Bull Natl Grassl Res Inst 49:17–24
39. Karperien A (2013) FracLac for ImageJ, version 2.5. Charles Sturt University. https://imagej.net/ij/plugins/fraclac/FLHelp/Introduction.htm. Accessed 10 July 2026.
40. Dubuc B, Zucker SW, Tricot C, Quiniou JF, Wehbi D, Berry MV (1989) Evaluating the fractal dimension of surfaces. Proc R Soc Lond A Math Phys Sci 425(1868):113–127. https://doi.org/10.1098/rspa.1989.0101
41. Ahammer H, Helige C, Dohr G, Weiss-Fuchs U, Juch H (2008) Fractal dimension of the choriocarcinoma cell invasion front. Physica D 237(4):446–453. https://doi.org/10.1016/j.physd.2007.09.016
42. Anguiano E, Pancorbo M, Aguilar M (1993) Fractal characterization by frequency analysis. I. Surfaces. J Microsc 172(3):223–232. https://doi.org/10.1111/j.1365-2818.1993.tb03416.x
43. Higuchi T (1988) Approach to an irregular time series on the basis of the fractal theory. Physica D 31(2):277–283. https://doi.org/10.1016/0167-2789(88)90081-4
44. Mayrhofer-Reinhartshuber M, Ahammer H (2016) Pyramidal fractal dimension for high resolution images. Chaos 26(7), 073109. https://doi.org/10.1063/1.4958709
45. Rousseeuw PJ, Leroy AM (2003) Robust regression and outlier detection. Wiley Series in Probability and Statistics. Wiley-Interscience, Hoboken, NJ. ISBN 978-0-471-48855-2. https://www.wiley.com/en-au/shop/general-introductory-statistics/robust-regression-and-outlier-detection-p-9780471488552. Accessed 10 July 2026

Chapter 6

Computational Pipeline to Infer Micro-, Meso-, and Macro-Scale Brain-State Transitions

Brandon R. Munn and James M. Shine

Abstract

Attractors are a framework for understanding and modeling the neural dynamics argued to underlie diverse cognitive phenomena, be it working memory, learning, neural oscillations, or arousal. Neural attractors have been argued to be present at the scale of single-cells, networks of cells, and between systems-scale regions of the whole-brain. Here, we describe a methodology inspired by statistical mechanics to calculate low-dimensional effective energy landscapes from neural data providing insight into the probability of brain-state transitions, which allow us to infer the brain's approximate attractor structure.

Keywords State transitions, Attractors, Neural dynamics, Energy landscape, Brain flexibility

1 Introduction

The brain is a complex dynamic system whose function emerges from coordinated activity across multiple spatial and temporal scales [1]. From moment-to-moment fluctuations in microscale neuronal activity, to long-range interactions between distributed macroscale cortical regions, neural dynamics constantly evolve as the brain processes, integrates, and responds to endogenous and exogenous stimuli. Understanding how such diverse patterns of neural activity are structured, and how they change over time and spatial scales, is essential to advancing our models of brain function.

One powerful way to track neural activity is using dynamical systems theory and related analyses [2], which characterises neural activity by treating it as a trajectory through a space of potential brain-states. A brain-state can be defined as a vector of activity across a set of neural elements determined by the recording technique (e.g., neurons, electrode channels, voxels, or parcels) at a single point in time. Importantly, this conceptual framework is scale-invariant, consistent across species, and modality-agnostic, and it has be applied to spike trains [3], calcium imaging [4], Magnetoencephalography (MEG)

Antonio Di Ieva and Jennilee M. Davidson (eds.), *Computational Neuroscience*, Methods in Molecular Biology, vol. 3062, https://doi.org/10.1007/978-1-0716-5440-8_6,

[5], electroencephalography (EEG) [6], and functional magnetic resonance imaging (fMRI) [7–9].

While neural activity is highly dynamic, its variability is not random. Recurrent motifs, brain-states that are revisited repeatedly, suggest the presence of attractors in neural dynamics. Attractors are stable configurations of neural activity that have been proposed to underlie diverse cognitive functions including working memory [10, 11], learning [12], oscillations [13, 14], adaptive behavior[15, 16], and the state of consciousness [6, 9]. Conversely, the ability to flexibly transition between brain-states is essential for adaptive behavior.

To quantify the stability and transitions of brain-states dynamics, we adopt a statistical mechanics–inspired approach that transforms empirical observations of state change into a low-dimensional energy landscape [17]. In this framework, the "effective energy" of a brain-state transition is defined as the inverse probability of observing a given displacement in brain-state over time. Note this energy is a statistical measure and has no direct relationship to the concept of metabolic energy (e.g., via ATP conversion). This transformation allows us to visualize and quantify how probable, and hence, how energetically stable, different brain-state transitions are. Analogous to a ball rolling along a landscape under the force of gravity, a shallow plain generates variable and flexible dynamics, with many accessible transitions between states, while a deep valley indicates a strong attractor and repeated state motifs.

Crucially, the shape of this energy landscape is not fixed. It can change dynamically over time, modulated by internal neuromodulatory tone, cognitive engagement, or external stimuli, and its deformation can signal fundamental shifts in brain function. Disruptions to the energy landscape have been associated with pathological states including loss of consciousness, and it may provide further insight for neurodegeneration and brain pathology, underscoring its relevance to both healthy and disordered brain function.

In this chapter, we present practical methodologies for estimating both temporal and spatial signatures of neural dynamics from empirical and synthetic data. We focus on the energy landscape analysis to estimate the time-varying range of brain-state displacements and transform these into a low-dimensional energy landscape. The energy landscape provides a compact, interpretable signature of brain organization and its alterations.

This method is presented along with MATLAB-code for the varying steps. Together, these tools provide a unified computational framework to study the multiscale organization of the brain, with broad applicability across modalities and species.

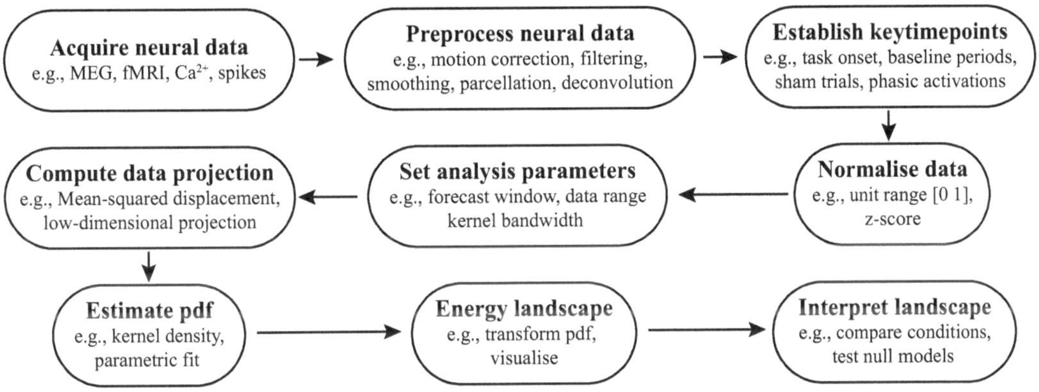

Fig. 1 Pipeline for calculating effective energy landscape of brain-state transitions

2 Materials

2.1 Software and Hardware

1. A standard desktop or laptop computer.
2. MATLAB (R2020a or later recommended) with Signal Processing Toolbox. (optional) Parallel Computing Toolbox for large datasets.

2.2 Data Preprocessing

Neural activity time-series of dimensions: [N×T] where N is the number of variables (e.g., channels, regions, or units) and T is the number of timepoints. This can be derived from calcium imaging, fMRI (e.g., BOLD signal from parcels), EEG/MEG, or spiking activity. Standard and modality specific processing should be applied (e.g., motion correction, artefact removal, segmentation/parcellation, smoothing, filtering) and modality specific preprocessing (deconvolution, spike-sorting, heart rate/eyeblink/muscle activity removal, referencing, source-localization) should be applied.

2.3 Analysis Parameters

1. data: matrix of shape [N × T].
2. taus: vector of time-lags in timepoints, e.g., taus = 1:20.
3. kernel_bandwidth: for density estimation (e.g., 0.1×std of data).
4. (Optional) scaling constants for energy transformation (analogous to Boltzmann factor, typically set to 1).

2.4 Null Data (Optional)

1. White-noise.
2. Brownian motion.
3. Oscillatory sinusoidal signals.

The energy landscape pipeline from recording and preprocessing the neural data through to interpretation can be seen in Fig. 1.

3 Method of Estimating the Energy Landscape

3.1 Load Data

Having collected and preprocessed your neural data following standard modality steps, begin by loading your data into MATLAB ensuring variables are organized across the rows and time is over columns (i.e., size N × T):

```
data = load('your_data.mat'); % Ensure dimensions
are [N × T]
```

(Optional) You may want to normalize your data, especially if contrasting across datasets or conditions. For instance, apply z-scoring across time for each channel or parcel:

```
data = zscore(data, 0, 2); % Zero mean, unit
variance across time
```

Alternatively, you may normalize to unit range if zero is meaningful:

```
data = normalize(data, 'range');
```

3.2 Parameter Defining the Forecast Window ($\{\tau\}$)

The energy landscape is estimated over a set of forecast windows, $\{\tau_1, \tau_2, \ldots\}$, which define the temporal offsets used to compute displacement. These should reflect the range of the neural dynamics you are interested in.

Case 1—Known range: If a particular range is known (e.g., 1–20 successive timepoints):

```
taus = 1:20;
```

Or, for coarser sampling:

```
taus = 200:200:1000; %e.g., a forecast window of
every 200ms over 1second
```

Case 2—Unknown range: In this case, we suggest using the autocorrelation of your data to guide the forecast lag selection. For example:

```
autocorr(data(1,:), NumLags=50); %Note use a
sufficiently long lag to see changes in dynamics
```

Using this approach, you may choose forecast lags based on peaks, troughs, or the first zero-crossing in the autocorrelation function (e.g., a standard choice of forecast window is double the time for the first zero crossing).

3.3 Parameter Defining the Displacement Space

The energy landscape is calculated over a range of Mean-Squared Displacement (MSD) values. As above choosing this range of interest can be determined whether there is a specific space of interest or more broadly to capture the spread of activity.

Note, these parameters can (and should) be refined after initial testing.

If a maximum MSD value is known, define:

```
maxMSD =; % Set maximum expected MSD
ds = linspace(0, maxMSD, 100);
```

If the maximum MSD is unknown, estimate it from the data:

```
example_tau = taus(end);
example_disp = data(:, 1+example_tau:end) - data(:,
1:end-example_tau);
example_msd = mean(example_disp.^2, 1);
maxMSD = 1.2 * max(example_msd); % Adding edge
buffer
ds = linspace(0, maxMSD, 100);
```

Note in both examples, we have set a 100 datapoint sampling from the pdf range, this can be modified.

3.4 Selecting Timepoints of Interest

The next choice in calculating the energy landscape, is the range of dynamics of interest. If you are calculating within a recording you may want to restrict the MSD calculation to specific timepoints alternatively if contrasting energy landscapes across different behaviors or recordings, you may want to use all timepoints.

If you are using all valid timepoints:

```
T = size(data,2);
tdash = 1:T - max(taus); %Where tau is the forecast
window chosen
```

Alternatively, you may want to focus on the change of neural dynamics following given timepoints (e.g., trial onsets or change-points, stimulus times, dynamic features):

```
tdash = [t1, t2,..., tn]; % Ensure at least ~100
data points for reliable PDF estimation
```

Note: If computing global landscapes collapsed across regions, you have N × t samples. For region-wise analysis, ensure sufficient samples per region.

3.5 Compute the Mean Squared Displacement (MSD)

The MSD quantifies how far neural activity travels in state space over a given forecast window. The MSD is calculated as the average change over each of the N variables,

$$MSD_{t,t\prime} = \langle |\boldsymbol{x}_{t\prime+t} - \boldsymbol{x}_{t\prime}|^2 \rangle_N,$$

where $\boldsymbol{x}$ is the brain-state, $\langle\ \rangle_N$ is the mean across N parcels, and it is calculated for timepoints t' defined earlier.

```
% Initialise
msd_forecast = cell(1, numel(taus));
% Loop over forecast windows
for idx = 1:numel(taus)
tau = taus(idx);
displacements = data(:, 1+tau:end) - data(:,
1:end-tau);
msd = mean(displacements.^2, 1); % Average across N
msd_forecast{idx} = msd;
end
```

Comment: By summarizing change via MSD (rather than full brain-state vectors), this method retains continuous signal information while reducing the dimensionality (required for probability estimation), while avoiding alternative approaches that require state-space discretization, which can limit sampling efficiency in high-dimensional systems.

3.6 Estimate MSD Probability Density Function

Estimate the probability density for a given MSD at each forecast window using either parametric or non-parametric methods.

In most cases, you will not know what family the resulting distribution will follow. In this case, we suggest a kernel density estimation approach to estimate the pdf from the provided data. Here, kernel density estimation applies kernel smoothing to each data-point for probability density estimation, i.e., it is a non-parametric method to estimate the probability density function based on kernels as weights. We, typically, use a Gaussian kernel density estimation $\mathrm{P}(MSD, t) = \frac{1}{BWn}\sum_{i=1}^{n} K(\frac{MSD_{t,i}}{BW})$, where $K(u) = \frac{1}{2\sqrt{\pi}}e^{-\frac{1}{2}u^2}$, here, BW is the bandwidth parameter. BW defines the resolution at which MSD states are distinguishable, small BW will lead to a fine-grained landscape (many shallow minima), whereas, a large BW will lead to a coarse landscape with few, deeper basins. Note histogram-based estimation is also acceptable and very interpretable; however, it will be bin-size sensitive and less smooth.

```
% Initialise
pdfs = nan(numel(taus), numel(ds)); % Storage for
PDF values
%Kernel bandwidth
BW = 1;
```

Note: Bandwidth selection is critical. Compare the estimated PDF to the empirical distribution (e.g., via overlayed scatter plots) to validate the fit. Perhaps try the standard deviation of the data or Silverman's rule (MATLAB's default selection criteria) as a good starting point.

```
for idx = 1:numel(taus)
msd = msd_forecast{idx};
msd = msd(tdash); % Restrict to selected timepoints
% Estimate PDF using kernel density estimation
pd = fitdist(msd(:),'Kernel', 'bandwidth', BW); %
Adjust bandwidth as
% needed
pdfs(idx,:) = pdf(pd,ds);
end
```

Alternatively, if you have an idea of what the probability distribution should follow you can directly fit this distribution.

For example, if you expect, it to be normally distributed, you can replace the distribution fitting line with the explicit estimation:

```
pd = fitdist(dat,'Normal'); %fit a normal
distribution
```

3.7 Computing the Effective Energy Landscape

To quantify the likelihood of brain-state transitions, we convert likelihood probabilities to an energy of transition. Importantly, the term "energy" here is used in reference to its definition in statistical physics, and hence, does not represent the biological use of the term, which instead stands for the metabolic energy used by the brain to maintain or change neural activity. Specifically, we define the energy landscape, E, as the natural logarithm of the inverse probability of observing a given BOLD mean-squared displacement (MSD) at a given time-lag T, $\mathrm{P}(MSD, t)$, calculated as $\mathrm{E} = \ln\left(\frac{1}{\mathrm{P}(MSD, t)}\right) = -\ln(\mathrm{P}(MSD, t))$. This is derived from the Boltzmann relationship in in statistical mechanics the energy of a given state, E_σ, and its probability are related $P(\sigma) = \frac{1}{Z}e^{-\frac{E_\sigma}{T}}$, where Z is the normalization function and T is a scaling factor equivalent to temperature in thermodynamics. In our analysis, $\sum_\sigma P_\sigma = 1 \rightarrow Z = 1$ by construction and we can trivially set $T = 1$ for the observed data (however, it may be an interesting parameter to explore).

```
% Initialise
NRG = nan(size(pdfs);
% Loop over forecast windows
for idx = 1:numel(taus)
NRG(idx,:) = -log(pdfs(idx,:) + eps);
% Where we have added eps to avoid log(0).
end
```

This creates an energy landscape matrix (NRG) of dimensions Forecast windows (τ) × displacement space (ds).

Note, if you will be contrasting absolute energies across differing recordings you may want to convert your probability density functions (units of 1/MSD) into a probability mass. This will make the energy depend on the probability of each bin and remove the unit-dependence.

```
dx = ds(2) - ds(1);
Pmass = pdfs * dx;
NRG = -log(Pmass + eps);
```

3.8 Visualizing the Energy Landscape

Having calculated the energy landscape, we can now plot the energy landscape as a mesh or heatmap:

```
[X, Y] = meshgrid(ds, taus);
figure;
mesh(X, Y, NRG, 'EdgeColor', [0.5, 0.5, 0.5]);
xlabel('MSD'); ylabel('Forecast Window (\tau)');
zlabel('Energy');
title('Temporal Energy Landscape');
colormap turbo; colorbar;
```

3.9 Interpreting the Landscape

In this framework, a highly probable relative change in activity (as quantified by the MSD) corresponds to a relatively low energy transition, whereas a surprising transition will possess the most energy. Rare transitions may reflect potentially disruptive or transient dynamics. A landscape that is highly uniform, suggests highly flexible dynamics. Alternatively, low-energy transitions, indicates repeated motifs and stable dynamics, which would be observed if the system was in an attractor.

Examples of this analysis on neural fMRI recordings can be seen in Fig. 2, where following activation of neuromodulatory regions control the effective likelihood of differing brain-state transitions.

3.10 Comparing Landscape Conditions

In most cases, you will be calculating multiple energy landscapes for varying conditions. In this case, a standard approach is to normalize the differing plots relative to an energy landscape, such as by subtracting a baseline setting.

In the case of subtracting two energy landscapes, such as a task state (NRG2 estimated using data P_2) relative to a pretrial rest period (NRG1 estimated using pdf P_1):

```
delta_NRG = NRG2 - NRG1;
```

Note that this is equivalent to:

$$\Delta E = \ln\left(\frac{P_1}{P_2}\right)$$

Which is a log-ratio of probability density change, highlighting where transitions become more or less surprising, across settings.

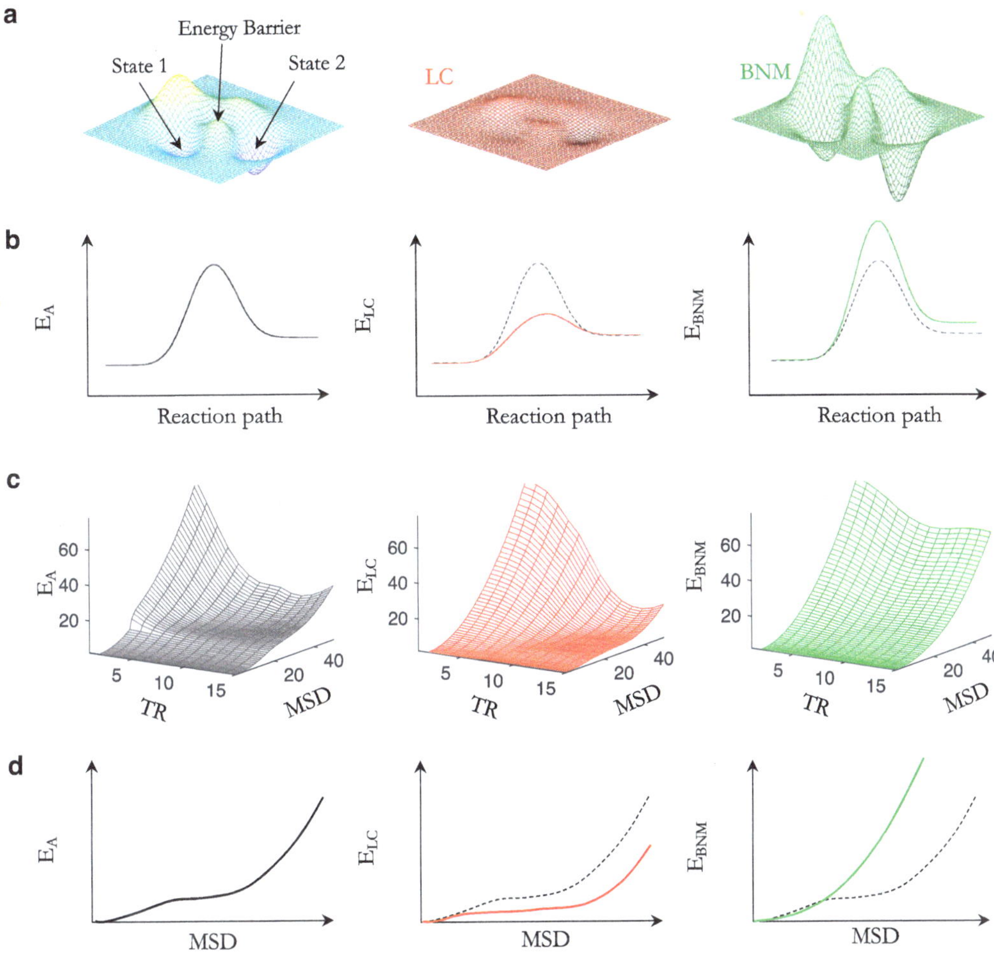

Fig. 2 Intuition and an empirical example of energy landscapes calculated on neural data following phasic activations of the adrenergic pontine locus coeruleus (LC) and cholinergic forebrain basal nucleus of Meynert (BNM). (**a**) An example energy landscape, which defines the energy required to move between different brain-states: flattening the energy landscape increases state transitions (red); whereas accentuating the energy landscape leads to attractor states (green). (**b**) The topography of the energy landscape is analogous to the activation energy (E_A) that must be overcome to convert one state to another. (**c**) Empirical BOLD energy landscape as a function of mean-squared displacement (MSD) and TR of the baseline activity (E_A, black) and after phasic bursts in adrenergic LC (E_{LC}, red) and BNM (E_{BNM}, green). (**d**) Empirical activation energy as a function of MSD averaged across lags t=[10:15] TR during base baseline activity (E_A, Left) and following phasic bursts in LC (E_{LC}, red) and BNM (E_{BNM}, green). Relative to the baseline energy landscape, phasic bursts in LC lead to a flattening or reduction of the energy landscape, whereas peaks in BNM lead to an accentuation of the well by raising the energy landscape. This figure is adapted from Munn et al., Nat Commun, 2021 with permission

3.11 Null Models

In other cases, you may want to assess the significance of observed changes against null models. Here are two examples nulls either data-derived or theoretical null models.

Random timepoints:

Using a data-derived approach, the energy landscape following predefined triggers, can be contrasted relative to random trigger timepoints. Here, the analysis would be repeated but the time indices would be randomly permuted.

```
t_null = randperm(T - max(taus), numel(t));
```

Random data:

A secondary approach would be contrasting changes relative to random data such as white noise (or more richer data such as Brownian motion)

```
data_null = randn(N, T);
```

Repeating the energy landscape calculation with null inputs and compare against your observed result to assess significantly large deviations.

4 Further Resources & Acknowledgments

Further resources and instructions along with example code and neural data to run this analysis can be found at https://github.com/Bmunn/BSI. B.R.M was supported by the Australian Research Council Discovery Early Career Researcher Award (DE260101935) & Dementia Australia Research Foundation Project Grant; J.M.S. was supported by the National Health and Medical Research Council (GNT1193857).

References

1. Munn BR, Müller EJ, Favre-Bulle I et al. (2024) Multiscale organization of neuronal activity unifies scale-dependent theories of brain function. Cell 187:7303–7313.e15. https://doi.org/10.1016/j.cell.2024.10.004
2. John YJ, Sawyer KS, Srinivasan K et al (2022) It's about time: linking dynamical systems with human neuroimaging to understand the brain. Netw Neurosci 1–20. https://doi.org/10.1162/netn_a_00230
3. Munn BR, Müller EJ, Medel V et al. (2023) Neuronal connected burst cascades bridge macroscale adaptive signatures across arousal states. Nat Commun 14: 6846. https://doi.org/10.1038/s41467-023-42465-2
4. Favre-Bulle IA, Muller E, Lee C et al. (2025) Brain-wide impacts of sedation on spontaneous activity and auditory processing in larval zebrafish. J Neurosci 45:e0204242025. https://doi.org/10.1523/JNEUROSCI.0204-24.2025
5. Krzemiński D, Masuda N, Hamandi K et al. (2020) Energy landscape of resting magnetoencephalography reveals fronto-parietal network impairments in epilepsy. Netw Neurosci 4:374–396. https://doi.org/10.1162/netn_a_00125
6. Müller EJ, Munn BR, Redinbaugh MJ et al. (2023) The non-specific matrix thalamus facilitates the cortical information processing modes relevant for conscious awareness. Cell Rep 42:112844. https://doi.org/10.1016/j.celrep.2023.112844
7. Taylor NL, Whyte CJ, Munn BR et al. (2024) Causal evidence for cholinergic stabilization of attractor landscape dynamics. Cell Rep 43:114359. https://doi.org/10.1016/j.celrep.2024.114359
8. Taylor NL, D'Souza A, Munn BR et al. (2022) Structural connections between the noradrenergic and cholinergic system shape the dynamics of functional brain networks. NeuroImage 260:119455. https://doi.org/10.1016/j.neuroimage.2022.119455

9. Munn BR, Müller EJ, Wainstein G, Shine JM (2021) The ascending arousal system shapes neural dynamics to mediate awareness of cognitive states. Nat Commun 12: 6016. https://doi.org/10.1038/s41467-021-26268-x
10. Amit DJ, Amit DJ (1989) Modeling brain function: the world of attractor neural networks. Cambridge University Press
11. Wang X-J (2001) Synaptic reverberation underlying mnemonic persistent activity. Trends Neurosci 24:455–463
12. Hopfield JJ (1982) Neural networks and physical systems with emergent collective computational abilities. Proc Natl Acad Sci 79:2554–2558
13. Miller P (2016) Dynamical systems, attractors, and neural circuits. F1000Res 5:F1000-Faculty
14. Robinson PA, Rennie CJ, Rowe DL et al. (2003) Neurophysical Modeling of Brain Dynamics. Neuropsychopharmacology 28:S74–S79. https://doi.org/10.1038/sj.npp.1300143
15. Fiore VG, Dolan RJ, Strausfeld NJ, Hirth F (2015) Evolutionarily conserved mechanisms for the selection and maintenance of behavioural activity. Philos Trans R Soc B Biol Sci 370:20150053
16. Wainstein G, Whyte CJ, Martens KAE et al (2025) Gain neuromodulation mediates task-relevant perceptual switches: evidence from pupillometry, fMRI, and RNN Modelling. eLife 13
17. Stephens GJ, Mora T, Tkačik G, Bialek W (2013) Statistical thermodynamics of natural images. Phys Rev Lett 110:018701. https://doi.org/10.1103/PhysRevLett.110.018701

Chapter 7

Computational Spatial Modeling of Extrachromosomal DNA to Decipher Cancer Evolution

Magnus Haughey and Imran Noorani

Abstract

Imaging techniques such as DNA fluorescent *in situ* hybridization (FISH) and nascent RNAscope enable detection and quantification of extrachromosomal DNA (ecDNA) in tumor cells. Here, we describe a workflow of ecDNA detection and spatial characterization using DNA FISH, and integration of these data with a spatial evolutionary model to estimate past evolution in individual patients' tumors, such as brain cancers. This analysis leverages the spatial patterns of ecDNA measured across multiple spatial regions of a solid tumor to predict the competitive advantage conferred by oncogenic ecDNA to their host tumor cell, and the number of these ecDNAs present around the time of the beginning of clonal expansion of the tumor, informing us on the origins of cancer.

Keywords Extrachromosomal DNA, Fluorescent in situ hybridization, Multi-region sampling, Evolutionary modeling

1 Introduction

Glioblastoma (GBM) is the most common intrinsic brain cancer in adults, carrying a poor prognosis of only 14 months median survival [1]. The outcomes for patients with this type of brain cancer have not substantially improved in the last few decades, owing largely due to the failure in advanced clinical trials of targeted therapies and immunotherapies. One of the reasons for failure of targeted therapy in glioblastoma is the presence of extrachromosomal DNA (ecDNA), which is present in up to 60% of these cancers [2, 3]. ecDNA is circular DNA that exists outside the native chromosomes. In GBM, as well as other types of solid cancer, these circular DNA molecules typically carry oncogenes such as epidermal growth factor receptor (*EGFR*). ecDNA molecules are segregated randomly during mitosis [4], a mechanism that can rapidly generate both extremely high oncogene copy numbers, and large cell-to-cell variation in oncogene expression, promoting tumor growth. Furthermore, this unique property of

Antonio Di Ieva and Jennilee M. Davidson (eds.), *Computational Neuroscience*, Methods in Molecular Biology, vol. 3062, https://doi.org/10.1007/978-1-0716-5440-8_7,

uneven segregation during cell mitosis makes ecDNA-driven tumors more adaptable in the face of changing environmental pressures, allowing for rapid changes in copy number of these oncogenes as illustrated by experiments, in which *EGFR* inhibitors are given to GBM cell lines, triggering a rapid decline in *EGFR* copy number due to loss of ecDNA [4, 5].

Although, in recent years, ecDNA has been noted to be a frequent event in several solid tumors, not only in GBM but other brain cancers such as medulloblastoma [6], methods to decipher the evolutionary contributions of these unique DNA molecules to cancer development have been lacking. A new method called SPECIES has been developed for this purpose. Specifically, this computational model leverages the single cell copy number counts of ecDNA-amplified oncogenes in cancer tissue slides (using DNA FISH or nascent RNAscope) to predict the evolutionary trajectories of ecDNA. By simulating the ecDNA copy number distributions across multiple regions of a cancer, such as GBM, SPECIES informs us on the selection pressure for the presence of this ecDNA (*i.e.*, whether it contributes to cancer growth, or is a neutral feature, or even a hindrance to tumor growth), and provides information as to how many copies of ecDNA were present at the time of clonal expansion of the tumor, thereby giving information regarding the origins of an individual tumor. A recent study using SPECIES determined that the type of oncogene on extrachromosomal DNA can dictate the evolutionary trajectory of a GBM tumor, as shown, for example, in differences between *EGFR*-driven and *PDGFRA*-driven GBM, and the fact that *EGFR* amplification on ecDNA is frequently an initiating event for this type of cancer growth [7, 8].

Here, we provide a step-by-step protocol describing how investigators perform quantification of extrachromosomal DNA copy numbers using DNA FISH and microscopy, and how the SPECIES computational model can be implemented to simulate the data and understand the evolutionary properties of extrachromosomal DNA.

2 Materials

2.1 DNA FISH Staining

1. Agilent Technologies Protocol Kit (Histology FISH Accessory Kit—K5799; RRID:SCR_013575).
2. Abbott Molecular Diagnostic FISH probe kit (RRID:SCR_008392).
3. Probes of interest, e.g., *EGFR* FISH probe (Vysis/Abbott), VYSIS EGFR/CEP7 FISH PROBE KIT #81666 (Abbott Molecular Inc.).
4. Slide scanner eg VS200 Olympus Slide Scanner.

2.2 Computational Requirements for Running SPECIES

1. The scripts contained in the SPECIES repository were tested using Apple clang version 17.0.0 and Python v3.11.0.
2. Prior to running the SPECIES scripts, download and install the following Python packages by entering the following command in the command line:

```
pip3 install numpy pandas seaborn argparse
```

3 Methods

3.1 DNA FISH Staining

Fluorescent *in situ* hybridization (FISH) provides single nucleus detail on the copy number of the ecDNA of interest, allowing for high-resolution computational modeling of ecDNA dynamics. We recommend performing a complementary method to further confirm ecDNA presence for a given tumor, such as whole genome sequencing with Amplicon Architect analysis to detect structural variants characteristic of ecDNA [9, 10]. A typical method for performing DNA FISH is presented below [8].

1. FISH is performed on 4 αM Formalin Fixed Paraffin Embedded (FFPE) tissue sections using the Agilent Technologies Protocol (Histology FISH Accessory Kit—K5799; RRID:SCR_013575) and the Abbott Molecular Diagnostic FISH probe protocol (RRID:SCR_008392).
2. Dewax FFPE sections in Xylene for 5 min then rehydrate in 100%, 80%, and 70% ethanol. Next, wash sections twice with Agilent Technologies wash buffer.
3. Incubate FFPE sections at 98°C for 10 min in pretreatment solution in a coplin jar water bath. Remove the jar from the water bath and leave to slowly cool for 15 min.
4. Wash the FFPE sections twice with wash buffer. Apply stock pepsin to the slide for 10 min at 37°C. Wash the FFPE slides a further two times with wash buffer, before dehydration using 70%, 80%, and 100% ethanol, followed by probe hybridization.
5. Apply gene specific probes containing chromosome specific centromere enumeration probes (CEPs), for example, against *EGFR* (Vysis/Abbott) and *MYC* (Vysis/Abbott), directly to the tissue sections, sealing the coverslip with rubber solution glue.
6. Denature the probes on the tissue at 75°C for 7 min. Incubate the slide overnight in a moist box at 37°C for 16 h.
7. Wash slides for 10 min at 73°C with 0.4× SSC containing 0.3% Igepal (Sigma), and then for 10 min at room temperature with

2× SSC containing 0.1% Igepal. Leave slides to air dry, before counter staining with Vectashield mounting medium containing DAPI.

3.2 DNA FISH Microscopy and Quantification

1. Images of the stained tissue sections are captured using the VS200 Olympus Slide Scanner (or similar digital microscope) at 40× magnification, using manufacturer's instructions (Fig. 1).
2. Typically, at least 3 independent regions per section are chosen for imaging for a particular sample. Alternatively, if a tissue microarray is being used for multiple samples, the whole microarray is imaged for each sample. For a cancer study, we recommend a minimum of 500 nuclei per sample in order to enhance the reliability of ecDNA copy number distributions and computational modeling. For downstream use of the SPECIES model for computational modeling of ecDNA dynamics, a minimum of two independent tumor regions is needed for quantification—*e.g.*, with glioblastoma, we recommend one tumor core region and one tumor margin region. An appropriately trained pathologist is essential for providing guidance of selection of these regions using the corresponding H&E-stained (Hematoxylin & Eosin) tumor slides.
3. Quantification of FISH foci per nucleus can be performed using a microscopy analysis software, such as QuPath software (this is freely available). DNA FISH microscopy images can be directly uploaded onto QuPath. Selection of the regions of interest is next performed using image selection tools on the software.

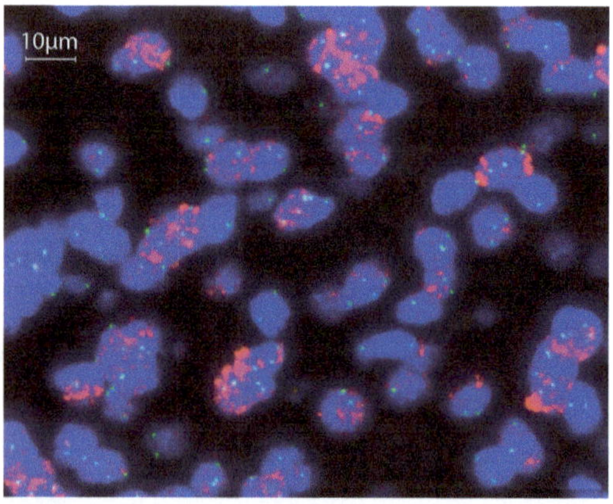

Fig. 1 Example DNA FISH-stained section of glioblastoma tissue, showing abundant *EGFR*-amplifying ecDNA. Each red spot corresponds to one copy of *EGFR*, and here it can be seen that such *EGFR* genes can localize in large clusters (termed "hubs" of many copies of ecDNA). *Blue* DAPI, *green* centromere 7, *red EGFR*

4. Quantification is performed in a supervised fashion using the Cell Detection and Subcellular Cell Detection features, with the "includeClusters" function to account for large ecDNA clusters when present (this is a frequent observation in some tumors such as glioblastoma and other brain cancers). Parameters need to be appropriately optimized for each image to be quantified, given that FISH spots may appear larger or smaller for individual sections due to inherent batch-to-batch variation between FISH sections and microscopy capture.
5. The number of spots for each nucleus is then used to plot a histogram of ecDNA copy number (corresponding to FISH spot counts) per region. An example of this is shown in Fig. 2.

3.3 Downloading SPECIES

1. The following instructions assume the use of a Unix environment (e.g., MacOS, Linux).
2. All files relating to the SPECIES simulation platform can be downloaded by entering either

```
git clone
git@github.com:MagnusHaughey/spatial_ecDNA_
patterns.git
```
or
```
git clone
https://github.com/MagnusHaughey/spatial_ecDNA_
patterns.git
```
in the command line. This will download the SPECIES files into a new directory called `./spatial_ecDNA_patterns/` in your current working directory.
3. Alternatively, these files can be downloaded at https://github.com/MagnusHaughey/spatial_ecDNA_patterns. On the main landing page for the SPECIES git repository, select the green "Code" dropdown menu and choose "Download ZIP." Move and unzip the downloaded ZIP archive in your preferred location.

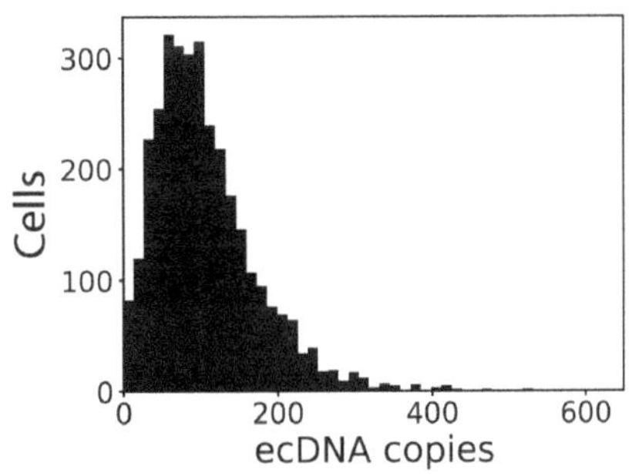

Fig. 2 Example single-cell ecDNA copy number distribution for the core region of the tumor of patient 17NE1105A (GB-UK cohort, see [8, 11])

3.4 Running a single SPECIES simulation

1. Having downloaded SPECIES, navigate to the directory in which the files are located using the command line (*see* **Note 1**).
2. Compile the SPECIES main simulation script using the following command:

```
g++ -o ./spatial_ecDNA_patterns_2D ./spatial_ecDNA_
patterns_2D.cpp
```

3. Simulate a single ecDNA-driven tumor by executing the main SPECIES simulation script:

```
./spatial_ecDNA_patterns_2D [--verbose] [-n N] [-q
Q] [-k K] [-s S] [-x X]
```

   For information on the simulation flags and arguments, *see* **Note 2**. For example, the command

```
./spatial_ecDNA_patterns_2D -n 10000 -q 5 -k 1 -s
0.1 -x 10
```

   will perform a simulation of a tumor with a maximum size of 10,000 cells, a cell pushing strength of 5, an initial ecDNA copy number of 1, an ecDNA selective advantage of 0.1, and a random seed of 10.
4. The resulting output data will be deposited in a new directory named `./results/`, in a subdirectory containing the specific parameters of the simulation. For the example in Step 3, the simulated tumor data would be saved in a file named `tissue.csv`, located at `./results/n=10000_k=1_q=5_s=0.1/seed=10/`.
5. Each line of the `tissue.csv` file is in the format: `x, y, #ecDNA`.

3.5 Visualizing Output Data from SPECIES

1. Included with the SPECIES files is a python script (named `plot_spatial_data.py`) that enables simple and fast visualization of the SPECIES simulated tumors. An output `tissue.csv` file can be plotted using the following command:

```
python3 ./plot_spatial_data.py [--path PATH]
[--plotCopyNumber]
```

   Where the `[--path PATH]` flag specifies the file path (either relative or absolute) to the `tissue.csv` file to be visualized. The `[--plotCopyNumber]` flag is optional (*i.e.*, can be included or left out) and toggles between a simple visualization of ecDNA-negative and ecDNA-positive cells in the tumor (without flag), and a more detailed visualization indicating specific ecDNA abundance per each cell (with flag) (Fig. 3).
2. Continuing with the example from Sect. 3.4, one could visualize the resulting `tissue.csv` file using the command

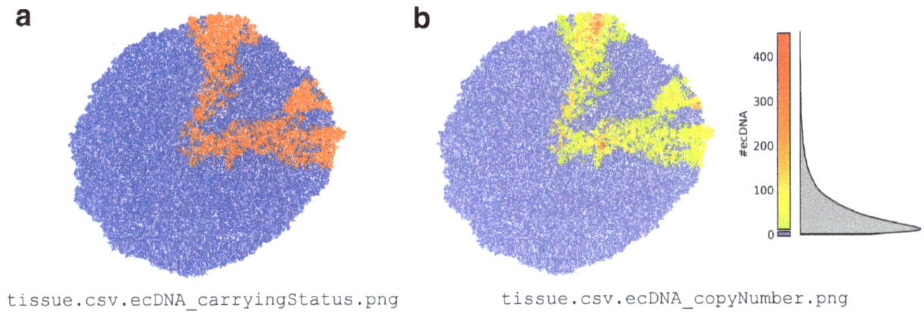

Fig. 3 Example of visualizations generated using `plot_spatial_data.py`, for an ecDNA-driven tumor simulated with SPECIES using parameters $(n, k, s, q) = (10000,\ 1,\ 0.1,\ 1)$. Visualized tumor (**a**) without and (**b**) with the `--plotCopyNumber` flag

```
python3 ./plot_spatial_data.py --path ./results/n=
10000_k=1_q=5_s=0.1/seed=10/tissue.csv
```

This will generate an image of the simulated tumor in a file named `tissue.csv.ecDNA_carryingStatus.png` in the same directory as the original `tissue.csv` file. If the `[--plotCopyNumber]` flag is used, the output file will be named `tissue.csv.ecDNA_copyNumber.png`.

3.6 Running Multiple SPECIES Simulations

1. To run multiple SPECIES simulations, the `batch_run.sh` and `batch_run_parameters.txt` files can be used. Edit the lines in `batch_run_parameters.txt` to specify the number of simulation repeats, final tumor size, and values of input parameters *k*, *s*, and *q*, then execute the `batch_run.sh` file in the command line using the command

```
bash ./batch_run.sh
```

3.7 Parameter Inference Using FISH-Stained Human Glioblastoma Samples

1. The SPECIES file repository contains a subdirectory named `./GB_UK_ecDNA_copy_number_distributions/` which contains multi-region single-cell ecDNA abundance data, estimated from DNA FISH experiments. These data correspond to untreated, *IDH* wild-type human glioblastomas, collectively referred to as the GB-UK cohort (see [8, 11]).
2. Data for each patient tumor is organized into subdirectories of `./GB_UK_ecDNA_copy_number_distributions/` with titles in the format `/[patient ID]_[FISH_targeted_oncogene]/`, *e.g.*, `/A1_EGFR/`.
3. Within each patient subdirectory, data files corresponding to each spatial region of the tumor can be found. These are named `core.dat`, `margin.dat`, and `leading_edge.dat` (the latter available only for certain patient tumors) corresponding, respectively, to tissue samples extracted from the tumor core, infiltrating margin and extreme leading edge region. Each file contains a list of values, corresponding to the estimated

number of ecDNAs, with each row corresponding to a single cell (*see* **Note 3**).

4. To perform parameter fitting (approximate Bayesian computation (ABC) with rejection sampling) of the SPECIES spatial model to multi-region ecDNA data, run the following command:

```
bash ./ABC_full_algorithm.sh
```

   This algorithm executes randomly initiated SPECIES simulations, extracts multi-region samples from the resulting simulated tumors (*see* **Note 4**), and compares the corresponding spatial ecDNA data between the patient and simulated tumors using the Wasserstein distance (*see* **Note 5**), allowing the user to estimate the best-fit SPECIES parameter values for s, k, and q for each given patient tumor. Uniform prior distributions are assumed for the Bayesian parameter estimation. These distributions can be specified by editing the variables in the file named `ABC_input_parameters.txt`. Data generated by this script is written to an output file in a new directory named `./ABC_raw_results/` (*see* **Note 6**).

5. Summarize the resulting raw parameter fitting data by executing the following command:

```
python3 ./ABC_summarize.py [--epsilon_core a]
[--epsilon_margin b] [--epsilon_leadingEdge c]
[--posterior_size d]
```

Each flag is optional, *see* **Note 7** for further details. This will summarize the raw ABC data and output processed data for each patient in a new directory named `./ABC_processed_results/*patient_ID*/`, including plots of the posterior distributions and estimates of best-fit parameter values for k, s, and q.

4 Notes

1. This can be done either by right-clicking on the parent directory containing all SPECIES files (called `/spatial_ecDNA_patterns/` by default) in the File Explorer and selecting "New Terminal at Folder," or by opening a new Terminal window and navigating to the `/spatial_ecDNA_patterns/` directory using the `cd` command (e.g., `cd ~/Downloads/spatial_ecDNA_patterns/`).
2. Flags for the SPECIES simulator correspond do:

Flag	Description	Optional?	Range	Notes
`--verbose`	Output simulation details in real time	Yes		
`-n`	Maximum size of simulated tumor	No	> 0	SPECIES will terminate when the simulated tumor has reached the size of `n` cells
`-k`	Number of ecDNA copies in first tumor cell	No	≥0	
`-s`	Competitive advantage of ecDNA-positive cells	No	Any floating point	The replication rate of each ecDNA-negative cell in the simulated tumor is equal to 1, whilst that of ecDNA-positive cells is equal to (1+`s`).
`-q`	Strength of cell pushing	No	≥1	$q = 1$ corresponds to short range cell displacement leading to boundary-driven tumor growth. Larger values of `q` lead to longer-ranging cell displacement, with sufficiently large values (e.g., $q = 1000$) leading to exponential tumor growth.
`-x`	Simulation random seed	No	Any integer	Two independent executions of SPECIES with the same random seed will result in the exact same sequence of events (i.e., cell births, deaths, and ecDNA segregations), giving the exact same simulated final tumor (assuming all other input parameters are also unchanged).

3. For each patient tumor region, the "Num spots estimated" function in QuPath provides a proxy measurement for the number of ecDNA particles in each cell present in the slide. Since this is an estimate, the number can take non-integer values. It is recommended simply to round these values to the nearest integer for downstream analysis. Furthermore, as estimates may be unreliable for cells with low numbers of ecDNA (since lowly abundant ecDNA is difficult to distinguish from chromosomal DNA in imaging data) we recommend filtering out cells which are estimated to have, *e.g.*, ≤3 ecDNA copies.
4. During ABC parameter fitting, random instances of SPECIES are executed. Each simulated tumor is compared to a patient tumor by extracting circular sub-regions of the simulated

tumor, for each spatial region, *i.e.*, core, margin, and leading edge. Each circular extracted subregion is defined by its central coordinates on the lattice (for the core sample, these coordinates are always the center point of the tumor, defined as its "center of mass") and the radius of the circular region. For each spatial region, and for each patient tumor, the radius is dynamically determined such that the number of cells is equivalent between patient and simulated tumor region.

5. We employ the Wasserstein distance [12, 13] to quantify the similarity between the ecDNA copy number distributions for patient and simulation derived tumors for each spatial region (core, margin, and leading edge). The Wasserstein distance is a commonly used metric to measure the distance between probability distributions.
6. The ABC algorithm outputs all variables for each random SPECIES simulation, in order to enable full reconstruction of each random tumor simulated during the parameter inference. Each line in the resulting raw output file corresponds to a single SPECIES-patient comparison, and is of the format:
   patient_ID,s,k,q,x,c_d,c_r,m_d,m_x,m_y,m_r,le_d,le_x-,le_y,le_r
   Each variable name is described below:
7. Optional flags for `ABC_summarize.py` correspond to:

Variable	**Description**
`patient_ID`	ID of patient tumor being compared
`s`	Simulated tumor parameter *s*
`k`	Simulated tumor parameter *k*
`q`	Simulated tumor parameter *q*
`x`	Simulated tumor random seed
`c_d`	Core region Wasserstein distance
`c_r`	Simulated tumor core region radius
`m_d`	Margin region Wasserstein distance
`m_x`	Simulated tumor margin region x-coordinate
`m_y`	Simulated tumor margin region y-coordinate
`m_r`	Simulated tumor margin region radius
`le_d`	Leading edge region Wasserstein distance
`le_x`	Simulated tumor leading edge region x-coordinate
`le_y`	Simulated tumor leading edge region y-coordinate
`le_r`	Simulated tumor leading edge region radius

Flag	Description	Default
`--epsilon_core`	Maximum permissible percentile for margin distance	0.25
`--epsilon_margin`	Maximum permissible percentile for leading edge distance	0.25
`--epsilon_leadingEdge`	Maximum permissible percentile for leading edge distance	0.25
`--posterior_size`	Size of posterior sample set (percentage of total number of particles for given patient)	0.05

References

1. Stupp R, Mason WP, vanden Bent MJ et al (2005) Radiotherapy plus concomitant and adjuvant temozolomide for glioblastoma. N Engl J Med 352(10):987–996. https://doi.org/10.1056/NEJMoa043330
2. Kim H, Nguyen N-P, Turner K et al (2020) Extrachromosomal DNA is associated with oncogene amplification and poor outcome across multiple cancers. Nature Genet 52(9):891–897. https://doi.org/10.1038/s41588-020-0678-2
3. Bailey C, Pich O, Thol K et al (2024) Origins and impact of extrachromosomal DNA. Nature 635(8037):193–200. https://doi.org/10.1038/s41586-024-08107-3
4. Lange JT, Rose JC, Chen CY et al (2022) The evolutionary dynamics of extrachromosomal DNA in human cancers. Nature Genet 54(10):1527–1533. https://doi.org/10.1038/s41588-022-01177-x
5. Nathanson DA, Gini B, Mottahedeh J et al (2014) Targeted therapy resistance mediated by dynamic regulation of extrachromosomal mutant EGFR DNA. Science 343(6166):72–76. https://doi.org/10.1126/science.1241328
6. Chapman OS, Luebeck J, Sridhar S et al (2023) Circular extrachromosomal DNA promotes tumor heterogeneity in high-risk medulloblastoma. Nature Genet 55(12):2189–2199. https://doi.org/10.1038/s41588-023-01551-3
7. Haughey M, Noorani I, Swanton C et al (2025) Extrachromosomal DNA: shaping the evolutionary dynamics of cancer. Trends in Cancer 11(9):901–916. https://doi.org/10.1016/j.trecan.2025.06.004
8. Noorani I, Haughey M, Luebeck J et al (2025) Extrachromosomal DNA–driven oncogene spatial heterogeneity and evolution in glioblastoma. Cancer Discov 15(10):2078–2095. https://doi.org/10.1158/2159-8290.CD-24-1555
9. Deshpande V, Luebeck J, Nguyen N-PD et al (2019) Exploring the landscape of focal amplifications in cancer using AmpliconArchitect. Nat Commun 10(1):392. https://doi.org/10.1038/s41467-018-08200-y
10. Noorani I, Luebeck J, Rowan A et al (2024) Oncogenic extrachromosomal DNA identification using whole-genome sequencing from formalin-fixed glioblastomas. Ann Oncol 35(6):570–573. https://doi.org/10.1016/j.annonc.2024.03.008
11. Noorani I, Sidlauskas K, Pellow S et al (2023) Clinical impact of anti-inflammatory microglia and macrophage phenotypes at glioblastoma margins. Brain Commun 5(3):fcad176. https://doi.org/10.1093/braincomms/fcad176
12. Bassetti F, Bodini A, Regazzini E (2006) On minimum Kantorovich distance estimators. Stat Probab Lett 76(12):1298–1302. https://doi.org/10.1016/j.spl.2006.02.001
13. Bernton E, Jacob PE, Gerber M et al (2019) Approximate Bayesian computation with the Wasserstein distance. J R Stat Soc Ser B Stat Methodol 81(2):235–269. https://doi.org/10.1111/rssb.12312

Check for updates

Chapter 8

Network Analysis Pipeline for Neuronal 2D and 3D Organoid Multielectrode Recordings

Timothy P.H. Sit and Susanna B. Mierau

Abstract

Multielectrode recordings from 2D or 3D cultures are commonly used to assess neuronal activity including action potential firing and burst rates. Yet, multielectrode recordings, which carry parallel time series, can also be used to infer functional connectivity and to compare network features from neuronal networks within neuronal 2D cultures or 3D organoids. This chapter provides a protocol for using the microelectrode array network analysis pipeline (MEA-NAP). With MEA-NAP, users with limited computational experience can easily evaluate network activity in their microelectrode array (MEA) recordings. Users with more computational experience can customize this open-source tool. Analysis of the functional connectivity and network features in MEA-NAP can provide insight into the development of neural circuits and how disruption of neuronal function, for example, in neurologic and psychiatric disorders, alters network function.

Keywords Microelectrode array, Multielectrode, Networks, Microscale, Development, In vitro, Neuronal cultures, Brain circuits, Functional connectivity, Network topology

1 Introduction

Multielectrode recordings from 2D neuronal or 3D organoid cultures can easily be performed with microelectrode array (MEA) systems [1–3]. The MEA chips on which the 2D neurons are cultured, or 3D organoids are placed, are available in different electrode numbers and configurations and can be used to sample neuronal network activity. This chapter focuses on the network analysis of MEA data from standard-density MEA systems, ranging from 16 to 64 electrodes, using the microelectrode array network analysis pipeline (MEA-NAP) [4]. First, we describe the computing requirements and inputs to MEA-NAP. Second, steps for preparing the MEA data and selection of parameters in MEA-NAP are discussed. Examples are provided for a variety of use cases. This chapter is intended to be complementary to the extensive documentation provided with MEA-NAP on our GitHub page (see Sect. 2). In particular, this chapter focuses on how

Antonio Di Ieva and Jennilee M. Davidson (eds.), *Computational Neuroscience*, Methods in Molecular Biology, vol. 3062,
https://doi.org/10.1007/978-1-0716-5440-8_8,

new users might utilize MEA-NAP to answer different scientific questions from their MEA data.

MEA-NAP is an open-source tool for comparing neuronal and network activity [4]. MEA-NAP has multiple steps including: (1) Spike Detection, (2) Comparing Neuronal Activity, (3) Inferring Functional Connectivity, and (4) Comparing Network Activity (Fig. 1). It also has the optional step of (5) Statistical Comparisons.

The input to MEA-NAP (Fig. 2A) includes the raw voltage timeseries from the MEA recordings and a comma separated values (.csv) file that includes the list of filenames and their age (or time) and genotype (or other group) names (Fig. 2B). The output includes both tabular data and folders containing figures for assessing the neuronal and network activity detected. MEA-NAP will complete Steps 1–4 or Steps 1–5 autonomously when run (Fig. 2C). However, users can also restart the analysis at Steps 2–5 using a prior output folder, for example, if one wanted to try different parameters or use a subset of the spiked detected files. MEA-NAP incorporates leading spike detection and network analysis methods, including adapting resources commonly used at larger spatial scales, such as the Brain Connectivity Toolbox [5], for microscale network analysis. A complete list of references and sources are included in the paper describing MEA-NAP [4] and in our documentation on GitHub (see Next Section).

2 Software and Hardware

MATLAB (MathWorks) R2021b version or later is required along with a valid license and the following MATLAB toolboxes: Signal Processing Toolbox, Parallel Computing Toolbox, Statistics and Machine Learning Toolbox, and Wavelet Toolbox. The pipeline is compatible with Windows, Mac, and Linux operating systems (OS); however, users of older Mac OS may need to enable XCode in the Security and Privacy setting if they encounter an error (*see* Sect. 4). Regardless of operating system, a computer with at least 16 GB RAM is needed to run MEA-NAP; however, the analysis will run faster on computers with 32 GB or higher RAM. We have also had success running MEA-NAP on high performance computing clusters at more than one university. This can be particularly helpful for analyzing large datasets (e.g., more than 200 files), although not required.

The up-to-date MEA-NAP tool is publicly available at our GitHub site: https://github.com/SAND-Lab/MEA-NAP. There is extensive documentation available on the GitHub, which can be conveniently viewed: https://analysis-pipeline.readthedocs.io/en/latest/. The version of MEA-NAP that was released with MEA-NAP's publication [4] is also available on a permanent repository: https://doi.org/10.7910/DVN/Z14LWA.

Fig. 1 Overview of the microelectrode array network analysis pipeline (MEA-NAP). Diagram illustrates the five steps in MEA-NAP, a semiautomated, open-source platform for comparing neuronal activity, functional connectivity, network topology, and dynamics in neuronal 2D and 3D organoid cultures

After downloading and launching the MEA-NAP GUI in MATLAB, there is an option at the bottom of the GUI to "Test the pipeline." Selecting this option will automatically download two sample files from the permanent repository and run MEA-NAP on those two files. This can be used to confirm that installation was successful and to identify any missing toolboxes.

MEA-NAP is currently set up to provide data conversion for MEA recordings from Multi-channel Systems (MCS) 60-electrode single-well MEAs and Axion Biosystems 16- and 64-electrode multiwell plates. Users of other MEA systems will need to convert their data to the correct format prior to using MEA-NAP.

3 Methods

3.1 Recording Microelectrode Array (MEA) Data for Network Analysis

1. **Choice of number of electrodes:** MEA-NAP was initially designed for 60-electrode MEA recordings from MCS single-well MEA system and then expanded for use for the 64-electrodes-per-well multiwell MEA recordings from Axion Biosystems. Many of the network features work best for 60- to 64-electrode MEA recordings, particularly the features, which rely on modular decomposition. However, MEA-NAP also works well with 16-electrodes-per-well multiwell MEA recordings from Axion Biosystems. For advanced network features, including modular decomposition or clustering analysis, 16-electrodes may provide an insufficient number of nodes for this type of network analysis. Thus, if users are interested in all the MEA-NAP features, the 60- or 64-electrode MEAs are recommended. For simpler network features and dimensionality reduction techniques, the 16-electrode MEAs are typically sufficient and facilitate higher throughput recordings.
2. **Choice of acquisition rate:** A minimum of 12.5 kHz for the Axion Biosystems and 25 kHz for the MCS MEA systems is recommended to provide sufficient sampling for template-based spike detection. The acquisition rate should be written down at the time of the recording.
3. **Length of recording:** For most applications, 10 min per recording is a good start for assessing network activity in the

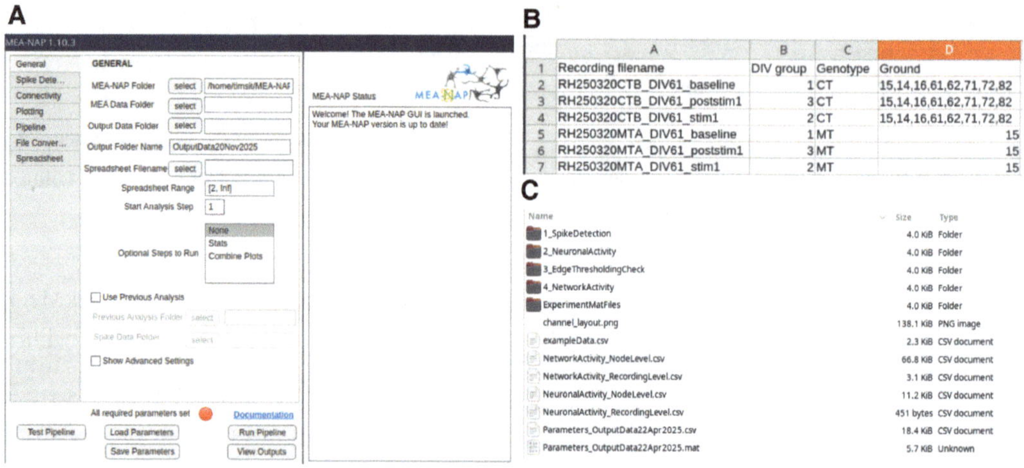

Fig. 2 MEA-NAP graphical user interface (GUI), inputs, and outputs. (**A**) Image of the MEA-NAP GUI. (**B**) Example csv file with the 3 required columns specifying the recording filename, days-in-vitro (DIV), and genotype (or other condition) group for each recording, and an optional column specifying, which electrodes are to be excluded in spike detection analysis. (**C**) Example output folder from MEA-NAP

MEA recordings from 2D and 3D neuronal cultures. Shorter recordings may not be sufficient if the neuronal activity is sparse, infrequent, or there is limited functional connectivity. Recordings longer than 10 min may need to be split and analyzed in 10-minute-long chunks, depending on the processing capabilities of the computer used to run MEA-NAP.

4. **Save the raw voltage-data:** MEA-NAP requires the raw voltage time series to perform spike detection and to compare spike detection methods. If your data only includes the spike times detected by the acquisition software, then it is not possible to use the MEA-NAP tools to compare and validate choice of spike detection methods or to remove electrical artifacts. Users with robust spike detected data and knowledge of MATLAB can format their data for MEA-NAP and start with Step 2 (Neuronal Activity). Short, but descriptive, recording filenames are ideal for use in MEA-NAP. For example, a filename "CTL250505_P1_DIV28" could be used for a set of cultures assigned the code CTL initiated on 2025-05-05 in MEA multiwell plate number 1 and recorded on days-in-vitro (DIV) 28. Long filenames should be shortened and other punctuation removed before converting the files to .mat format for MEA-NAP (Sect. 3.2).

5. **Visually inspect, and ideally photograph, each well after MEA recordings to confirm electrodes are covered by living neuronal tissue:** Note any electrodes that were grounded (or were not covered by cells), as these electrodes should be specified for exclusion in MEA-NAP.

3.2 Converting MEA Data for MEA-NAP

1. Run MEA-NAP in MATLAB by typing MEApipeline in the command window and pressing the return key. This will launch the graphical user interface (GUI). Screenshots of the GUI are available in the MEA-NAP documentation on GitHub. Select the tab for "File Conversion." This will allow you to select the folder where the raw voltage data are located on your computer. You can currently select ".raw from Axion Biosystems," ".raw from Multi-channel Systems," or ".h5 from Multi-channel Systems." The file conversion for the Axion Biosystems will "unpack" the recordings from the multi-well plates creating a .mat file for each MEA well adding an "_A1," for example, to distinguish the wells from the same plate. For the Multi-channel Systems (MCS) recording files, please follow the instructions on the MEA-NAP documentation on GitHub to convert either .mcd (MCRack) or .msrd (MCExperimenter) files using the appropriate free MCS tools to convert the files to .raw or .h5, respectively.
2. A short but descriptive name should be chosen for the .csv "batch" file (also called "spreadsheet" in the GUI), which will provide the list of filenames for MEA-NAP. The file conversion will generate a .csv file that includes all the filenames of files converted in that folder. You can then edit the .csv to add the age and group names. If your filenames are formatted with the DIV[number], this can be automatically populated by the file conversion in the batch file.

3.3 Preparing the Converted MEA Data for MEA-NAP

1. **What is the scientific question that you wish to address with MEA-NAP?** This is the first and foremost question when preparing the .csv "batch" file as an input to MEA-NAP with your converted data and with all of the subsequent parameter selection. Depending on the filenames included in the .csv "batch" file and how the "Age" and "Genotype" group columns are assigned, MEA-NAP will generate informative plots and comparisons that may look at specific aspects of a subset of the data from a single experiment, all the data from an experiment or multiple experiments. For example, if you have weekly recordings from control (CTL) and a disease-model (DIS) organoids. Your "Age" groups might be DIVs (e.g., 150, 200, 250) and your "Genotype" may be "CTL" and "DIS." The Age group must start with a number, but does not need to be an age. Note, if you put actual age in this column (e.g., 149, 150, 151) versus the "Age" group (e.g., 150), the plots will create 3 "Age" groups instead of 1. The "Genotype" must start with a letter (no numbers) and should be short (3-5 characters) to create informative plots. In another example, you want to compare the features of organoids generated in different batches, here the "Genotype" group might be "B1"

and "B2," for the first and second batches. You may need to combine the .csv files if the recordings from the two batches were converted separately. When MEA-NAP is run, all of the .mat files in the .csv file must be in the same folder. In a third example, you want to compare the effect of a drug, you might label the "Age" groups "1, 5, 10," where "1" is the pre-drug MEA recording and "5" and "10" are 5 and 10 min after the drug was added. Here, the "Genotype" group could be named for individual networks (e.g., "A5" for well A5) or for category (e.g., "CTL," "DIS"). To make different comparisons in the same dataset, one can rerun MEA-NAP starting from Step 2 using different .csv "batch" files and/or subsequent parameters.

2. Ensure all the MEA recordings .mat files that you want analyzed in the .csv "batch" file are in the same folder (and not divided into subfolders) before running MEA-NAP.
3. Complete the "Age" and "Genotype" columns for each file in the .csv "batch" file.
4. Add the coordinates of any electrodes that were grounded (e.g., "15" for many MCS 60-electrode MEA chips), or should be excluded from MEA-NAP analysis, in the "Ground" column. This is especially important if the 2D neuronal or 3D organoid culture does not cover all the electrodes on the MEA, or if the cells over some electrodes only contain cellular debris. The impedance of living cells is different from dead cells/tissue or the culture media or recording solution alone. Thus, electrodes without living cells may also have higher voltage variations.

3.4 Running MEA-NAP Using the Default Parameters

1. Follow the "Guide for New Users" in the MEA-NAP documentation on GitHub for explanation of the specific fields in the "General" tab of the MEA-NAP GUI.
2. These fields in the "General" tab include the locations of the raw MEA voltage data .mat files and the .csv "batch" file (spreadsheet) and the name and location of where MEA-NAP's folder of outputs should be saved. These fields are required.
3. Avoid long folder names for the output folder.
4. If you have recorded the same cultured network at different time points (e.g., DIV) and the filenames are the same for each cultured network except for the number following "DIV," the optional "combine plots" function will generate additional figures with network plots from each DIV. This is useful for large datasets but requires a uniform naming system across multiple timepoints. If statistical comparisons are desired, this option can be selected as an option.

5. The last required field is on the second tab "Spike Detection" where the user must choose the appropriate "Channel Layout" if the data is not 64-electrode MEA data from Axion Biosystems. Other current options include "Axion16" for 16-electrode MEA data and MCS60 for 60-electrode MEA data recorded with an MCS MEA2100 system (choose "MCS60old" if data was recorded on the MEA1600 system). There is also a "custom" option where advanced MATLAB users can design their own layout for use in MEA-NAP.
6. When the "Channel Layout" is changed, the sampling frequency and units for the voltage trace will be updated with new defaults for that layout. If your data were acquired with a different sampling frequency, for example, that needs to be updated.
7. The GUI light will turn from red to green when all the required parameters are completed. At this point, we recommend "Save parameters" be selected before running. This way if there was an error in any of the parameters you selected, you can restart MEA-NAP and load the parameters file without having to reenter all of the information.
8. Run MEA-NAP when ready. Do not close MATLAB or let the computer go into sleep or off mode while it is running. If this happens, you can restart MEA-NAP at later steps if Step 1 was completed.

3.5 Selecting Spike Detection Parameters Using MEA-NAP

1. Accurate spike detection underlies the robustness of comparing action potential (spike) firing and burst rates and functional connectivity and network features. For 2D cultures with good signal-to-noise ratios across electrodes, we recommend the default settings. The continuous wavelet transform (CWT) template-based spike detection using a wavelet called "bior1.5" identifies action potentials based on their characteristic morphology. Thus, in 2D neuronal cultures, it provides both high sensitivity and specificity using default parameters in MEA-NAP. It is superior, in both sensitivity and specificity, to threshold based methods based on standard deviation, where even in good signal-to-noise conditions, the threshold method shows (1) rate dependence and (2) false positives in inactive or minimal activity electrodes [6].
2. MEA recordings performed before and after application of 500 nM–1 uM of tetrodotoxin (TTX) provide a gold standard for validating choice of spike detection parameters. See an example for organoid recordings [4]. Note that TTX blocks the pore of open voltage-gated sodium channels. Thus, a cell must fire in the presence of TTX before it can be blocked by TTX. Thus, depending on the frequency of spontaneous activity, or how

quickly the TTX permeates into 3D organoid tissue, it may take ~10 min to fully block activity in some cultures. Also note that the threshold-based methods based on standard deviation, or median absolute deviation, will often detect some "spikes" from electrodes with no action potential because, particularly for lower thresholds, there will by definition be some voltage defections (even in TTX) that fall below the threshold set. For advanced MATLAB users, setting an absolute voltage threshold based on the pre-TTX recording can help resolve this issue.

3. In the absence of TTX data, MEA-NAP provides the option to select multiple wavelets and/or thresholds (based on the median absolute deviation) and then to compare the spike detection using plots from the output folder labeled "1B_SpikeDetectionChecks."
4. For 3D organoid data, the neurons can be densely packed near the electrodes leading to a small signal relative to the background activity (this includes voltage fluctuations from neurons and groups of neurons near enough to the electrode to affect the background activity but not close enough to distinguish action potentials and the recording limits of the MEA system). The amplitude of spikes detected will vary based on the number and level of maturity of the neurons near the electrode and the electrode properties and MEA system's ability to accurately estimate the signal (where the voltage signals are shown as plus and minus change from the estimated baseline relative to the reference electrode). Thus, while the default parameters using the template-based method provide robust spike detection for some organoid recordings. Other organoids and/or MEA chip/systems may require using threshold-based methods or a combination of methods.
5. After evaluating spike detection using the MEA-NAP outputs in the folder "1B_SpikeDetectionChecks," MEA-NAP can be rerun starting at Step 2 with the desired spike detection method for downstream analysis (neuronal and network activity) entered as the "Spike Method for Analysis" on the "Spike Detection" tab and unchecking the box to "Detect Spikes." In contrast, if a user wants to try different spike detection methods, MEA-NAP can be rerun from Step 1 with the new parameters.
6. The MEA-NAP output folders "Step 2 Neuronal Activity" and "Step 4 Network Activity" can also be used to understand how the choice of spike detection parameters affects the network properties. In many cases, the spike detection method will affect the mean firing rates and may affect the mean or maximum strength of connectivity (edge weight), while other network features such as the number of connections (node

degree) or network density may be unaffected. Where the spike detection method alters the neuronal or network activity comparisons between age or group, careful validation of the spike detection method is required.

3.6 Selecting Connectivity Parameters

1. For MEA-NAP's "Step 3 Inferring Functional Connectivity," the spike time tiling coefficient (STTC) lag sets the window for inferring correlated activity in the pair-wise comparisons of electrodes' activity [7]. MEA-NAP can compare the choice of STTC lag and, thus, the defaults are set to 10, 25, and 50 ms. Comparisons of graph metrics by STTC lag can be viewed in the "Step 4" output folder. Running MEA-NAP with multiple lags increases the computational time for Steps 3 and 4 and the number of figures created in the output folders. Thus, for large datasets, or where the user does not wish to compare lags, we recommend changing this option to "[10]" alone for most datasets. Longer lags may be suitable for more immature neurons with infrequent activity. However, longer lags may decrease the sensitivity for true connections (edges) because MEA-NAP uses probabilistic thresholding to determine significant functional connections.
2. We recommend using "weighted" adjacency matrices as differences in the strength of connectivity (edge weights) are seen as neurons develop and are often affected in neurological and psychiatric disorders at the micro- and macroscale.
3. The graphical outputs for MEA-NAP's "Step 3" including "Probability Detection Checks." These figures illustrate how the mean threshold for significant connections changes as the number of iterations for the probabilistic thresholding increases. Typically, these values become more consistent by 180–200 iterations (also called "repeats"). If these values are still unstable in the output figures, MEA-NAP could be re-run starting at Step 3 with a larger number of iterations.
4. Using the probabilistic thresholding, the default for determining significant connections is above the 95th percentile (entered in the GUI as $p<0.05$). If a more stringent definition of significant edges (high specificity, lower sensitivity for true edges) is desired, this value can be changed to the 99th percentile (entered 0.01), for example.
5. The effect of STTC lag choice can also be compared for the nodal roles assigned by the Node Cartography, which uses the entire dataset to determine the relevant boundaries for hub and non-hubs, for example. If a user changes the default STTC lag, the corresponding lags should be changed for the Node Cartography as well.

3.7 Selecting Plotting Parameters

1. All of the neuronal and network features calculated by MEA-NAP are saved in tabular format in the main output folder. However, a major feature of MEA-NAP is the automated generation and organization of figures in the output folder. This allows users with limited experience with MATLAB to quickly visualize their data at both the individual network and entire experiment level, as well as examine the age and group comparisons. It also facilitates convenient sharing of the output figures with other team members and collaborators who may not have access to MATLAB. Thus, choice of image format for the figures is important and depends on intended use.
2. For MEA-NAP users interested in assessing whether there is network activity in their MEA data or users trying out different parameters, selecting a widely compatible figure format such as .png is desirable. These figures can easily be viewed on most computers and through web-based cloud services (e.g., uploading an output folder to Dropbox to share with collaborators).
3. For users who want to use the figures in scientific posters, presentations, grants, or manuscripts, .svg provides files that can be edited in common software such as PowerPoint, Adobe Illustrator, or Affinity Designer, for example. This is particularly important where resizing the plots make text, line thickness or other aspects of figure unreadable. Users must take great care not to alter the data itself in this format.
4. For users with MATLAB experience who want to make changes to how the figures are plotted, selecting .fig (MATLAB figure format) will allow the figures to be opened and edited in MATLAB. Advanced MATLAB users could also do this using the .mat files saved in the output folder.
5. Multiple output figure formats can be selected using the control button. However, each additional figure format (particularly .svg) will add to both the computational time and the size of the output folder. For example, a dataset with ~240 files produced a 50 GB output folder with .png and .svg figures.

3.8 Selecting Advanced Features

1. On the "General" tab of MEA-NAP, there is a box that can be checked "Show Advanced Settings." Checking this box reveals more tabs that are labeled with the appropriate corresponding section (e.g., "Advanced Spike Detection"). New MEA-NAP users are advised to use the non-advanced settings first and then, if needed, explore the advanced settings.
2. The advanced features may be particularly useful for users who wish to compare single-electrode and/or network burst

properties at different temporal scales. The default setting in MEA-NAP is to use the ISI_N method [8], which automatically sets the threshold for spikes within and outside of bursts using the distribution of interspike intervals (ISIs). This automated method works well with a bimodal distribution of ISIs where there is clearly bursting (clustering of action potentials in short time periods) compared to action potentials occurring outside of bursts. However, there are a number of cases where this automated detection fails, e.g., for noisy and/or irregular activity [8]. In these cases MEA-NAP uses 70 ms as the threshold, which worked for most 2D and 3D neuronal culture recordings. Bursting is fractal in nature, meaning there can be bursts within bursts within bursts [9]. There are also large network bursts (bursting that occurs across multiple electrodes) that last for seconds that are visible in the raster plots of the firing rates by electrode produced in the "Step 2 Neuronal Activity" outputs that users may seek to quantify. By adjusting either the number of action potentials, the number of electrodes, and/or the threshold for detecting bursts, users can explore how their independent variables impact bursting at different temporal scales.

3. The advanced features also offer the opportunity to adjust how the plots are scaled, colors, and even which metrics that are calculated. If the latter is changed, it is important for the user to understand, which network metrics are dependent on other network metrics to be calculated (e.g., network metrics that rely on the modular decomposition).

4 Notes

1. We recommend regularly updating MEA-NAP from our GitHub page to take advantage of recent updates and patches.
2. Issues with MEA-NAP can be raised on our GitHub page.
3. For users of older Mac OS, please follow the instructions on our GitHub Read-the-docs documentation for enabling XCode in the Privacy and Security settings. If this error is encountered, follow the instructions in the documentation and then restart MATLAB. For some Mac OS, users will need to repeat this process–waiting for the error to be produced by the first MEA-NAP run–after restarting the computer. The XCode allows MATLAB to "call" C (programming language) in order to perform the spike time tiling coefficient (STTC) calculations for inferring functional connectivity faster.

Acknowledgments

We would like to thank our coauthors [4] on the creation of MEA-NAP and the many members of the Synaptic and Network Development (SAND) research group, collaborators, and users around the world who have tested MEA-NAP on MEA data and provided valuable feedback.

References

1. Schröter M, Paulsen O, Bullmore ET (2017) Micro-connectomics: probing the organization of neuronal networks at the cellular scale. Nat Rev Neurosci 18:131–146
2. Giandomenico SL, Mierau SB, Gibbons GM, Wenger LMD, Masullo L, Sit T, Sutcliffe M, Boulanger J, Tripodi M, Derivery E, Paulsen O, Lakatos A, Lancaster MA (2019) Cerebral organoids at the air-liquid interface generate diverse nerve tracts with functional output. Nat Neurosci 22:669–679
3. Szebényi K, Wenger LMD, Sun Y, Dunn AWE, Limegrover CA, Gibbons GM, Conci E, Paulsen O, Mierau SB, Balmus G, Lakatos A (2021) Human ALS/FTD brain organoid slice cultures display distinct early astrocyte and targetable neuronal pathology. Nat Neurosci 24:1542–1554
4. Sit TPH, Feord RC, Dunn AWE, Chabros J, Oluigbo D, Smith HH, Burn L, Chang E, Boschi A, Yuan Y, Gibbons GM, Khayat-Khoei M, De Angelis F, Hemberg E, Hemberg M, Lancaster MA, Lakatos A, Eglen SJ, Paulsen O, Mierau SB (2024) MEA-NAP: a flexible network analysis pipeline for neuronal 2D and 3D organoid multielectrode recordings. Cell Rep Methods 4:100901
5. Rubinov M, Sporns O (2010) Complex network measures of brain connectivity: uses and interpretations. Neuroimage 52:1059–69
6. Dunn AWE, Sit T, Chabros J, Szebényi K, Gibbons G, Limegrover C, Wenger L, Chiaradia I, Giandomenico SL, Lancaster MA, Lakatos A, Eglen SJ, Paulsen O, Mierau SB (2020) Comparing spike detection in 2D murine cortical culture and 3D human cerebral organoid microelectrode array (MEA) recordings. Federation of European Neuroscience Societies (FENS) Forum Abstract, e-Poster presentation
7. Cutts CS, Eglen SJ (2014) Detecting pairwise correlations in spike trains: an objective comparison of methods and application to the study of retinal waves. J Neurosci 34:14288–303
8. Bakkum DJ, Radivojevic M, Frey U, Franke F, Hierlemann A, Takahashi H (2014) Parameters for burst detection. Front Comput Neurosci 7:193
9. Mazzoni A, Broccard FD, Garcia-Perez E, Bonifazi P, Ruaro ME, Torre V (2007) On the dynamics of the spontaneous activity in neuronal networks. PLoS One 2:e439

Chapter 9

A Computational Protocol for Whole Brain Histology Imaging

Chao-Ying Huang and Li-An Chu

Abstract

Recent advancements in tissue clearing and multiplex labeling have enabled the rapid acquisition of high-resolution, whole-brain 3D imagery [3]. However, the resulting datasets are often massive in size and structurally complex, creating significant bottlenecks in data storage, processing, and analysis. Here, we propose a robust, end-to-end pipeline designed to manage and analyze these large-scale volumes efficiently. This protocol utilizes a chunk-based data structure (OME-Zarr) to drastically lower hardware memory requirements [2], enabling the processing of terabyte-scale data on standard workstations. We design a workflow that integrates artifact correction, automated deep learning segmentation, and atlas registration. Furthermore, the pipeline facilitates dynamic, multiscale 3D visualization via a web-based interface (Neuroglancer) [7], allowing for the remote exploration of massive datasets without the need for complete data transfer.

Keywords Expansion microscopy, Light sheet, OME-Zarr, Deep learning, Brain registration, Neuroglancer, Big data, Image analysis

1 Introduction

Understanding the complex cytoarchitecture of the brain requires high-resolution, three-dimensional (3D) imaging of intact structures. Recent advances in tissue clearing techniques and light sheet fluorescence microscopy (LSFM) have enabled the rapid acquisition of cellular-resolution data from whole mouse brains [1]. However, the resulting datasets are massive often exceeding hundreds of gigabytes to terabytes per sample presenting significant bottlenecks in data storage, processing, and analysis. Standard image processing workflows, which traditionally rely on loading entire image volumes into Random Access Memory (RAM), fail catastrophically when confronted with these data magnitudes.

To address the challenge of data accessibility, this protocol leverages the OME-Zarr (NGFF) format [2]. Unlike legacy formats (e.g., TIFF stacks) that require sequential loading, OME-Zarr utilizes an N-dimensional array structure stored in "chunks." This architecture allows

Antonio Di Ieva and Jennilee M. Davidson (eds.), *Computational Neuroscience*, Methods in Molecular Biology, vol. 3062,
https://doi.org/10.1007/978-1-0716-5440-8_9,

computational tools to load only the specific spatial region required for a task, drastically reducing memory overhead. Furthermore, OME-Zarr supports multiscale resolution pyramids—similar to the technology used in satellite mapping allowing users to seamlessly zoom from a whole-brain overview to cellular details without latency. When paired with parallel processing libraries like Dask [3], this format enables the processing of terabyte-scale datasets on standard workstation hardware.

Quantitative analysis of these volumes requires precise anatomical mapping. We describe a dual approach to brain registration to accommodate different experimental needs. For high-precision mouse brain analysis, we utilize the BIRDs (Bi-channel image registration and deep-learning segmentation) [4], which is a tool optimized for aligning LSFM data specifically to the Allen Mouse Brain Common Coordinate Framework (CCFv3) [5]. Alternatively, for projects involving other species (e.g., Drosophila or Rattus) or requiring specific template flexibility, we employ the CMTK (Computational Morphometry Toolkit) [6], a versatile algorithm capable of robust affine and warping registrations across diverse biological templates.

Finally, the sheer size of these datasets often makes traditional visualization methods impractical. We overcome this by deploying **Neuroglancer** [7], a WebGL-based viewer that streams volumetric data directly to a web browser. By hosting the OME-Zarr pyramid via a local server, Neuroglancer allows for the instant visualization of raw anatomical data, segmentation masks, and registered atlas annotations in a 3D space. Crucially, this setup could enable visualization form any device with internet access; by using tunneling services, researchers can securely share full-resolution, interactive views of their data with collaborators worldwide via a simple URL, facilitating a level of data accessibility and transparency previously unattainable in large-scale microscopy.

2 Materials

2.1 Hardware

2.1.1 Standard Image Processing Workstation

1. Central Processing Unit (CPU): Intel or AMD (12–16 cores)
2. Graphics Processing Unit (GPU): Nvidia RTX 3080 or RTX A4000
3. Random Access Memory (RAM): 64–128 GB
4. Storage: 1 TB M.2 Solid State Drive (SSD)
5. Power Supply Unit (PSU): 800 W Gold
6. Local Area Network (LAN): 2.5 Gbps Ethernet port (*see* **Notes 1 and 2**)

2.1.2 High-Performance Image Processing Workstation

1. CPU: Intel or AMD (32–64 cores)
2. GPU: Nvidia RTX A6000; RAM: 256–512 GB
3. Storage: 4 TB M.2 SSD
4. PSU: 1200 W Platinum
5. LAN: 10 Gbps (Ethernet) or 25 Gbps (SFP) (*see* **Note 1**)

2.1.3 Standard Data Storage Server

1. CPU: Intel or AMD (4 cores)
2. RAM: 32 GB
3. OS Drive: 2 × 1 TB M.2 SSD (configured in RAID 1)
4. Data Storage: 50 TB Hard Disk Drive (HDD) array configured in RAID 6 (*see* **Notes 3 and 4**)

2.1.4 High-Performance Data Storage Server

1. CPU: Intel or AMD (16 cores)
2. RAM: 32 GB
3. OS Drive: 2 × 1 TB M.2 SSD (configured in RAID 1)
4. Data Storage: 200 TB HDD array configured in RAID 6

2.2 Software

1. **Fiji (ImageJ)** [8]**:** Install with the following plugins: BigStitcher [9], BIRDs [4], CMTK [6] GUI, BigDataViewer [10], Weka [11], and Labkit [12].
2. **Python Environment:** Python 3.10 configured with PyTorch, MONAI, Zarr, Dask, Neuroglancer, and Unicorn.
3. **Deep Learning Frameworks:** nnUNet [13] (self-adapting framework).

3 Methods

Mouse brain samples are obtained via tissue clearing and hydrogel expansion. Image samples from the superior to inferior lobe in the horizontal plane using light sheet microscopy to generate 16-bit single-channel images with a spatial resolution of 1.82 μm × 1.82 μm × 4 μm.

3.1 Data Conversion and Preprocessing

1. **Stitching:** Import raw image tiles into Fiji using the BigStitcher plugin. Calculate pairwise shifts using phase correlation, perform global optimization, and fuse the dataset into a single volume [9] (*see* **Note 5**).
2. **Destriping:** Apply stripe artifact correction to the fused image series using the Pystripe Python module with default wavelet decomposition parameters. Save the output as a 2D TIFF series (total volume approximately 200–250 GB) [6, 14] (*see* **Note 6**).

3. **Stitching:** Stitch each 2D array data on the same z-axis into one z-slice [14].
4. **Conversion:** Convert the image file format to OME-Zarr using dask modules to load and write the dataset (*see* **Note 7**).
5. **Chunking:** Set the storage chunk size to 128×128×128 or 64×128×128 (*see* **Note 8**).
6. **Compression:** Set the compressor of the chunk to the Blosc compressor with clevel 5 [15] (*see* **Note 9**).
7. Pyramid Generation: Generate 5 levels of downsampled volume for the OME-Zarr pyramid.
8. Resizing: Resize the volume using dask.map_overlap and skimage. transform.resize() modules [3] (*see* **Note 10**).

3.2 Signals Segmentation

1. **For high signal to noise ratio (SNR) signals:** Use the Labkit and Weka machine learning approach utilizing pixel-wise random forest classification [12]. Train on one slice, inference on the remaining slices, and manually quality control the results to correct false positives or negatives.
2. **For low SNR/complex features:** Use a Deep Learning approach (e.g., 3D UNet with MONAI [16] or nnUNet [13]) (Fig. 1).
3. **Data Preparation:**
   - If no pre-trained model exists: Crop 48 continuous slices of 1600×1600 ROIs containing target signals.

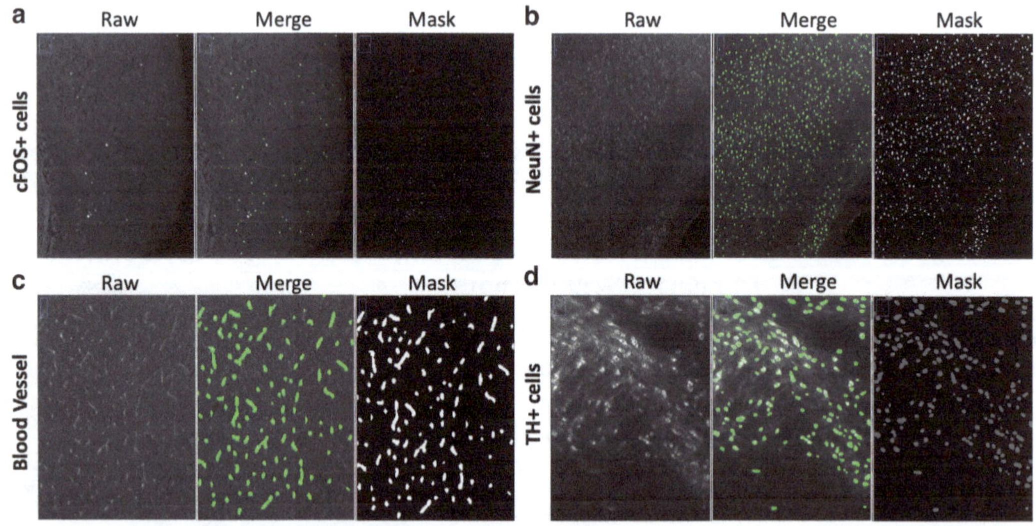

Fig. 1 Representative raw images and nnU-Net–based segmentation results. Example raw images (Raw), overlay of predictions (Merge), and segmentation masks (Mask) generated by nnU-Net for multiple tissue targets. *a* anti-cFOS–positive neuronal nuclei. *b* anti-NeuN–positive neuronal nuclei. *c* lectin-labeled blood vessels. *d* anti-TH–positive dopaminergic neuron soma

- If a pre-trained model exists: Prepare 16 continuous slices of 1600 × 1600 ROIs.
- Annotate them using the same method as the high SNR signals.

4. **Image Preprocessing:**
   - Apply Z-Score normalization to raw intensity values (typically 100–4000 grayscale).
   - For **spherical morphology:** Apply a light 3D Gaussian filter (5 × 5 × 3) (*see* **Note 11**).
   - For **tubular morphology:** Apply gamma correction with an intensity of 1.3 (*see* **Note 11**).
   - Crop images into 3D patches. Ensure patch size is smaller than the Zarr chunk size (*see* **Note 12**).
5. **Augmentation (MONAI pipeline):** Apply random affine rotation, random flip, random contrast adjustment, random intensity shifting, random bias field, and random zoom (*see* **Note 13**) (Skip this step if using nnUNet).
6. **Training:** Train or fine-tune the model using the Adam optimizer. Use a weighted loss function combining binary cross-entropy loss and Dice loss. Evaluate performance based on the F1 score. Expect training times of 2–10 h depending on GPU specifications.
7. **Post-processing:** Stitch the inferred patches using an overlap strategy, caching overlapped planes from previous patches.
8. Save the resulting mask into an OME-Zarr file for overlay visualization.

3.3 Brain Registration

1. Open Fiji and import the dataset via File -> Import -> HDF5/H5/Zarr/OME-NGFF. Select a resolution tier smaller than 10 GB (e.g., approx. 2000 × 2000 × 1000, or tier 3 downsampled by 8×) (*see* **Note 14**).
2. Perform registration using either BIRDs (Fig. 2) or the CMTK GUI (*see* **Note 15**).
3. Upsample and convert the resulting annotation file to match the original image dimensions and mask for overlay in Neuroglancer.

3.4 Quantitative Signal Analysis

1. **Spherical Structures:**
   - Use dask.map_overlap with skimage functions to calculate distance transforms and apply a local maxima filter to identify centroids.
   - Apply a watershed algorithm to separate adjacent binary spheres if required.
   - Map the center point coordinates to the annotation chunk to identify the brain region (Fig. 3).

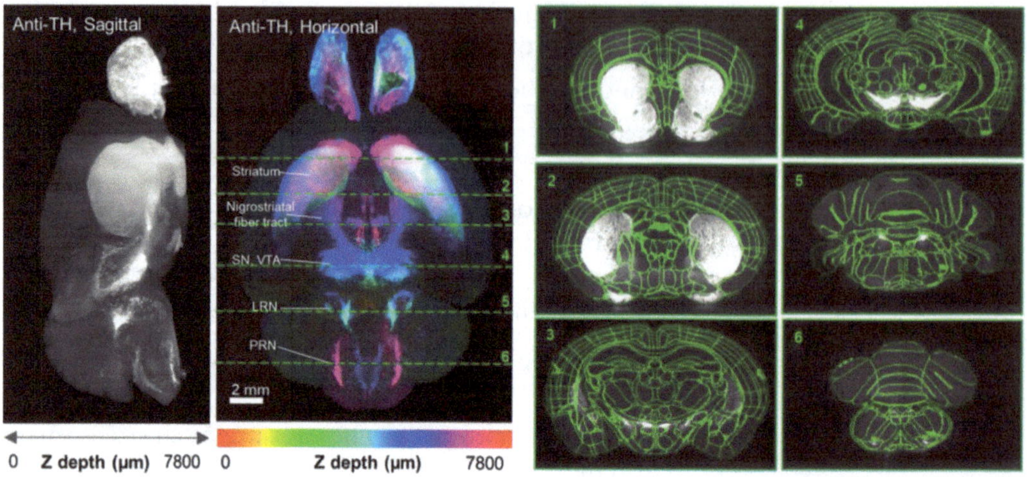

Fig. 2 Quantitative registration of whole-brain immunolabeling. Horizontal and sagittal views of Tyrosine Hydroxylase (TH) expression are shown in 3D (left). To validate spatial accuracy, cross-sectional views (right) show the experimental signal overlaying the standard brain atlas (green contours), confirming precise registration in regions including the lateral reticular nucleus (LRN) and pontine reticular nucleus (PRN). Reproduced from Lin et al., 2024, *eLife*, https://doi.org/10.7554/eLife.93212.4, used under the CC BY 4.0 license [4, 17]

2. **Tubular Structures:**
   - Use dask.map_overlap with skimage functions to calculate distance transforms and identify the widest segments.
   - Extract the structure skeleton using skimage.morphology.skeletonize().
   - Calculate bifurcation points using a 3×3 convolution kernel.
   - Map skeleton coordinates and bifurcations to the annotation chunk.
3. Generate a report containing region-specific data and accumulative tier results based on structure hierarchy.

3.5 Neuroglancer Visualization (Fig. 4)

1. Host a uvicorn data service server via Python in the data directory.
2. Use the Neuroglancer Python module to host the service on the local machine.
3. Input localhost:port into the Neuroglancer GUI. Use the "plus" interface to overlay multiple data layers.
4. Configure port forwarding or Cloudflare tunneling to allow external access (*see* **Note 16**).

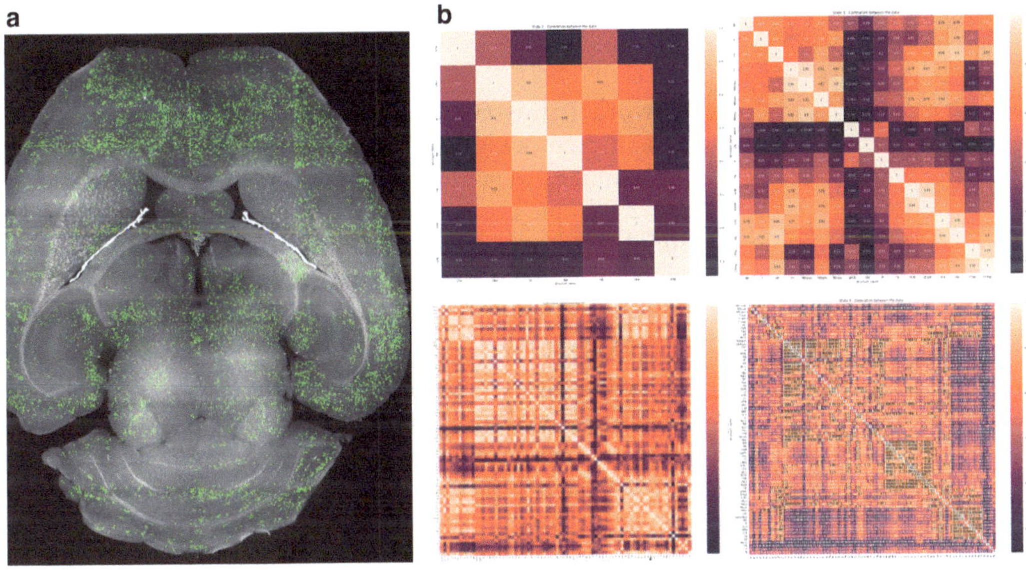

Fig. 3 Whole-brain analysis after 3D cell-center mapping of cFOS signals. *a* Representative coronal slice from the whole-brain dataset. Green dots denote detected cFOS-positive cell centers after 3D cell-center mapping. *b* Whole-brain activity correlation matrices generated using different parcellation resolutions, illustrating how brain-wide functional relationships vary with the degree of regional subdivision

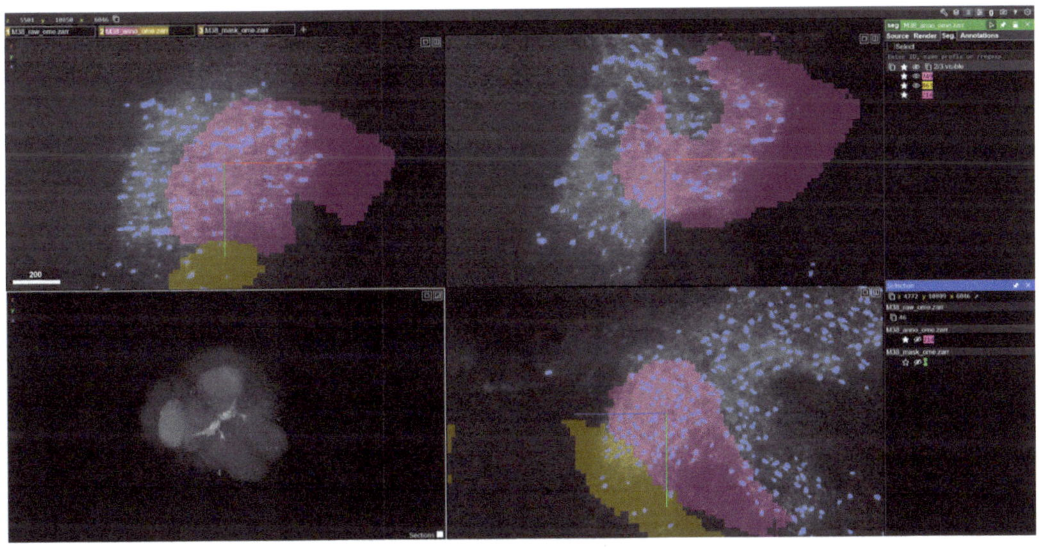

Fig. 4 Representative Neuroglancer visualization of whole-brain data. Displayed is a Neuroglancer view showing the overlay of the raw fluorescence image (white), detected cell masks (blue), and associated brain-region annotations (purple and yellow). This visualization enables simultaneous inspection of cellular positions relative to anatomical boundaries

4 Notes

1. **Network Infrastructure:** It is preferred that the data storage and preprocessing devices reside on the same local LAN to minimize latency. We recommend a transfer speed of at least 2.5 Gbps. For a reasonable budget, a 10 Gbps Ethernet setup is preferred; speeds exceeding 10 Gbps typically require fiber optic cabling and Small Form-factor Pluggable (SFP) interface, which drastically increases infrastructure and maintenance costs.
2. **Operating System Selection:** We recommend Windows as the primary operating system because specific commercial image processing systems (e.g., Imaris) may not fully support Linux. If Linux-exclusive tools are required, we prefer installing Windows Subsystem for Linux (WSL) and Docker. A more advanced alternative is a dual-boot system or utilizing two separate machines (Windows and Linux) connected via high-speed LAN.
3. **OS Drive Redundancy:** We utilize two 1 TB M.2 SSDs configured in **RAID 1 (Mirroring)** for the operating system. This ensures that if one boot drive fails, the system remains operational without data loss or downtime, which is critical for servers running long-term storage or web services.
4. **RAID Configuration:** For the data storage array, we strictly recommend **RAID 6**.
   - **Rationale:** RAID 6 provides dual-parity, meaning the array can withstand the simultaneous failure of **two**hard drives without data loss. This is crucial for large arrays (50TB+) where the rebuild time for a replaced drive can take days, during which a second drive failure is a statistically significant risk.
   - **Alternatives:**
   - **RAID 5:** Offers single-drive parity. We do *not* recommend this for large drives because if a second drive fails during the rebuild process (a common occurrence due to stress on the drives), all data is lost.
   - **RAID 10:** Stripes data across mirrored pairs and offers faster write speeds than RAID 6. However, it cuts usable storage capacity by 50%, making it significantly more expensive for high-capacity storage needs.
5. **Stitching Strategy:** We utilize **BigStitcher** (https://github.com/JaneliaSciComp/BigStitcher) for its ability to handle large, multitile datasets efficiently using virtual stacks. If the

sample was cleared and imaged effectively, phase correlation is usually sufficient for calculating pairwise shifts between tiles [4].

6. **Artifact Removal:** Light sheet microscopy frequently results in horizontal stripe artifacts due to light absorption or scattering by tissue structures. We utilize **Pystripe** (https://github.com/chunglabmit/pystripe) to filter these artifacts. Pystripe employs combined wavelet and FFT filtering to selectively remove the specific frequency components of the stripes without degrading the biological signal [4].
7. **Data Management:** We highly recommend making a copy of the original volume during the conversion process. Leveraging the multicore capabilities of the system via dask is recommended for efficient loading and writing.
8. **Chunking Strategy:** Small chunk sizes result in the generation of thousands of small files, which can create file management overhead and slow down I/O operations. We find that dimensions of $128 \times 128 \times 128$ or $64 \times 128 \times 128$ provide an optimal balance.
9. **Compression Strategy:** We prioritize the **Blosc** compressor (clevel 5) because it is optimized for fast binary data access, significantly reducing the "time-to-view" latency in Neuroglancer.
   - **Alternatives:**
   - **Gzip:** If maximum software compatibility or higher compression ratios are required (at the cost of slower read/write speeds), **Gzip** (level 5 or 6) is a robust alternative supported by virtually all analysis tools.
   - **Zstd (Zstandard):** For users seeking a modern balance between high compression ratios and speed, **Zstd** is an excellent choice, though one must ensure that all downstream tools (specifically specific older versions of visualization plugins) support the codec.
10. **Memory Overhead:** While map_overlap is effective for resizing, we have observed a 5× to 10× increase in RAM usage during this operation. Ensure sufficient memory is available before execution.
11. **Morphology Filters:** Preprocessing filters should be chosen based on target morphology. A 3D Gaussian filter helps remove high-frequency imaging noise common in spherical targets, whereas gamma correction highlights tubular structures.
12. **Patch vs. Chunk Size:** It is critical that the training crop/patch size is smaller than the OME-Zarr chunk size. Zarr loads data by the chunk; if the patch size exceeds the chunk

size (or misaligns significantly), the system will load all overlapping chunks into RAM, leading to inefficient data loading and potential memory overflows.

13. **Deep Learning Implementation:** For users developing custom 3D UNet models or implementing specific augmentation pipelines (e.g., RandomAffine, RandomElasticDeformation), we refer to the official **MONAI** tutorials and transformation documentation (https://github.com/Project-MONAI/tutorials). These resources provide verified implementations for medical image-specific deep learning tasks.
14. **Registration Tiers:** Loading the full-resolution volume into Fiji is often impossible due to memory constraints. Downsampled tiers (Tier 3 in our case) provide sufficient anatomical detail for registration without crashing the software.
15. **Tool Selection:**
    - **BIRDS:** Limited to mouse brain registration on the Allen Brain Atlas but generally yields higher accuracy for this specific task. (https://github.com/bleach1by1/BIRDS_plugin) [4].
    - **CMTK GUI:** More versatile, capable of registering mouse, rat, and fly brains (including mushroom bodies), provided a properly annotated template is available (https://github.com/jefferis/fiji-cmtk-gui?tab=readme-ov-file) [6].
    - Follow the installation and usage instructions provided on their respective GitHub and Wiki pages.
16. **Remote Access and Security:** For sharing visualization with outside collaborators, we recommend obtaining a dedicated domain name (e.g., mylab.com) and utilizing a Cloudflare tunnel service to mask the raw IP address of the local server. Alternatively, port forwarding can be used, but strict institutional network security protocols must be followed when exposing ports.

References

1. Huisken J, Swoger J, Del Bene F, Wittbrodt J, Stelzer EH (2004) Optical sectioning deep inside live embryos by selective plane illumination microscopy. Science 305:1007–1009
2. Moore J et al. (2021) OME-NGFF: a next-generation file format for expanding bioimaging data-access strategies. Nat Methods 18:1496–1498
3. Rocklin M (2015) Dask: parallel computation with blocked algorithms and task scheduling. In: SciPy 126–132. https://proceedings.scipy.org/articles/Majora-7b98e3ed-013.pdf
4. Tyson AL et al. (2021) A deep learning algorithm for 3D cell detection in whole mouse brain image datasets. PLoS Comput Biol 17:e1009074
5. Wang Q et al. (2020) The Allen mouse brain common coordinate framework: a 3D reference atlas. Cell 181(936–953):e920
6. Rohlfing T, Maurer Jr CR, O'Dell WG, Zhong J (2004) Modeling liver motion and deformation during the respiratory cycle using intensity-based nonrigid registration of gated MR images. Med Phys 31:427–432

7. Maitin-Shepard J (2016) Neuroglancer: WebGL-based viewer for volumetric data. GitHub Repository. https://github.com/google/neuroglancer. Accessed 20 Nov 2023
8. Schindelin J et al. (2012) Fiji: an open-source platform for biological-image analysis. Nat Methods 9:676–682
9. Horl D et al. (2019) BigStitcher: reconstructing high-resolution image datasets of cleared and expanded samples. Nat Methods 16:870–874
10. Pietzsch T, Saalfeld S, Preibisch S, Tomancak P (2015) BigDataViewer: visualization and processing for large image data sets. Nat Methods 12:481–483
11. Arganda-Carreras I et al. (2017) Trainable Weka Segmentation: a machine learning tool for microscopy pixel classification. Bioinformatics 33:2424–2426
12. Arzt M et al (2022) LABKIT: Labeling and segmentation toolkit for big image data. Front Comput Sci 4:777728. https://www.frontiersin.org/journals/computer-science/articles/10.3389/fcomp.2022.777728/full
13. Isensee F, Jaeger PF, Kohl SAA, Petersen J, Maier-Hein KH (2021) nnU-Net: a self-configuring method for deep learning-based biomedical image segmentation. Nat Methods 18:203–211
14. Swaney J et al (2019) Scalable image processing techniques for quantitative analysis of volumetric biological images from light-sheet microscopy. bioRxiv 576595. https://www.biorxiv.org/content/10.1101/576595v1
15. Alted F et al (2023) Blosc: a blocking, shuffle and lossless compression library. https://www.blosc.org
16. Cardoso MJ et al (2022) MONAI: An open-source framework for deep learning in healthcare. https://arxiv.org/abs/2211.02701
17. Lin YH et al (2024) Revealing intact neuronal circuitry in centimeter-sized formalin-fixed paraffin-embedded brain. Elife 13:RP93212. https://elifesciences.org/articles/93212

Part II

Computational Clinical Neurosciences

Chapter 10

Computational Pipeline in Neuroradiomics

Ghasem Azemi and Antonio Di Ieva

Abstract

Neuroradiomics is a field in neuroimaging that extracts large-scale, quantitative features from central and peripheral nervous systems' images using computational techniques. These features capture intensity distributions, textures, shapes, and spatial patterns. By replacing subjective interpretation with algorithm-driven analysis, neuroradiomics improves objectivity, reproducibility, and consistency, while revealing associations that aid diagnosis, prognosis, and monitoring of neurological and neurosurgical diseases. This chapter presents a step-by-step implementation of the neuroradiomics pipeline, with recommended tools and illustrative Python scripts and command-line snippets provided. The workflow begins with dataset preparation, imaging formats, and computational environments, followed by preprocessing procedures to standardize and normalize neuroimaging data. Segmentation methods are then introduced to define regions of interest, after which radiomic features are extracted using dedicated Python libraries. Subsequent sections demonstrate feature selection, machine learning model development, and performance evaluation, all implemented in reproducible Python environments. Throughout, emphasis is placed on transparency, reproducibility, and practical guidance to support both research and clinical applications. An example implementation is provided to demonstrate practical application and illustrate Python-based workflow implementation with code snippets. By the end of this chapter, readers will have a comprehensive and practical framework for implementing neuroradiomics workflows that would enable them to design, adapt, and apply these methods across diverse neurological applications.

Keywords Neuroradiomics, Radiomic features, Neuroimaging, Preprocessing, Segmentation, PyRadiomics, Machine learning, Predictive modeling

1 Introduction

Neuroradiomics is a rapidly advancing domain within neuroimaging that focuses on the extraction of large-scale, quantitative features from images, such as Magnetic Resonance Imaging (MRI), Computed Tomography (CT), and Positron Emission Tomography (PET), of the central and peripheral nervous systems using computational and algorithmic techniques. These features capture diverse aspects of the image, including intensity distributions, texture patterns, structural shape, and spatial organization, often revealing subtle imaging biomarkers that are beyond the resolution of human perception.

Antonio Di Ieva and Jennilee M. Davidson (eds.), *Computational Neuroscience*, Methods in Molecular Biology, vol. 3062, https://doi.org/10.1007/978-1-0716-5440-8_10,

Beyond feature extraction, neuroradiomics offers broader advantages in terms of objectivity, reproducibility, and data richness. By shifting from subjective visual interpretation to algorithm-driven analysis, it reduces inter-observer variability and promotes consistency across institutions. Moreover, by uncovering hidden patterns and complex associations within imaging data, neuroradiomics supports a more nuanced, data-driven understanding of brain structure and pathology, ultimately aiding in the diagnosis, prognosis, and monitoring of neurological disorders [1, 2].

Neuroradiomics has demonstrated substantial clinical utility across a range of neurological and neurosurgical conditions. In neuro-oncology, it enables non-invasive characterization of brain tumors, supports tumor classification and grading, assesses treatment response, and predicts clinical outcomes [1–7]. It also facilitates the identification of tumor heterogeneity and the integration of imaging with genomics, i.e., radiogenomics [1, 8]. Similarly, in neurodegenerative diseases such as Alzheimer's disease, Parkinson's disease, and amyotrophic lateral sclerosis (ALS), radiomic features can detect subtle patterns of atrophy, microstructural alterations, and metabolic abnormalities, aiding in early diagnosis, differential classification, and disease monitoring [9]. In cerebrovascular neuroimaging, radiomics have been shown to usefully characterize brain aneurysms, arteriovenous malformations and cavernomas [10, 11], with many applications in stroke imaging as well [12]. One of the most transformative aspects of neuroradiomics is its ability to function as a virtual biopsy—providing detailed insights into tissue pathology without surgical intervention [13]. These capabilities have profound implications for personalized medicine, where image-derived biomarkers can inform precision diagnostics, prognostication, and treatment planning tailored to individual patients.

A key strength of neuroradiomics lies in its ability to integrate multimodal data. In many clinical and research settings, radiomic features derived from structural MRI can be combined with complementary information from PET, CT, ultrasound, or advanced MRI sequences such as diffusion-weighted imaging (DWI) or perfusion imaging. Moreover, neuroradiomic data can be fused with clinical, genomic, or histopathological variables to create more comprehensive predictive models. This multimodal integration supports a systems-level understanding of neurological disease and enhances the accuracy and generalizability of diagnostic and prognostic tools.

A typical neuroradiomics workflow is illustrated in Fig. 1. It begins with the acquisition of structural or functional brain imaging data, commonly MRI, CT, or PET, and involves several key stages that transform raw neuroimaging data into clinically actionable insights. These stages include image preprocessing, region-of-interest (ROI) delineation, quantitative feature extraction, feature selection and dimensionality reduction, followed by classification or statistical analysis to derive clinically relevant information.

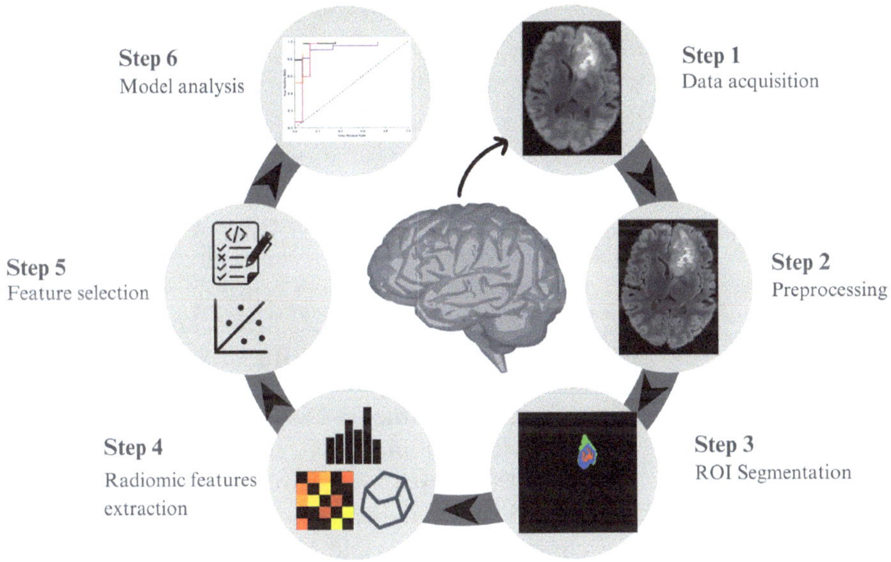

Fig. 1 Overview of a typical neuroradiomics workflow, from brain imaging acquisition through preprocessing, delineation of ROI, feature extraction, and selection, to classification or statistical analysis

The computational workflow for neuroradiomics can be implemented using various programming environments, with MATLAB and Python being among the most used in neuroimaging research. MATLAB provides a mature, user-friendly environment with extensive toolboxes for image processing, signal analysis, and statistical modeling, making it suitable for prototyping and visualization. However, Python has become the preferred platform for neuroradiomics due to its open-source nature, flexibility, and rich ecosystem of libraries such as *nibabel*, *pyradiomics*, *scikit-learn*, and *pytorch*. Python's versatility and strong community support enable reproducible, scalable, and machine learning–integrated analyses. Accordingly, this chapter focuses on implementing the neuroradiomics pipeline using Python-based tools.

2 Materials

Prepare and organize all neuroimaging datasets, metadata, and computational environments. Ensure that hardware and software resources are appropriately configured before starting the workflow.

2.1 Neuroimaging Datasets

1. After proper approvals from the ethics committees, prepare and incorporate private or institution-specific (in-house) datasets obtained from clinical archives, research cohorts, or multicenter collaborations where available. Ensure that these datasets include detailed metadata—such as patient demographics, clinical profiles, and labelled outcomes (e.g., disease

classifications, treatment responses, or survival data)—to enable training, validation, and interpretation of radiomic models in clinically relevant contexts.

2. For neuroradiomics studies—particularly those focused on developing new computational methodologies, identifying novel imaging biomarkers, or validating automated analysis tools—publicly available datasets can be used. These datasets offer standardized imaging, annotations, and accompanying clinical metadata that support reproducibility, benchmarking, and cross-study comparisons. Below are several widely used neuroimaging datasets: (*see* **Note 1** for links to the datasets)

   2.1. ADNI (Alzheimer's Disease Neuroimaging Initiative) offers longitudinal MRI and PET scans from individuals with normal cognition, mild cognitive impairment (MCI), and Alzheimer's disease, along with extensive clinical, cognitive, and biomarker data [14].

   2.2. BraTS (Brain Tumor Segmentation) dataset provides multimodal MRI scans of glioma patients, including T1, T1-contrast-enhanced (T1ce), T2, and FLAIR images, along with expert-annotated tumor subregion masks [15].

   2.3. HCP (Human Connectome Project) provides high-resolution structural and functional MRI (fMRI) data from healthy young adults, aimed at mapping brain connectivity and architecture [16].

   2.4. OpenNeuro: It is an open platform offering diverse neuroimaging datasets, including fMRI, electroencephalogram (EEG), and multimodal studies, often organized using the BIDS (Brain Imaging Data Structure) standard [17].

   2.5. TCIA (The Cancer Imaging Archive) hosts various neuro-oncology datasets, including pre- and post-treatment MRIs, with corresponding clinical and genomic data in some cases [18].

2.2 Data Standards and Formats for Private Datasets

1. Store imaging data in standardized formats to ensure compatibility, interoperability, and reproducibility across neuroradiomics workflows. Private imaging datasets are commonly acquired in the DICOM (Digital Imaging and Communications in Medicine) format, which is the clinical standard for storing image data along with detailed metadata. For research and computational workflows, the NIfTI (Neuroimaging Informatics Technology Initiative) format is widely adopted due to its simplicity, consistency, and broad software support.

2. (Optional): Organize imaging datasets and associated metadata according to the BIDS. Follow its conventions for directory structure, file naming, and metadata documentation to enable seamless data sharing, interoperability with neuroimaging

software, and automated preprocessing pipelines (*see* **Note 2** for a BIDS example).

3. Prepare annotation files such as tumor or lesion masks in NIfTI format to ensure spatial correspondence with imaging data. Record clinical and outcome metadata in structured formats (e.g.,*.json*, *.csv*) to maintain linkage with imaging data.

2.3 Computational Resources and Hardware

1. Prepare a workstation with sufficient resources to manage high-resolution 3D and 4D neuroimaging data. A multi-core CPU ($\geq$8 cores) and at least 64 GB of RAM are recommended to ensure smooth execution of preprocessing and analysis workflows.
2. Include a dedicated GPU if accelerated processing is required, particularly for deep learning or computationally intensive tasks (*see* **Note 3**). While not essential for all radiomics studies, GPU resources can significantly reduce processing time.
3. For projects involving very large cohorts or advanced machine learning (ML) models, consider scaling to high-performance computing clusters or cloud-based platforms. These resources are not required for standard analyses but can support faster and more efficient processing when needed.

2.4 Programming Environments and Python Libraries

1. Install Python using the Anaconda distribution (*see* **Note 4**), which provides a convenient package manager and preconfigured scientific libraries.
2. Create a dedicated conda environment for neuroradiomics tasks, e.g., *neuroradio_env*, to ensure reproducibility and prevent conflicts between dependencies.
3. Prepare the environment by installing the core libraries required for handling neuroimaging data, preprocessing, feature extraction, and model development:
   - 3.1. *nibabel* and *nilearn*: Essential for loading, manipulating, and visualizing neuroimaging data in NIfTI format.
   - 3.2. *simpleitk* and *antspyx*: Provide powerful tools for image registration, resampling, and spatial normalization.
   - 3.3. *pyradiomics*: The most widely used library for radiomic feature extraction, offering a standardized and extensible framework in compliance with the Image Biomarker Standardization Initiative (IBSI).
4. Include ML and deep learning libraries to support model development and evaluation:
   - 4.1. *scikit-learn*: Provides a wide range of classical ML algorithms, hyperparameter optimization tools, and evaluation metrics.

4.2. *pytorch* or *tensorflow*: Commonly employed for deep learning workflows such as automated segmentation or classification.

4.3. *pandas* and *numPy*: For structured data handling and numerical computation.

5. Install *matplotlib* and *seaborn* utilities for generating figures, visualizing features, and summarizing results.

Use the commands below to create and configure the *neuroradio_env* environment, following the steps outlined in items 1–5 above. The commands can be run directly from a terminal on macOS or Linux, from an Anaconda Prompt on a Windows machine, or alternatively saved in a *.sh* file and executed on a macOS or Linux system.

```
# 1. Create and activate a dedicated environment
conda create -n neuroradio_env python=3.10
conda activate neuroradio_env

# 2. Install neuroimaging and radiomic feature extraction libraries
conda install -c conda-forge nibabel nilearn simpleitk antspyx
pip install pyradiomics

# 3. Install machine learning and deep learning libraries
conda install scikit-learn
# (Optional) Install deep learning frameworks
conda install pytorch torchvision torchaudio -c pytorch
# OR (if you prefer TensorFlow instead of PyTorch)
pip install tensorflow

# 4. Install utilities for data handling and visualisation
conda install pandas numpy matplotlib seaborn
```

2.5 Software Tools for Medical Image Analysis

1. Install and configure the specialized neuroimaging software packages required for your specific workflow (*see* **Note 4**), as described in Sect. 3 . These tools support key operations such as data preprocessing, image registration, brain extraction, segmentation, and visualization.

2. The following software tools are commonly integrated into neuroradiomics workflows. Depending on the specific pipeline, not all these tools may be required (*see* **Note 5**):

   2.1. *3D Slicer*: Provides advanced capabilities for medical image visualization, segmentation, and surgical planning, with support for multimodality fusion [19].

   2.2. *ANTs* (Advanced Normalization Tools): Provides state-of-the-art spatial registration, template construction, and image-driven morphometric analysis [20].

2.3. *DeepMedic*: A convolutional neural network (CNN)–based framework designed for automated volumetric medical image segmentation [21].

2.4. *fMRIPrep*: A robust and standardized pipeline for preprocessing fMRI and MRI data, ensuring reproducibility and consistency across participants [22].

2.5. *FreeSurfer*: Widely used for cortical and subcortical segmentation, reconstruction of brain surfaces, and morphometric analysis [23].

2.6. *FSL* (FMRIB Software Library): Offers a suite of tools for structural and functional MRI analysis, including motion correction, brain extraction, and statistical modelling [24, 25].

2.7. *nnU-Net*: A self-configuring deep learning framework that automatically adapts its architecture and training settings to the dataset at hand for segmentation tasks [26, 27].

2.8. *ImageJ*: General-purpose image processing application used for interactive inspection and manual measurements [28, 29].

2.9. *ITK-SNAP*: Facilitates semiautomated segmentation and 3D visualization, particularly useful for interactive annotation and verification of segmented regions [30].

2.10. *SynthStrip*: Deep learning–based skull-stripping tool that is particularly effective for MRI data acquired in inconsistent or low-quality imaging settings [31].

3 Methods

3.1 Preprocessing Implementation

Preprocess all neuroimaging data to improve image quality, harmonize intensity and spatial properties, and ensure readiness for feature extraction and analysis. In this section, only the preprocessing of MRI scans is discussed. Preprocessing of MRIs can be performed either by executing the shell commands or by using automated pipelines such as *fMRIPrep*. The following steps are recommended:

1. Convert all raw DICOM files to NIfTI format using dedicated converters such as *dicom2nifti* or *dcm2niix*, as the conversion consolidates scans into a single file (*.nii* or.*nii.gz*) or paired.*nii* and *.json* files for streamlined processing.
2. Store scans for each subject in structured directories to facilitate reproducibility and downstream processing.
3. For each MRI sequence of each subject, perform the following preprocessing steps:

3.1. Bias field correction: Correct intensity non-uniformities caused by scanner inhomogeneities using, for example, the N4 [32] algorithm in *ANTs* software package.

3.2. Spatial normalization and registration: Align individual scans to a standard anatomical template (e.g., MNI152—Montreal Neurological Institute-152 template—an average of 152 healthy brain MRIs) using tools such as *FSL FLIRT* (FMRIB's Linear Image Registration Tool) and *FNIRT* (FMRIB's Non-linear Image Registration Tool) [24, 33], or *ANTs*. Perform linear registration for global alignment and non-linear registration for detailed anatomical matching.

3.3. Skull stripping (brain extraction): Remove non-brain tissues using tools such as *FSL BET* [24], *FreeSurfer*, or *SynthStrip*. Verify that only brain parenchyma is retained for analysis.

3.4. Use the following shell script, beginning with the comment "# preprocessing of a T1-weighted MRI using ANTs and FSL", to implement the preprocessing steps listed above for all T1-weighted scans in a specified

```
# preprocessing of a T1-weighted MRI using ANTs and FSL

subject_id="sub-01"  # Set subject ID (e.g., sub-01)
input_image="${subject_id}_T1w.nii.gz"

# set paths
f="/sampleBIDSdataset/${subject_id}/anat"
template="/templates/MNI/MNI152/mni_icbm152_t1_tal_nlin_sym_09c.nii.gz"
# step 1: Bias field correction (no resampling)
N4BiasFieldCorrection -d 3 -i "$f/${input_image}" -s 3 \
-c [100x50x20,0.000001] -o "$f/${subject_id}_T1w_BFC.nii.gz"

# step 2: Affine normalisation to MNI template using FSL
flirt -in "$f/${subject_id}_T1w_BFC.nii.gz" -ref "$template" \
-out "$f/${subject_id}_T1w_BFC_normalised.nii.gz" \
-omat "$f/${subject_id}_T1w_BFC_trans.mat"

# step 3: Skull stripping using ANTs
antsBrainExtraction.sh \
-d 3 \
-a "$f/${subject_id}_T1w_BFC_normalised.nii.gz" \
-e "/ANTs_ANSR_Brain_Templates/Oasis/MICCAI2012-Multi-Atlas-Challenge-Data/
  T_template0.nii.gz" \
-m "/ANTs_ANSR_Brain_Templates/Oasis/MICCAI2012-Multi-Atlas-Challenge-Data/
  T_template0_BrainCerebellumProbabilityMask.nii.gz" \
-o "$f/${subject_id}_T1w_BFC_normalised_"
```

directory. The script integrates *ANTs* software package for bias field correction and skull stripping, and *FSL* for spatial normalization.

4. For BIDS-formatted datasets, *fMRIPrep* can be used to implement standardized preprocessing with minimal manual intervention (*see* **Note 6**). *fMRIPrep* integrates *ANTs*, *FSL*, and *FreeSurfer* to generate high-quality anatomical and functional outputs.

3.2 Segmentation of Regions of Interest

Segment regions of interest (ROIs) such as brain tumors (or any kind of other lesion) or anatomical structures to prepare for radiomic feature extraction. Accurate segmentation ensures that extracted features correspond to meaningful biological regions, supporting reliable downstream analyses. Implement segmentation using manual, automated, or hybrid approaches as follows:

1. Delineate ROIs manually when high precision is required, such as tumor or lesion boundaries.
   1.1. Use interactive software tools such as *ImageJ*, *ITK-SNAP*, or *3D Slicer*.
   1.2. Have segmentation performed by expert radiologists or neuroscientists to ensure anatomical accuracy.
2. Use automated tools to reduce time and effort while improving consistency across subjects. Automated methods are particularly suitable for large-scale or multicenter studies. Recommended tools include (*see* **Note 4**):
   2.1. *DeepMedic*: Implement this 3D convolutional neural network for brain lesion segmentation, including tumor and stroke lesions. It has been validated extensively on neuroimaging datasets.
   2.2. *nnU-Net*: Apply this self-configuring deep learning framework for a wide range of segmentation tasks. It automatically adapts to the dataset and has demonstrated high performance in clinical imaging applications.

3.3 Feature Extraction in Neuroradiomics

Extract radiomic features from preprocessed images within the defined ROIs. Accurate preprocessing and ROI delineation are essential to ensure reproducibility and biological interpretability of the extracted features [34]. For Python users, *pyradiomics* provides an open-source library for standardized extraction of radiomic features from medical images, enabling reproducible and comprehensive quantification of tissue characteristics (*see* **Note 7** for other available software platforms). Comprehensive details on all radiomic features extracted using *pyradiomics*, including the mathematical formulas used for their derivation, can be found in its documentation (*see*

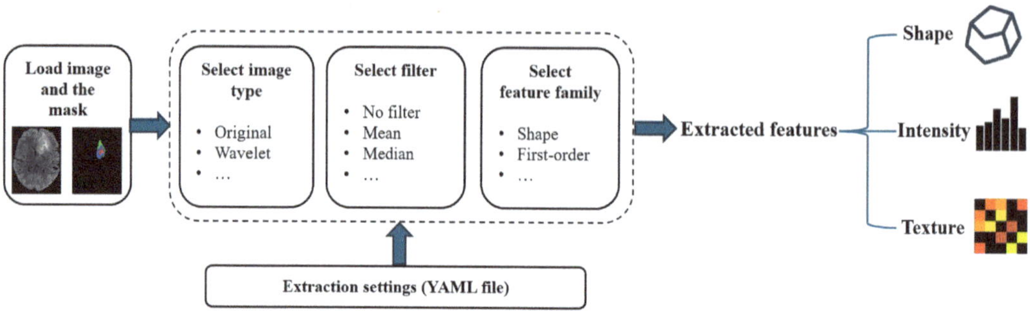

Fig. 2 Configuration workflow for *pyradiomics* feature extraction. Select the image type, choose an image filter, and specify feature families. Settings are stored in a YAML (*.yaml*) file to ensure standardized and reproducible extraction of radiomic features, whether executed in Python or via the command line

Note 4 for a link to the *pyradiomics* documentation). Fig. 2 illustrates the typical configuration workflow for extracting radiomic features using *pyradiomics*.

Follow the steps below to perform feature extraction.

1. Choose the radiomic feature types to extract based on study objectives. Common categories include:
   1.1. Shape Features: Capture geometric properties of the ROI in three dimensions, including volume, surface area, compactness, elongation, and sphericity.
   1.2. Intensity-Based (first-order) Features: Describe the statistical distribution of voxel intensities within the ROI, including mean, median, variance, skewness, kurtosis, and entropy.
   1.3. Texture Features: Quantify spatial relationships and patterns of voxel intensities within the ROI to assess tissue heterogeneity. Common texture feature families include:
      1.3.1. GLCM (Gray Level Co-occurrence Matrix): Measure frequency of specific pairs of intensity values at defined spatial distances and directions, capturing contrast, correlation, and homogeneity.
      1.3.2. GLRLM (Gray Level Run Length Matrix): Measure the length of consecutive voxels with the same intensity in a particular direction, describing texture granularity.
      1.3.3. GLSZM (Gray Level Size Zone Matrix): Quantify the size and distribution of connected regions with the same intensity, reflecting texture uniformity and heterogeneity.
      1.3.4. GLDM (Gray Level Dependence Matrix): Measure dependence between a voxel and its neighboring

voxels with similar intensity, indicating texture complexity and smoothness.

2. Specify the image type (e.g., original, Laplacian of Gaussian [LoG], wavelet) and the image filters to apply (e.g., mean, median).
3. Define the settings in a configuration YAML file (*.yaml*). A sample configuration file, beginning with the comment "# This is the parameters file used", is shown below. The configuration specifies the target label, the image types to be included, and the selected feature classes.This YAML file can be used in the snippet provided below, beginning with the command "import os", to extract features from different filtered versions of the original scan, where each filter is applied sequentially to the image, and the radiomics features are computed for each filtered output.

```
# This is the parameters file used (in the example provided in Section 3.8) ↩
  to extract radiomics features from MRI sequences.
setting:
  label: 2 # label corresponding to the edema region in the BraTS tumour ↩
    segmentation nifti files

# image types to use
imageType:
  Original: {}
  LoG:
    sigma: [2.0, 3.0, 4.0, 5.0]
  Wavelet: {}

# feature classes: which families must be calculated.
featureClass:
  shape: # extract all shape features
  firstorder:
  glcm: # extract only the listed features from the glcm
    - 'Autocorrelation'
    - 'JointAverage'
    - 'ClusterProminence'
    - 'ClusterShade'
    - 'ClusterTendency'
  glrlm:
  glszm:
  gldm:
```

4. Execute the feature extraction pipeline and verify that output feature tables include all selected categories.
5. For large datasets or computationally intensive workflows, use the *pyradiomics-cuda* (*see* **Note 4**) extension to leverage CUDA-based GPU acceleration.

```
import os
import SimpleITK as sitk
from radiomics import featureextractor

# load input image and mask using sitk
# load pyradiomics settings from yaml file in the current directory
params_path = os.path.join(os.getcwd(), 'Params_edema.yaml') # <-- may need
  editing
extractor = featureextractor.RadiomicsFeatureExtractor(params_path)

# define filters to apply
filters = {
    "AdditiveGaussianNoise": sitk.AdditiveGaussianNoiseImageFilter(),
    "BinomialBlur": sitk.BinomialBlurImageFilter(),
    "Mean": sitk.MeanImageFilter(),
    "Median": sitk.MedianImageFilter(),
}

# apply each filter and extract features
for name, filt in filters.items():
    filtered_image = filt.Execute(image)
    features = extractor.execute(filtered_image, mask)
```

3.4 Feature Selection and Dimensionality Reduction

Implement feature selection or dimensionality reduction to manage high-dimensional radiomic feature sets and improve model interpretability, performance, and computational efficiency. Follow these steps:

1. For feature Selection, retain a subset of the original features that are most predictive of the target outcome without altering their representation. Common methods include:
   1.1. SelectKBest (SKB) [35]: Filter-based method that selects the top K features based on a univariate statistical test, such as ANOVA F-value or mutual information. Use *sklearn.feature_selection.SelectKBest* in Python.
   1.2. Boruta [36]: Wrapper-based algorithm that evaluates feature importance using random forest models and compares each feature to shadow features created by permuting the original values. Features outperforming shadows are retained. Use the *boruta_py* package with *scikit-learn* random forest classifiers.
   1.3. Relief-F [37]: Filter-based method that measures how well each feature distinguishes between similar instances from different classes, effective for multiclass problems and feature interactions. Implement using libraries such as *skrebate.*
2. For dimensionality Reduction, use PCA (Principal Component Analysis) to transform the original high-dimensional feature space into a lower-dimensional representation by creating new

composite variables that preserve data variability [38]. PCA is an unsupervised method projecting data onto orthogonal components that capture maximum variance. Use *sklearn.decomposition.PCA* to implement it in Python.

3.5 Data Balancing

Address class imbalance in neuroradiomics datasets to improve model generalization and predictive performance. Imbalanced datasets often occur when one class (e.g., healthy controls) has many more samples than another class (e.g., patients with a specific pathology). Implement data balancing using one of the following steps prior to model training to ensure balanced class representation.

1. RUS (Random Under-Sampling):
   - 1.1. Reduce the size of the majority class by randomly removing samples to match the number of minority class instances.
   - 1.2. Use Python libraries such as *imbalanced-learn* (*imblearn.under_sampling.RandomUnderSampler*) to implement RUS efficiently.
2. SMOTE (Synthetic Minority Over-Sampling Technique):
   - 2.1. Generate synthetic examples for the minority class using nearest-neighbor interpolation.
   - 2.2. Implement SMOTE in Python using *imbalanced-learn* (*imblearn.over_sampling.SMOTE*).

3.6 Model Analysis

Develop and evaluate predictive models using radiomic features to classify or predict clinical outcomes. Follow these steps:

1. Choose an appropriate ML model based on the study objectives. Common models include Support Vector Machine (SVM), Random Forest (RF), and Logistic Regression (LR), with SVM and RF often used for both classification and regression tasks, and LR primarily applied to binary or multinomial classification problems.
2. Implement ML models in Python using *scikit-learn*.
3. Divide the dataset into training, validation, and testing subsets to ensure unbiased evaluation.
4. Apply K-fold cross-validation to assess model stability and generalization.
5. In Python, use the *train_test_split* function from *scikit-learn* or *StratifiedKFold* for stratified K-fold cross-validation.
6. Fit the chosen model on the training set, adjusting hyperparameters using the validation set. Apply hyperparameter

optimization using grid search or randomized search (*GridSearchCV* or *RandomizedSearchCV* in *scikit-learn*).

7. Use performance metrics appropriate for the task to assess the model using the test set or cross-validation results. Typical performance metrics include:
   7.1. Classification task: accuracy, area under the receiver operating characteristic curve (AUC), sensitivity, specificity, and F1-score.
   7.2. Regression task: mean squared error (MSE), root mean squared error (RMSE), and coefficient of determination (R^2).
   7.3. In Python, compute metrics using the sklearn.metrics module (e.g., *accuracy_score, roc_auc_score*).
8. Save trained models, preprocessing pipelines, and evaluation results for reproducibility and future analyses using *pickle* in Python.

3.7 Explainability (Optional but Recommended)

After completing model training and evaluation, interpret how radiomic features contribute to predictions. Apply explainability techniques to gain insight into model behavior, improve transparency, and support clinical trust in neuroradiomics workflows. Among available explainability approaches, SHapley Additive exPlanations (SHAP) is widely used. SHAP assigns each feature an importance value for individual predictions, based on principles from cooperative game theory.

1. Implement SHAP in Python using the *shap* library, which integrates with models from *scikit-learn.*
2. Train your model as usual, then pass the fitted model and dataset to SHAP explainer classes (e.g., *shap.TreeExplainer* for tree-based models).
3. Use visualization functions such as *summary_plot* to interpret and present results.
4. Describe how model decisions are linked to specific image features to improve transparency and support clinical trust in neuroradiomics workflows.

3.8 Example: Radiomic Fingerprinting of Peritumoral Edema for HGG vs. LGG classification

As an example, the workflow presented earlier in this chapter is demonstrated using the BraTS 2023 dataset to classify low-grade glioma (LGG) versus high-grade glioma (HGG), by extracting radiomic features from contrast-enhanced T1-weighted (T1-C) MRI scans. The step-by-step process is outlined below.

1. Download the MRI scans and corresponding ground-truth segmentation masks from the BraTS 2023 dataset (*see* **Note 1**). The dataset includes 325 HGG and 77 LGG cases. All

scans are already preprocessed, eliminating the need for additional preprocessing steps. The accompanying masks provide detailed delineations of tumor subregions—including the enhancing tumor, tumor core, necrosis, and peritumoral edema—so no further ROI segmentation is required.

2. Use the provided masks to isolate the tumor and peritumoral edema regions for analysis. An example MRI scan, together with its BraTS segmentation maps and the extracted edema mask are shown in Fig. 3.
3. Structure the dataset files and folders in a consistent way to ensure compatibility with the scripts and feature extraction pipeline. Place the downloaded dataset in a dedicated directory, and keep MRI scans, segmentation masks, configuration files and scripts in clearly labeled subfolders. The folder structure should resemble the example shown in Fig. 4.
4. Prepare the *Params_edema.yaml* file to extract radiomic features from the original image, wavelet sub-band images, and LoG filtered images. Configure wavelet transformations to include all combinations of high- and low-pass filters, and apply LoG filters with sigma values of 2, 3, 4, and 5. The YAML file contents were presented earlier in Sect. 3.3.

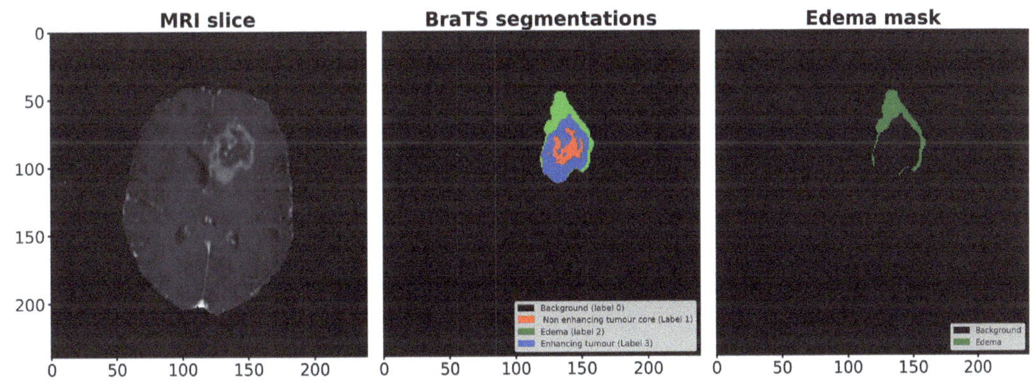

Fig. 3 Example MRI from the BraTS 2023 dataset showing tumor subregion segmentations and the extracted peritumoral edema mask used for feature extraction Sect. 3.8

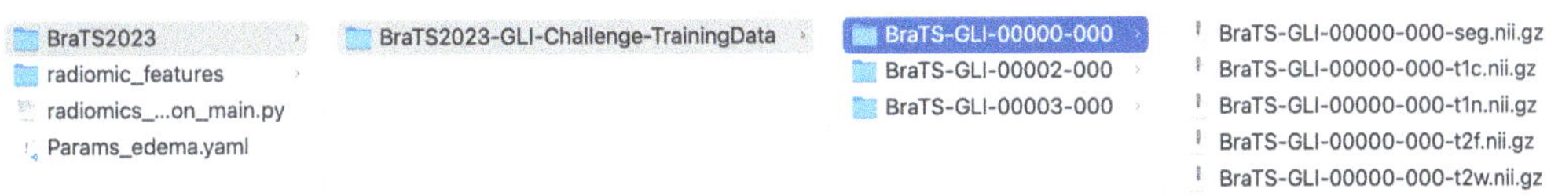

Fig. 4 Example directory structure for organizing the BraTS2023 dataset, including subject-level MRI scans, segmentation masks, and configuration files required for radiomic analyses

5. Use *pyradiomics* to extract radiomic features from the peritumoral edema region on T1-c images of all glioma (GLI) cases. Note that in the BraTS segmentation masks provided,

```
# settings
results_folder = 'radiomics_features'
params = os.path.join(os.getcwd(), 'Params_edema.yaml')
results_csv_file = os.path.join(results_folder,
  f'feature_vectors_edema_GLI_T1_C.csv')
data_folder = 'BraTS2023/BraTS2023-GLI-Challenge-TrainingData'
# feature extraction
data_list = []
subdirs = os.listdir(data_folder)
for subdir in os.listdir(data_folder):
    start_time = time.time()
    subject_folder = os.path.join(data_folder, subdir)
    mask = os.path.join(subject_folder, subdir + '-seg.nii.gz')
    image = os.path.join(subject_folder, subdir + '-t1c.nii.gz')
    print(f"extracting radiomic features from the T1_C sequence of {subdir}
      subject ..." )
    Image = sitk.ReadImage(image)
    Mask = sitk.ReadImage(mask)
    # check the Mask
    mask_array = sitk.GetArrayFromImage(Mask)
    if np.all(mask_array == 0):
        print(f"\tno segmentation exists for subject {subdir}; skipping
          analysis...")
        continue
    if not np.any(mask_array == 2):
        print(f"\tlabel 2 does not exist in the segmentation mask for subject
          {subdir}; skipping analysis...")
        continue
    try:
        extractor = featureextractor.RadiomicsFeatureExtractor(params)
        feature_vector = extractor.execute(Image, Mask)
    except (IndexError, MemoryError) as e:
        print(f"\tan error occurred for subject {subdir}: {e}")
        continue
    row_data = {'subject_id': subdir, 'label': 'GLI'}
    for feature_name, feature_value in feature_vector.items():
        if feature_name.startswith(('original', 'log', 'wavelet')):
            row_data[feature_name] = feature_value
    data_list.append(row_data)
    iteration_time = time.time() - start_time
    print(f"\ttime taken to process {subdir} subject: {iteration_time}
      seconds")
columns = ['subject_id'] + [feature_name for feature_name in
  feature_vector.keys() if feature_name.startswith(('original', 'log',
  'wavelet'))] + ['label']
df = pd.DataFrame(data_list, columns=columns)
df.to_csv(results_csv_file, index=False) # save the extracted features
```

the label for the peritumoral edema region is 2 (see Fig. 3); therefore, in the *Params_edema.yaml* file, the label is also set to 2 to ensure that features are extracted specifically from this region.

6. Use the following Python script, beginning with the comment "# settings", to perform the feature extraction. Note that the file paths are based on the directory structure shown in Fig. 4; for a different structure, paths must be modified accordingly.
7. Split the dataset into train and test using the K-fold methodology and save the splits for each fold. This step is particularly important when comparing the performance of different feature sets extracted from multiple imaging modalities or when benchmarking different ML models.
8. After creating and storing the splits, prepare the radiomic features for both training and testing sets to ensure consistency across folds.
9. Use the function *data_preparation* below, beginning with the command "def data_preparation", to scale the features using either standard scaling or min–max normalization, balance the training data with resampling methods such as SMOTE or RUS, and reduce dimensionality or select the most relevant features using one of the supported techniques (PCA, SelectKBest, Boruta, or Relief-F). Adapt the options to match the experimental design and dataset characteristics.

```
def data_preparation(X_train,y_train,X_test,y_test,scaler,sampler,feature_selection,no_of_features_to_select):
  # 1) scaling: options: standard and minmax
  if scaler is not None:
      scaler = StandardScaler() if scaler=='standard' else MinMaxScaler()
      X_train, X_test = scaler.fit_transform(X_train), scaler.transform(X_test)
  # 2) resampling: options: none, smote, and rus
  if sampler is not None:
      if sampler=='smote':
          X_train,y_train = SMOTE(random_state=42).fit_resample(X_train,y_train)
      elif sampler=='rus':
          X_train,y_train = RandomUnderSampler(random_state=42).fit_resample(X_train,y_train)
  print('distribution of classes in resampled dataset %s' % Counter(y_train))
  # 3) dimensionality reduction and feature selection: options: none, pca, skb, boruta, relieff
  if feature_selection is not None:
      if feature_selection=='pca':
          pca, comps = PCA(), range(1,50) # range used to search for the optimal number of components
          cv_scores=[np.mean(cross_val_score(pca.set_params(n_components=n),X_train,y_train,cv=3)) for n in comps]
          optimal_n = comps[np.argmax(cv_scores)]
          pca = PCA(n_components=optimal_n).fit( X_train)
          X_train, X_test = pca.transform(X_train), pca.transform(X_test)
      elif feature_selection=='skb':
          k_values = np.arange(1,50) # range used to search for the optimal number of components
          cv_scores=[np.mean(cross_val_score(RandomForestClassifier(),
                       SelectKBest(f_classif,k=K).fit_transform(X_train,y_train),y_train,cv=3)) for K in k_values]
          optimal_k = k_values[np.argmax(cv_scores)]
          k_best = SelectKBest(f_classif,k=optimal_k).fit( X_train,y_train)
          X_train, X_test = k_best.transform(X_train), k_best.transform(X_test)
      elif feature_selection=='boruta':
          rfc=RandomForestClassifier(n_jobs=-1,class_weight='balanced',max_depth=5)
          boruta=BorutaPy(rfc,n_estimators=500,random_state=1,max_iter=100,perc=90).fit(X_train,y_train)
          X_train,X_test=boruta.transform(X_train),boruta.transform(X_test)
      elif feature_selection=='relieff':
          relieff=ReliefF(n_features_to_keep=no_of_features_to_select,n_neighbors=2).fit(X_train,y_train)
          X_train,X_test=relieff.transform(X_train),relieff.transform(X_test)
  return X_train,y_train,X_test,y_test
```

10. Define the ML models and specify the parameter distributions for hyperparameter tuning.
11. Use the Python snippet below, beginning with the comment "# ML settings", to define the RF, SVM, and LR models along with their parameter search spaces. These parameter

distributions will later be used in randomized search or other optimization strategies to identify the best-performing models.

```
# ML settings
models = {
    'RandomForest': RandomForestClassifier(random_state=42),
    'SVM': SVC(max_iter=1000, random_state=42),
    'LR': LogisticRegression(max_iter=1000, random_state=42)
}
# define parameter distributions for each model
np.random.seed(42)
param_distributions = {
    'RandomForest': {
        'n_estimators': randint(100, 500),
        'max_depth': [None] + list(randint(10, 50).rvs(size=4)),
        'criterion': ['entropy', 'gini', 'log_loss'],
        'min_samples_split': [2, 3, 4, 5],
        'max_features': ['sqrt', 'log2', None, 0.5, 0.7]
    },
    'SVM': {
        'C': uniform(0.1, 100),
        'gamma': ['scale', 'auto', 0.1, 0.01],
        'kernel': ['linear', 'rbf', 'poly', 'sigmoid'],
        'degree': randint(2, 5),
        'coef0': uniform(0.0, 10.0)
    },
    'LR': {
        'C': uniform(0.1, 5),
        'penalty': ['l1', 'l2'],
        'solver': ['liblinear'],
        'max_iter': [200],
        'class_weight': [None, 'balanced']
    }
}
```

12. Train the selected model within each fold of the cross-validation process, using the training split, and evaluate it on the corresponding test split.
13. For each fold, generate and save the confusion matrix to capture classification performance.
14. After completing all folds, compute the average (or aggregate) confusion matrix across the K folds to provide a robust overall summary of model performance. This average confusion matrix offers a clearer representation of how well the model performs across the entire dataset rather than on a single split.

4 Notes

1. Access links for selected publicly available datasets are listed below.
   1.1. ADNI: https://adni.loni.usc.edu
   1.2. BraTS 2023: https://www.kaggle.com/datasets/bkb2024/brats-2023-training
   1.3. HCP: https://db.humanconnectome.org
   1.4. OpenNEURO: https://openneuro.org/
   1.5. TCIA: https://www.cancerimagingarchive.net/access-data/
2. The structure of a typical neuroimaging dataset organized according to the BIDS specification, including subject-level directories and key metadata files, is illustrated in Fig. 5.
3. The use of GPUs has become integral to modern neuroradiomics pipelines, particularly in computationally intensive

```
sampleBIDSdataset/
├── README
├── dataset_description.json          # REQUIRED: Dataset-level metadata (e.g., name, BIDS version, authors).
├── participants.tsv                  # REQUIRED: Participant info table (e.g., ID, age, sex, group).
├── participants.json                 # OPTIONAL: Metadata describing each column in participants.tsv.
├── sub-01/                           # Subject 01 data directory.
│   ├── anat/                         # Contains anatomical MRI scans.
│   │   ├── sub-01_T1w.nii.gz
│   │   ├── sub-01_T1w.json
│   │   └── sub-01_inplaneT2.nii.gz
│   ├── func/                         # Contains functional MRI scans.
│   │   ├── sub-01_task-rest_bold.nii.gz
│   │   ├── sub-01_task-rest_bold.json
│   │   ├── sub-01_task-rest_events.tsv
│   │   └── sub-01_task-rest_sbref.nii.gz
│   ├── dwi/                          # Contains diffusion MRI data.
│   │   ├── sub-01_dwi.nii.gz
│   │   ├── sub-01_dwi.bval
│   │   ├── sub-01_dwi.bvec
│   │   └── sub-01_dwi.json
│   ├── meg/                          # Contains MEG recordings.
│   │   ├── sub-01_task-rest_meg.fif
│   │   ├── sub-01_task-rest_channels.tsv
│   │   ├── sub-01_task-rest_coordsystem.json
│   │   ├── sub-01_task-rest_events.tsv
│   │   └── sub-01_task-rest_headshape.pos
│   ├── eeg/                          # Contains EEG recordings.
│   │   ├── sub-01_task-rest_eeg.vhdr
│   │   ├── sub-01_task-rest_eeg.eeg
│   │   ├── sub-01_task-rest_eeg.vmrk
│   │   ├── sub-01_task-rest_channels.tsv
│   │   ├── sub-01_task-rest_coordsystem.json
│   │   └── sub-01_task-rest_events.tsv
├── sub-02/                           # Subject 02 data directory (same structure as sub-01).

├── sourcedata/                       # OPTIONAL: Original raw data prior to BIDS formatting.
└── derivatives/                      # OPTIONAL: Outputs from preprocessing or analysis (e.g., fMRI, MEG).
```

Fig. 5 Example folder hierarchy of a neuroimaging dataset formatted according to the BIDS standard. The dataset includes subject-specific directories with anatomical (anat), functional (func), diffusion (dwi), MEG (meg, [magnetoencephalography]), and EEG (eeg) data, alongside essential metadata files such as dataset_description.json and participants.tsv stages such as automated segmentation and ROI identification, feature extraction, and classification. Deep learning-based segmentation tools for MRI and CT imaging rely heavily on GPU acceleration to perform both training and inference efficiently. Likewise, GPU-compatible radiomics, along with ML frameworks, use GPU processing to accelerate the extraction and analysis of high-dimensional radiomic features.

4. The software tools and libraries mentioned in this chapter, including those for preprocessing, segmentation, feature extraction, and analysis, are available at:
   4.1. 3D Slicer: https://www.slicer.org
   4.2. Anaconda distribution: https://www.anaconda.com/download
   4.3. ANTs: https://github.com/ANTsX/ANTs
   4.4. dcm2nifti: https://pypi.org/project/dicom2nifti/
   4.5. DeepMedic: https://github.com/deepmedic/deepmedic
   4.6. dicom2nifi: https://www.nitrc.org/projects/dcm2nii/
   4.7. fMRIPrep: https://fmriprep.org/en/stable/
   4.8. FreeSurfer: https://surfer.nmr.mgh.harvard.edu
   4.9. FSL: https://fsl.fmrib.ox.ac.uk/fsl
   4.10. nnU-Net: https://github.com/MIC-DKFZ/nnUNet
   4.11. ImageJ: https://imagej.net
   4.12. ITK-SNAP: https://www.itksnap.org/pmwiki/pmwiki.php
   4.13. pyradiomics: https://github.com/AIM-Harvard/pyradiomics; https://pyradiomics.readthedocs
   4.14. pyradiomics documentation: https://pyradiomics.readthedocs.io/en/latest/
   4.15. pyradiomics-cuda 1.0.3: https://pypi.org/project/pyradiomics-cuda/; https://github.com/jiaoyining/cuRadiomics
   4.16. pytorch: https://pytorch.org
   4.17. SPM: https://www.fil.ion.ucl.ac.uk/spm/
   4.18. SynthStrip: https://surfer.nmr.mgh.harvard.edu/fswiki/SynthStrip
   4.19. tensorflow: https://www.tensorflow.org

5. When selecting software, consider the imaging modality, preprocessing requirements, dataset quality, and available computational resources. Open-source platforms such as *fMRIPrep*, *ANTs*, and *nnU-Net* support reproducibility and

methodological transparency, while tools like *ITK-SNAP* and *3D Slicer* provide essential capabilities for visual inspection and manual adjustments. Whenever possible, use consistent software versions across participants and imaging sessions to ensure reproducible processing and analysis outcomes.

6. *fMRIPrep* runs smoothly on macOS and Linux-based systems. It requires input datasets to be organized in the BIDS format (*see* **Note 2**), with Docker installed (visit https://docs.docker.com/get-docker/) and a valid *FreeSurfer* license. Preprocessing across multiple subjects (e.g., sub-01, sub-02, sub-03) can be automated using the shell script provided below, beginning with the comment "#sample run".

```
#sample run: ./fmriprep_preprocessing_MultiSubj.sh sub-01 sub-02 sub-03

# user inputs:
bids_root_dir=./sampleBIDSdataset # path to the BIDS dataset (edit as needed)
nthreads=4
mem=20 # GB
container=docker
mem_mb=$(((mem*1000)-5000)) # reduce some memory for buffer space

export FS_LICENSE=./license.txt # path to the FreeSurfer licence (edit as needed)

# loop through each subject ID passed as argument
for subj in "$@"; do
    start_time=$(date +%s)
    echo "starting preprocessing for subject: $subj at $(date)"

    fmriprep-docker $bids_root_dir $bids_root_dir/derivatives \
        participant \
        --participant-label $subj \
        --skip-bids-validation \
        --md-only-boilerplate \
        --fs-license-file $FS_LICENSE \
        --fs-no-reconall \
        --output-spaces MNI152NLin2009cAsym:res-2 \
        --nthreads $nthreads \
        --stop-on-first-crash \
        --mem_mb $mem_mb

    end_time=$(date +%s)
    elapsed_time=$((end_time - start_time))

    echo "finished preprocessing for subject: $subj"
    echo "time taken: $((elapsed_time / 60)) min $((elapsed_time % 60)) sec"
done
```

7. Other open-source platforms for radiomic feature extraction include:

   7.1. *CaPTk* (Cancer Imaging Phenomics Toolkit): A cross-platform toolkit for cancer imaging, providing a suite of radiomic features, ML tools, and both GUI and command-line interfaces [39].

   7.2. *IBEX* (Imaging Biomarker Explorer): A MATLAB and C++-based platform enabling flexible, customizable pipelines for various imaging modalities [40].

   7.3. *RaCaT* (Radiomics Calculation Tool): A standardized feature computation tool supporting reproducible workflows across studies [41].

References

1. Beig N, Bera K, Tiwari P (2020) Introduction to radiomics and radiogenomics in neuro-oncology: implications and challenges. Neuro-Oncol Adv 2(Suppl 4):iv3–iv14
2. Yip SS, Aerts HJ (2016) Applications and limitations of radiomics. Phys Med Biol 61(13):R150
3. Malik N, Geraghty B, Dasgupta A, Maralani PJ, Sandhu M, Detsky J, Tseng C-L, Soliman H, Myrehaug S, Husain Z (2021) MRI radiomics to differentiate between low grade glioma and glioblastoma peritumoral region. J Neurooncol 155(2):181–191
4. Tan R, Sui C, Wang C, Zhu T (2024) MRI-based intratumoral and peritumoral radiomics for preoperative prediction of glioma grade: a multicenter study. Front Oncol 14:1401977
5. Azemi G, Di Ieva A (2025) Radiomic fingerprinting of the peritumoral edema in brain tumors. Cancers (Basel) 17(3):478. https://doi.org/10.3390/cancers17030478
6. Tabassum M, Suman AA, Suero Molina E, Pan E, Di Ieva A, Liu S (2023) Radiomics and machine learning in brain tumors and their habitat: a systematic review. Cancers (Basel) 15(15):3845. https://doi.org/10.3390/cancers15153845
7. Koong K, Preda V, Jian A, Liquet-Weiland B, Di Ieva A (2022) Application of artificial intelligence and radiomics in pituitary neuroendocrine and sellar tumors: a quantitative and qualitative synthesis. Neuroradiology 64(4):647–668. https://doi.org/10.1007/s00234-021-02845-1
8. Ak M, Toll S, Hein K, Colen R, Khatua S (2022) Evolving role and translation of radiomics and radiogenomics in adult and pediatric neuro-oncology. Am J Neuroradiol 43(6):792–801
9. Liu H, Zhang X, Liu Q (2025) A review of AI-based radiogenomics in neurodegenerative disease. Front Big Data 8:1515341
10. Veeturi SS, Poppenberg KE, Pinter NK, Jaikumar V, Levy EI, Siddiqui AH, Tutino VM (2025) Evaluation of radiogenomics for risk stratification of intracranial aneurysms: a pilot study. Neuroradiology 67(9):2425–2435. https://doi.org/10.1007/s00234-025-03702-1
11. Tenhoeve SA, Wardrop T, Smith A, Couldwell WT, Rennert RC (2025) Emerging role for radiomics in the management of intracranial arteriovenous and cavernous malformations. Expert Rev Neurother 25(10):1223–1233. https://doi.org/10.1080/14737175.2025.2548328
12. Zhang J, Zhu H, Wu H, Xie H, Lin D, Zhu L (2025) Radiomics and ischemic stroke research: bibliometric insights and visual trends (2004–2024). Front Neurol 16, 1606388. https://doi.org/10.3389/fneur.2025.1606388
13. Jiang C, Kong Z, Zhang Y, Liu S, Liu Z, Chen W, Liu P, Liu D, Wang Y, Lyu Y (2020) Conventional magnetic resonance imaging–based radiomic signature predicts telomerase reverse transcriptase promoter mutation status in grade II and III gliomas. Neuroradiology 62(7):803–813
14. Veitch DP, Weiner MW, Miller M, Aisen PS, Ashford MA, Beckett LA, Green RC, Harvey D, Jack JrCR, Jagust W (2024) The Alzheimer's Disease Neuroimaging Initiative in the era of Alzheimer's disease treatment: a review of ADNI studies from 2021 to 2022. Alzheimer's & Dementia 20(1):652–694
15. Bonato B, Nanni L, Bertoldo A (2025) Advancing precision: a comprehensive review of MRI segmentation datasets from brats

challenges (2012–2025). Sensors (Basel, Switzerland) 25(6):1838
16. Marcus DS, Harwell J, Olsen T, Hodge M, Glasser MF, Prior F, Jenkinson M, Laumann T, Curtiss SW, Van Essen DC (2011) Informatics and data mining tools and strategies for the human connectome project. Front Neuroinform 5:4
17. Markiewicz CJ, Gorgolewski KJ, Feingold F, Blair R, Halchenko YO, Miller E, Hardcastle N, Wexler J, Esteban O, Goncavles M (2021) The OpenNeuro resource for sharing of neuroscience data. Elife 10:e71774
18. Clark K, Vendt B, Smith K, Freymann J, Kirby J, Koppel P, Moore S, Phillips S, Maffitt D, Pringle M (2013) The Cancer Imaging Archive (TCIA): maintaining and operating a public information repository. J Digit Imag 26(6):1045–1057
19. Fedorov A, Beichel R, Kalpathy-Cramer J, Finet J, Fillion-Robin J-C, Pujol S, Bauer C, Jennings D, Fennessy F, Sonka M, Buatti J, Aylward S, Miller JV, Pieper S, Kikinis R (2012) 3D Slicer as an image computing platform for the Quantitative Imaging Network. MRI 30(9):1323–1341. https://doi.org/10.1016/j.mri.2012.05.001
20. Avants BB, Epstein CL, Grossman M, Gee JC (2008) Symmetric diffeomorphic image registration with cross-correlation: evaluating automated labeling of elderly and neurodegenerative brain. Med Image Anal 12(1):26–41
21. Kamnitsas K, Ledig C, Newcombe VF, Simpson JP, Kane AD, Menon DK, Rueckert D, Glocker B (2017) Efficient multi-scale 3D CNN with fully connected CRF for accurate brain lesion segmentation. Med Image Anal 36:61–78
22. Esteban O, Markiewicz CJ, Blair RW, Moodie CA, Isik AI, Erramuzpe A, Kent JD, Goncalves M, DuPre E, Snyder M (2019) fMRIPrep: a robust preprocessing pipeline for functional MRI. Nat Methods 16(1):111–116
23. Fischl B, Van Der Kouwe A, Destrieux C, Halgren E, Ségonne F, Salat DH, Busa E, Seidman LJ, Goldstein J, Kennedy D (2004) Automatically parcellating the human cerebral cortex. Cereb Cortex 14(1):11–22
24. Jenkinson M, Beckmann CF, Behrens TE, Woolrich MW, Smith SM (2012) Fsl. NeuroImage 62(2):782–790
25. Smith SM, Jenkinson M, Woolrich MW, Beckmann CF, Behrens TEJ, Johansen-Berg H, Bannister PR, De LM, Drobnjak I, Flitney DE, Niazy RK, Saunders J, Vickers J, Zhang YY, De SN, Brady JM, Matthews PM (2004) Advances in functional and structural MR image analysis and implementation as FSL. NeuroImage 23:208–219. https://doi.org/10.1016/j.neuroimage.2004.07.051
26. Isensee F, Jaeger PF, Kohl SAA, Petersen J, Maier-Hein KH (2021) nnU-Net: a self-configuring method for deep learning-based biomedical image segmentation. Nat Methods 18(2):203–211. https://doi.org/10.1038/s41592-020-01008-z
27. IsenseeFWaldTUlrichCBaumgartnerMRoySMaier-HeinKHJaegerPF2024nnU-Net revisited: a call for rigorous validation in 3D medical image segmentation. In: Linguraru MG, (Ed.) Medical Image Computing and Computer Assisted Intervention – MICCAI 2024. Lecture Notes in Computer Science. Springer, Cham, 15009. https://doi.org/10.1007/978-3-031-72114-4_47
28. Rueden CT, Schindelin J, Hiner MC, DeZonia BE, Walter AE, Arena ET, Eliceiri KW (2017) ImageJ2: ImageJ for the next generation of scientific image data. BMC Bioinform 18(1):529. https://doi.org/10.1186/s12859-017-1934-z
29. Schneider CA, Rasband WS, Eliceiri KW (2012) NIH Image to ImageJ: 25 years of image analysis. Nat Methods 9(7):671–675. https://doi.org/10.1038/nmeth.2089
30. Yushkevich P, Pashchinskiy A, Oguz I, Mohan S, Schmitt J, Stein J, Zukić D, Vicory J, McCormick M, Yushkevich N, Schwartz N, Gao Y, Gerig G (2019) User-guided segmentation of multi-modality medical imaging datasets with ITK-SNAP. Neuroinformatics 17:1–20. https://doi.org/10.1007/s12021-018-9385-x
31. Hoopes A, Mora JS, Dalca AV, Fischl B, Hoffmann M (2022) SynthStrip: skull-stripping for any brain image. NeuroImage 260:119474. https://doi.org/10.1016/j.neuroimage.2022.119474
32. Tustison NJ, Avants BB, Cook PA, Zheng Y, Egan A, Yushkevich PA, Gee JC (2010) N4ITK: improved N3 bias correction. IEEE Trans Med Imaging 29(6):1310–1320
33. Andersson J, Jenkinson M, Smith S (2007) Non-linear registration, aka Spatial normalisation FMRIB technical report TR07JA2. FMRIB 2(1):e21
34. Lambin P, Rios-Velazquez E, Leijenaar R, Carvalho S, van Stiphout RG, Granton P, Zegers CM, Gillies R, Boellard R, Dekker A, Aerts HJ (2012) Radiomics: extracting more information from medical images using advanced feature analysis. Eur J Cancer 48(4):441–446. https://doi.org/10.1016/j.ejca.2011.11.036
35. Guyon IM, Elisseeff A (2003) An introduction to variable and feature selection. J Mach Learn Res 3:1157–1182
36. Kursa MB, Rudnicki WR (2010) Feature selection with the Boruta Package. J Stat Softw

36(11):1–13. https://doi.org/10.18637/jss.v036.i11

37. Urbanowicz RJ, Meeker M, La Cava W, Olson RS, Moore JH (2018) Relief-based feature selection: introduction and review. J Biomed Inform 85:189–203. https://doi.org/10.1016/j.jbi.2018.07.014
38. Jolliffe IT, Cadima J (2016) Principal component analysis: a review and recent developments. Philos Trans R Soc A Math Phys Eng Sci 374(2065):20150202. https://doi.org/10.1098/rsta.2015.0202
39. Pati S, Singh A, Rathore S, Gastounioti A, Bergman M, Ngo P, Ha SM, Bounias D, Minock J, Murphy G, Li H, Bhattarai A, Wolf A, Sridaran P, Kalarot R, Akbari H, Sotiras A, Thakur SP, Verma R, Shinohara RT, Yushkevich P, Fan Y, Kontos D, Davatzikos C, Bakas S (2020) The Cancer Imaging Phenomics Toolkit (CaPTk): technical overview. BrainLes 11993:380–394. https://doi.org/10.1007/978-3-030-46643-5_38
40. Parmar C, Rios Velazquez E, Leijenaar R, Jermoumi M, Carvalho S, Mak RH, Mitra S, Shankar BU, Kikinis R, Haibe-Kains B, Lambin P, Aerts HJ (2014) Robust Radiomics feature quantification using semiautomatic volumetric segmentation. PLoS One 9(7):e102107. https://doi.org/10.1371/journal.pone.0102107
41. Pfaehler E, Zwanenburg A, de Jong JR, Boellaard R (2019) RaCaT: an open source and easy to use radiomics calculator tool. PLoS One 14(2):e0212223. https://doi.org/10.1371/journal.pone.0212223

Chapter 11

Computational Pipeline in Clinical Connectomics

Valentina Pacella and Sara Bertagnoli

Abstract

Understanding the neural basis of cognitive deficits following brain injury requires methods that can move beyond focal lesion analysis to capture the distributed nature of brain networks. This chapter presents a practical, step-by-step protocol for indirect disconnection mapping, a method that estimates white matter tract disruption from structural imaging alone, without requiring diffusion-weighted scans. By leveraging normative tractography data, this approach provides a means to infer the network-level consequences of lesions in both MRI and CT datasets. The procedure integrates widely used neuroimaging tools (MRIcron, BCBtoolkit, FSL, and SPM12) and supports both group-level and single-case analyses, including voxelwise statistical testing and Bayesian inference. Illustrated with the CT and MRI imaging of a patient, the workflow enables researchers and clinicians to map cognitive symptoms to patterns of structural disconnection, offering a reproducible and anatomically grounded framework for investigating and predicting brain–behavior relationships in clinical populations.

Keywords White matter disconnection, Lesion analysis, Stroke, Neuroimaging

1 Introduction

Since the earliest days of clinical neurology and neuropsychology, the study of brain lesions has offered unparalleled insight into the neural basis of cognition. Pioneering observations from the nineteenth century, such as Broca's and Wernicke's syndromes [1, 2], established the foundational principle that damage to specific brain regions can selectively impair distinct cognitive functions. These classical approaches relied on postmortem examinations and gross anatomical localization, yet they laid the groundwork for a causal framework in cognitive neuroscience: understanding function through dysfunction.

With the advent of neuroimaging, lesion studies transitioned from postmortem analysis to in vivo mapping of brain–behavior relationships [3]. While early methods focused on coarse lesion overlap, more recent decades saw the development of voxelwise lesion-symptom mapping (VLSM [4]), which allowed for statistically principled, high-resolution inferences linking damaged tissue to specific deficits

Antonio Di Ieva and Jennilee M. Davidson (eds.), *Computational Neuroscience*, Methods in Molecular Biology, vol. 3062,
https://doi.org/10.1007/978-1-0716-5440-8_11,

[5]. By moving beyond predefined regions of interest, the VLSM technique strengthens traditional lesion analysis and enables the identification of brain regions associated with symptoms that were not previously predicted. It also supports comparisons between different patient groups (or within the same group based on symptom severity) and introduces statistical procedures capable of estimating the probability that specific structures are linked to those symptoms. However, these approaches largely remained confined to the visible lesion and often assumed that damage exerted its effects locally.

Today, it is increasingly recognized that brain lesions disrupt not only gray matter nodes but also the white matter tracts that link them. In line with a network-based model of cognition, symptoms can arise from disconnection, a loss of communication between anatomically intact but functionally discoupled regions [6–8]. This perspective has reinvigorated lesion-based cognitive neuroscience, allowing researchers to move beyond focalist models and incorporate the principles of distributed processing [9–11]. Thus, lesion studies evolved from mapping damaged zones to probing emergent properties, including higher-level functions that arise from dynamic interactions among structurally linked but spatially distributed systems [12]. Investigating disconnection, therefore, means accessing the brain's hidden architecture: its white matter scaffold, which enables complex cognitive processes to emerge from coordinated activity across regions.

Indirect disconnection mapping offers a practical method to operationalize this network perspective. By projecting each lesion onto a normative tractography dataset derived from healthy controls, one can infer the probability that any given voxel lies along a disconnected pathway, even when diffusion-weighted imaging is unavailable [13–15]. This approach provides a feasible means of testing hypotheses about network-level dysfunction in stroke and other focal brain injuries.

Parallel advances in neuroimaging software and statistical modeling, such as Bayesian hypothesis testing [16], have further strengthened the analytical rigor of modern lesion studies. These tools allow for robust inferences even in small samples or single-case studies, where traditional frequentist approaches may lack power or fail to provide interpretable estimates of uncertainty.

In this chapter, we present a practical and accessible workflow for conducting indirect disconnection analysis using clinically available neuroimaging data. The protocol accommodates both Magnetic Resonance Imaging (MRI) and Computerized Tomography (CT) modalities and leverages open-source tools, including MRIcron [17], MRIcroGL [18], BCBtoolkit [14], FSL [19], SPM [20], and Python. We illustrate the procedure with an exemplar case: the T1-weighted MRI and the CT scan of a previously described patient [21]. Throughout, we focus on how to derive individual or group-level disconnectome maps, interpret them anatomically, and apply statistical inference, including single-case Bayesian testing, within a reproducible and neuroanatomically grounded framework. Moreover, in single-case applications, we show how disconnection profiles can be used not only to explain observed deficits but also to generate predictions about symptom expression based on the lesion's impact on white matter connectivity.

Overall, these methods allow researchers and clinicians to understand how focal injuries disrupt brain networks, revealing hidden dimensions of the brain-cognition organization that cannot be captured by structural damage alone.

2 Materials

2.1 Hardware

A Linux or macOS workstation is recommended for optimal compatibility with neuroimaging tools.

Windows users are advised to operate within a virtual machine to ensure proper functionality of Linux/OS-based software (e.g., BCBtoolkit and FSL).

2.2 Software

1. MRIcron/MRIcroGL: for manual lesion delineation. https://www.nitrc.org/projects/mricron
2. BCBtoolkit: for image normalization (MRI) and computation of indirect disconnection maps. http://www.brainconnectivitybehaviour.eu
3. SPM12/25: for normalization of CT scans. Requires MATLAB and the Clinical Toolbox [22, 23] extension. https://www.fil.ion.ucl.ac.uk/spm/software/spm12
4. MIPAV: for manual coregistration and spatial normalization of challenging CT datasets. https://mipav.cit.nih.gov
5. FSL: for imaging manipulation, voxel-wise statistical mapping using the randomise tool, and for visualization. https://fsl.fmrib.ox.ac.uk/fsl/fslwiki
6. Python: for statistical analysis. Recommended libraries are nibabel, numpy, scipy, and statsmodels.

2.3 Optional Resources for Anatomical Interpretation

1. Gray matter atlas, such as Harvard-Oxford Cortical and Subcortical Atlases [24], and Automated Anatomical Labeling (AAL) [25]. Available by default with the MRIcron and FSL installation; useful for interpreting statistical results and immediate compatibility.
2. White matter atlas, such as Rojkova [26] or the JHU [27–29], both provided with BCBtoolkit, useful for identifying specific tract-level disconnections.

3 Methods

3.1 Lesion Delineation in Native Space

In disconnection-based analyses, the lesion is the seed from which network disruption is inferred. Its accuracy, completeness, and anatomical plausibility directly shape the reliability of all downstream analyses [30, 31]. The patients' clinical characteristics (timing, lesion type, symptom stability) inherently affect how the lesion appears and how interpretable it is in relation to behavior. Patients in the acute phase of stroke are often ideal candidates for investigations of the brain's functional architecture, since their motor and neuropsychological symptoms can largely be attributed to the original function of the damaged region. At this stage, surrounding networks have not yet undergone significant functional reorganization [32]. However, collecting data from acute patients poses practical challenges: access to stroke units is often restricted, and the accuracy and specificity of neuropsychological assessments can be compromised by the patient's overall medical condition. For this reason, many behavioral lesion-mapping studies rely on data from chronic stroke patients, who are generally more accessible. Moreover, ischemic strokes typically produce focal damage with clearer anatomical boundaries, whereas hemorrhagic lesions are often more diffuse. Hemispheric restrictions are common when studying lateralized symptoms (e.g., anosognosia and neglect [33–37]), but atypical or bilateral cases should be anticipated and evaluated in light of the underlying symptom profile [38].

These factors influence how well the lesion reflects the underlying pathophysiology and its behavioral manifestations. Once patient inclusion criteria have been defined and the appropriate imaging data selected, the next step is to delineate the lesion in native space manually. The following section provides a step-by-step guide for performing this task using MRIcron, covering both T1-weighted MRI and non-contrast CT scans (*see* **Note 1**).

To convert your image from DICOM to NIfTI, select Import → Convert DICOM to NIfTI from the MRIcron/MRIcroGL toolbar.

Patient MRI:

1. Open the native T1-weighted scan.
2. Use the drawing tool to manually trace the lesion slice by slice on the axial plane (*see* **Note 2**).
3. Recursively save the lesion as a.voi file (e.g., patientA_lesion.voi) throughout the delineation.
4. If satisfied with the result, fill in the lesion with the bucket option to ensure a complete and continuous volume.

Patient CT:

1. Load the native CT scan.

2. Adjust contrast to improve lesion visibility (e.g., minimum = 0.0 and maximum = 80.0).
3. Manually trace the lesion (*see* **Note 2**), save as.voi.
4. Use the bucket tool to fill the lesion once you are confident in the tracing.

3.2 VOI to NIfTI Conversion

1. Navigate to Draw → Convert and select the option VOI to NIfTI. Load each.voi file individually. A NIfTI lesion file will be automatically generated in the same directory. If the image appears blank after conversion, this is expected and does not indicate an error.
2. As a sanity check, binarize the lesion map using the MRIcroGL tool. This ensures the output is a clean binary mask suitable for normalization and analysis. The binarization can also be executed in FSL's fslmaths utility. The following command can be used in batch processing:

```
fslmaths yourlesion.nii -bin yourlesion_binarised.
nii
```

3.3 Normalization to MNI Space

Patient T1-MRI:

1. Launch BCBtoolkit (*see* **Note 3**) either by double-clicking the executable or by running./BCBToolkit.* from a terminal. Open the "Normalisation" module.
2. Provide two separate folders: one containing only the patient's native T1-weighted scan and another containing only the corresponding lesion mask. Ensure both files share the exact same filename (except for the extension), as this is required for the transformation to be correctly applied.
3. If the lesion is unilateral, select the enantiomorphic normalization option. This method reconstructs the lesioned tissue by mirroring the corresponding region from the contralateral hemisphere, improving spatial normalization in cases of focal damage. For bilateral or extensive lesions, use the classical normalization option instead.
4. Check the box "Apply to Others" to ensure that the transformation computed from the T1 image is also applied to the lesion mask (and any other associated modalities, if needed). The output will include both the normalized T1 and lesion files in MNI space, ready for further analysis.

Patient CT:

Normalization of CT scans can be challenging due to low tissue contrast and artifacts (*see* **Note 4**). Begin with the automated procedure in SPM~~12~~, and if this fails, proceed with manual normalization using MIPAV.

1. Normalization using SPM: -
   - Select "Display" in the menu to set the origin on the anterior commissure by adjusting right, forward, and up. Select "reorient" and choose both the CT scan and the lesion. If needed, adjust the head orientation by editing pitch, roll, and yaw. Reorient both the CT scan and the lesion accordingly.
   - From the "Batch" tool in the menu, navigate to SPM → Tools → Clinical → CT Normalise. In Input file image, select your CT scan; in Input lesion, select your lesion.
2. If normalization fails due to low contrast or distorted anatomy, use landmark-based registration in MIPAV.
   - Open the patient's CT scan and the MNI152 1 mm template in MIPAV.
   - Use the Draw Point VOI tool to manually place anatomical landmarks on the CT scan and corresponding landmarks on the MNI152 template. The more points you place, the more accurate the normalization (e.g., ~100 [39]).
   - Save each set of landmarks: go to VOI → Select All VOIs, then VOI → Save VOI → Save VOI As, and choose to save as voxel coordinates. Repeat the procedure separately for the CT and the MNI152 images.
   - Perform the landmark-based warping by selecting Algorithms → Registration → Landmark → Least Squares. Make sure to register select CT_patient.nii to MNI152. Specify the MNI152 landmark file as the reference matrix directory.
   - Save the normalized CT: File → Save image as.
   - Apply the transformation matrix to the lesion by opening the lesion nifti as file. Load the same set of CT landmark coordinates used in the previous step (VOI → Open VOI). Repeat the same procedure used to normalize the CT image. Save the transformed lesion image from the output window. Lesion mask binarization might be necessary (see Sect. 3.2, Point 2).
3. Check alignment visually.

3.4 Indirect Disconnection Mapping

The mapping of indirect disconnections infers which distant white matter tracts are likely to be disconnected, based on normative diffusion tractography from healthy individuals. Implemented in software such as the BCBtoolkit (*see* **Note 5**), the disconnectome approach projects each patient's lesion onto precomputed streamlines and calculates the probability that a given voxel belongs to a disrupted

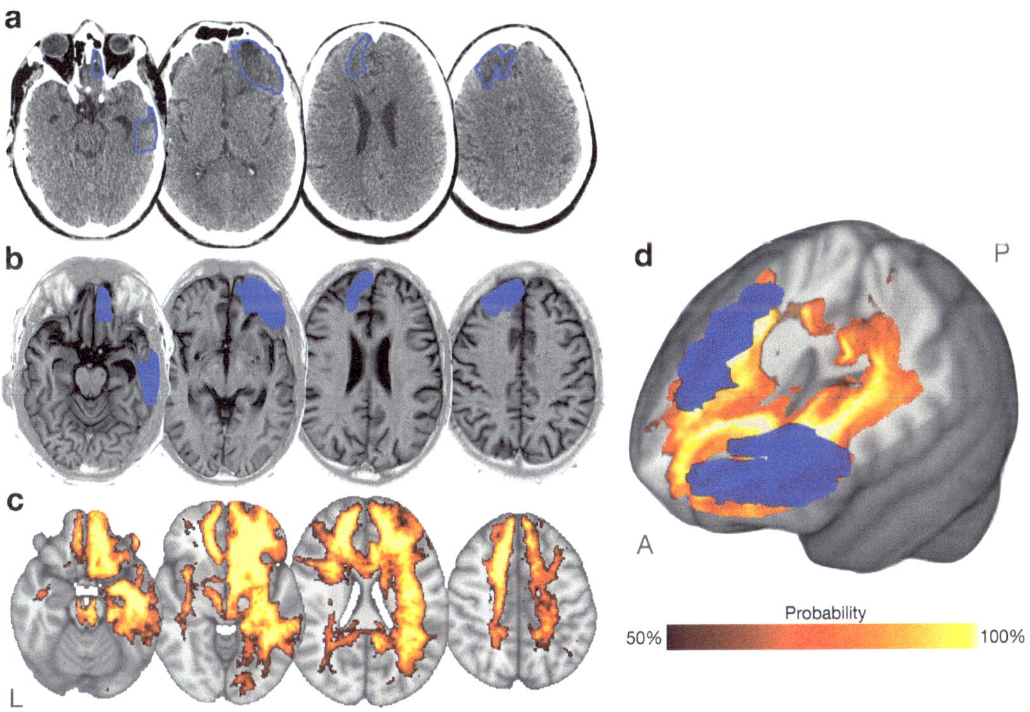

Fig. 1 Lesion delineation and indirect disconnection. (**a**) Representation of lesion delineation on the CT scan. (**b**) Illustration of the filled lesion on the T1. (**c**) The probability of disconnection as computed via the disconnectome maps tool. (**d**) 3D rendering of the lesion (blue) and white matter disconnections (red-yellow palette). The color bar represents the probability of disconnection, thresholded at 50%. L: left. A: anterior. P: posterior. This figure was generated from the example dataset

pathway. The resulting map reflects the likelihood of disconnection at each voxel, even in regions not visibly damaged on structural imaging. This allows researchers to move beyond local lesion anatomy and explore the network-level substrates of cognitive deficits. The following steps guide the user through generating these disconnectome maps from MNI-normalized lesion masks in the BCBToolkit.

1. Launch BCBtoolkit by double-clicking the executable or running./`BCBToolkit.*` from a terminal.
2. Select the Disconnectome maps tool. Provide the folder containing only the normalized lesion map (in MNI152 space, 1 mm resolution) derived from the CT (Fig. 1a) and MRI (Fig. 1b) scans and specify the output directory. If you are investigating a single case without group-level statistical comparison, set the probability threshold to 0.5, corresponding to at least 50% likelihood of disconnection. This threshold is supported by previous studies [40, 41] and reflects moderate-to-high certainty in white matter disruption based on normative tractography.

3. The BCBtoolkit will generate a voxelwise probabilistic disconnectome map, estimating the likelihood that each region is disconnected due to the lesion, based on data from healthy controls (Fig. 1c).

3.5 Group Disconnection Analysis

Group-level analysis enables the identification of brain regions where structural disconnections are statistically associated with variation in behavioral performance across a sample of patients (*see* **Note 6**). By regressing behavioral scores onto disconnection probability values across individuals, it is possible to isolate network components that contribute to specific cognitive or clinical symptoms.

To ensure meaningful and interpretable results, careful consideration must be given to patient inclusion and exclusion criteria. Comorbidities may influence neuropsychological profiles, and even stringent selection cannot always disentangle symptoms caused by the primary deficit from those related to additional conditions [30, 36]. Comparisons with a well-characterized control group can help distinguish brain regions causally linked to the deficit of interest from those that are merely more frequently lesioned [42].

It is equally important to control for timing: the interval between lesion onset and neuropsychological testing and between onset and imaging acquisition can influence results. Failing to account for these differences may conflate recovery dynamics with true anatomical effects. These variables should be matched across subjects or included as covariates in the model. To support methodological rigor in lesion studies, Moore and colleagues [31] provide a quality-control checklist that can serve as a reference during study design and reporting.

The following steps guide the user through the preparation of group-level input files and the execution of non-parametric permutation-based testing using FSL's randomise tool (*see* **Notes 7 and 8**). While this method is described for use with disconnectome maps, it can also be applied to normalized binary lesion masks using the same pipeline.

1. Gather behavioral data and prepare a CSV file with your dependent variable and covariates. Each column should contain one variable (e.g., Behavioral Score, Lesion Volume, Age, and Education), with one row per patient.
2. Create a 4D NIfTI file by stacking all MNI-normalized disconnectome maps into a single file. This can be done using FSL's fslmerge, navigating in your disconnectomes directory:

```
fslmerge -t output_4D.nii.gz *.nii.gz
```

3. Create the design matrix and contrast files, either using FSL's Glm_gui or manually, following standard FSL formatting and the formatting required by the chosen analysis design (*see* **Note 9**). Make sure to include all relevant nuisance covariates (e.g., lesion volume) to account for confounding effects.

4. Run FSL's randomise to perform voxelwise non-parametric permutation testing, specifying the design matrix (-d), the statistical contrasts (-c), the mask (-m), the number of permutations (-n), the type of correction (e.g., TFCE (-T)), and demeaning (-D):

```
randomise -i disconnectomes_4D.nii.gz -o
disco_results -d
yourdesignmatrix.mat -t yourdesigncontrasts.con -m
mask_MNI152.nii.gz -n numberofpermutations -T -D
```

3.6 Interpretation of Results

While visual inspection using standard atlases provides a first result's identification, a statistical quantification can help clarify the contribution of specific tracts, particularly when multiple pathways may be implicated in a given cognitive function. This section guides the user through basic visualization and advanced tract-level quantification, enabling the extraction of interpretable metrics that can be correlated with behavior or entered into multivariate models. These steps are especially useful for follow-up analyses aimed at modeling additive or interactive contributions of disconnected tracts (e.g., Pacella and colleagues [42]).

1. Once the statistical analysis is complete, open the corrected output maps (e.g., disco_results_tfce_corrp_tstat1.nii.gz, depending on your contrast of interest) in MRIcron or FSLeyes for visual inspection. To display only statistically significant clusters, the display threshold is set to 0.95 for corrected $p < 0.05$ and to 0.99 for corrected $p < 0.01$.
2. To identify the anatomical regions involved, overlap the thresholded results on a structural atlas: AAL/Harvard-Oxford atlas can be loaded in MRIcron/MRIcroGL: Use Draw → Descriptive to quantify overlap with known regions. The Rojkova atlas (as part of the BCBtoolkit) is particularly useful for the disco_results interpretation. Use the Tractotron tool available in the BCBToolkit to overlap your disco_results map with normative tracts maps. This tool outputs the probability that each tract is affected and the proportion of the tract disconnected.
3. In some cases, especially if multiple tracts are involved, you may want to go beyond visual inspection. For instance, you may want to quantify how much each tract is affected in each individual and relate this information to behavioral performance. This allows you to explore the specific contribution of each tract and to model additive or interaction effects between tracts.

   Start by selecting the tracts or regions of interest. For example, if you are using the Rojkova atlas, you can take each tract and threshold it to retain only the core regions (e.g., voxels present in at least 90% of the population). This ensures that the tract mask reflects a reliable portion of the anatomy across individuals. You can perform this thresholding with the following command in FSL:

```
fslmaths tract.nii.gz -thr 0.9 thresholded_tract.
nii.gz
```

Once the map is thresholded, it needs to be binarized so that only values of 0 or 1 remain:

```
fslmaths thresholded_tract.nii.gz -bin binary_mask.
nii.gz
```

This binary mask can now be applied to each patient's disconnectome map. By multiplying the disconnectome map by the binary tract mask, you isolate the values inside the tract. Then, using fslstats, you can extract the mean disconnection probability within that tract:

```
fslmaths patient_disconnectome.nii.gz -mas binary_
mask.nii.gz masked_disconnectome.nii.gz
fslstats masked_disconnectome.nii.gz -M
```

This mean value reflects how strongly the tract is affected in that specific patient.

If you are analyzing many patients and multiple tracts, this procedure can be automated using a shell script (https://github.com/vale-pak/disco_userguide). The script loops through all tract masks, applies them to each corresponding patient's map, and saves the mean values into a table (e.g., CSV format). The logic of the script is as follows:

- Define your folders: one containing the tract masks, one with patient disconnectome maps, and one where you want to save the results.
- Create output folders to store intermediate and final files.
- Loop through each tract mask
- Threshold at 90% to retain the core of the tract.
- Binarize the result.
- For each mask, find the corresponding patient map.
- Apply the binary mask to isolate the tract area.
- Compute the mean value inside that region using fslstats.
- Write the results into a CSV file for easy import into R, Python, or Excel.

3.6.1 Single-Case Disconnection Analysis

In addition to group-level comparisons, voxel-wise disconnection maps can be used to assess whether an individual patient with unique clinical manifestations shows significant disruption relative to a reference patient population. This section outlines how to statistically compare a single patient's disconnectome map to the disconnectomes of a control group of patients, voxel by voxel, using a Bayesian approach [16]. This is particularly useful for rare or highly specific deficits for which group-level analysis is not possible. This method builds on Bayesian estimation of the probability that a given voxel in the patient is significantly different from the same voxel in the control sample. It models the mean and variance of the control sample using the posterior predictive distribution and computes an estimated *p*-value for each voxel in the patient. Finally, a False Discovery Rate (FDR) correction is applied to account for multiple comparisons across the brain.

The required inputs are (i) a single disconnectome map for the patient (in MNI space, 1 mm); (ii) a folder containing multiple control group disconnectome maps (same resolution and alignment); and (iii) an output directory where results will be saved.

1. Load the patient's disconnectome map to extract the voxel data and the affine transformation.
   import nibabel as nib
   patient_img = nib.load('/path_to_patient/pat
2. Next, load the disconnectome maps of the controls (e.g., 27), ensuring that all images are aligned and in the same dimensions. These are stacked into a single 4D array, where each "time point" corresponds to one control (see Sect. 3.5).
   import numpy as np

```
control_data = []
for i in range(1, 27): # Assuming 27 controls
control_img = nib.load(f'/path_to_controls/{i:03d}.
nii.gz')
control_data.append(control_img.get_fdata())
control_data = np.array(control_data)
```

3. At each voxel, compare the patient's disconnectome value to the distribution of values across controls. For this, a Bayesian approach is used: the control mean and variance are estimated using a posterior predictive distribution, and a probability is computed that the patient's value comes from that distribution. The function below performs the Bayesian comparison and returns an estimated *p*-value and a credible interval.

```
from scipy import stats
def bayesian_voxelwise_test(patient_value, control_
values, iterations=10000):
control_values = np.asarray(control_values)
n = len(control_values)
s = np.std(control_values, ddof=1) if
np.std(control_values) > 0 else 1e-10
df = n - 1
theta_hat = ((n - 1) * s**2) / np.random.
chisquare(df, size=iterations)
```

```
z = np.random.normal(size=iterations)
mu_hat = np.mean(control_values) + (z *
np.sqrt(theta_hat / n))
z_ast = (patient_value - mu_hat) /
np.sqrt(theta_hat)
p_values = 2 * (1 - stats.norm.cdf(np.abs(z_ast)))
return np.mean(p_values)
```

Then, loop through all voxels and compute the estimated *p*-value for each one.

```
p_values = np.zeros(patient_data.shape)
for i in range(patient_data.shape[0]):
for j in range(patient_data.shape[1]):
for k in range(patient_data.shape[2]):
patient_val = patient_data[i, j, k]
control_vals = control_data[:, i, j, k]
p = bayesian_voxelwise_test(patient_val,
control_vals)
p_values[i, j, k] = p
```

This results in a full-brain 3D map of uncorrected *p*-values.

4. Once the voxelwise *p*-values are computed, the map is saved as an NIfTI file:

```
p_img = nib.Nifti1Image(p_values, affine=patient_img.
affine)
nib.save(p_img, '/output_directory/bayesian_p_
values.nii.gz')
```

5. Since the analysis includes tens of thousands of voxels, correction for multiple comparisons is essential. Here, FDR correction is applied:

```
from statsmodels.stats.multitest import
multipletests
# Flatten the 3D p-values into 1D array for
correction
flat_p_values = p_values.flatten()
# Apply FDR correction
_, corrected_flat_p, _, _ = multipletests(flat_p_
values, method='fdr_bh')
# Reshape the corrected values back into 3D
corrected_p = corrected_flat_p.reshape(p_values.shape)
# Save the corrected p-value map
corrected_img = nib.Nifti1Image(corrected_p, affine=
patient_img.affine)
nib.save(corrected_img, '/output_directory/
bayesian_corrected_p_values.nii.gz')
```

6. Two statistical maps are saved: uncorrected p-value map (e.g., bayesian_p_values.nii.gz) and FDR-corrected p-value map (e.g., bayesian_corrected_p_values.nii.gz). These maps can be opened in MRIcron, FSLeyes, or any other NIfTI-compatible

viewer. As usual, significant voxels are those below the conventional threshold (e.g., $p < 0.05$ for corrected values).

3.6.2 Single-Case Behavior Prediction

Beyond statistical inference at the single-case level, researchers and clinicians may also benefit from tools offering predictive modeling based on structural disconnection patterns. One such tool is the Disconnectome Symptoms Discoverer [43], a web-based platform that allows users to upload a patient's lesion mask and receive prospective predictions about likely cognitive outcomes. The tool leverages a large database of disconnectome maps and behavioral scores (N = 163 stroke patients), applying dimensionality reduction and machine learning to project each new patient into a low-dimensional morphospace of disconnection-symptom relationships. The output is a personalized cognitive profile prediction, estimating a range of behavioral domains.

1. Visit the interactive tool at Disconnectome Studio, which implements the Disconnectome Symptoms Discoverer (DSD) framework.
2. Upload the patient's binary lesion mask in MNI space via the platform's interface. The DSD framework will generate the disconnectome map and project the patient into a latent disconnectome morphospace, linking disconnection patterns with neuropsychological outcomes.
3. The tool provides a CSV file with personalized predictions of cognitive scores (e.g., language fluency, visuospatial skills) at ~1 year post-stroke. The report includes predicted scores, associated confidence metrics, and positioning within the morphospace of disconnection–symptom relationships. This enables data-driven inferences about probable cognitive outcomes in contexts where behavioral data might be limited.

4 Notes

1. For lesion delineation reliability, lesion masks should be traced independently by at least two trained raters in a double-blind procedure. This helps mitigate subjectivity in lesion definition. Discrepancies should be resolved through discussion, and if disagreement persists, a third expert should adjudicate. Consensus masks should only be finalized once agreement is reached on lesion boundaries.
2. If you make a mistake, use Ctrl (command in OS)+z to return back to one move. This can be applied only once.
3. Other normalization software is available, such as ANTs and SPM.
4. Always perform a thorough visual inspection of both raw and normalized anatomical images before proceeding with analysis.

This is especially critical for CT scans, where artifacts, low contrast, or variability in acquisition parameters may compromise automatic normalization procedures. Manual adjustments (e.g., reorientation, landmark registration) are often necessary for older or low-quality CT datasets.

5. Other indirect disconnection software is available, such as Structural Disconnection (SDC [44]) and the quickDisco tool, as part of the Functionnectome software [45]. The latter allows for the identification of disconnection according to tract type (association, commissural, projection).
6. Sample size caution. With N <30, results should be interpreted cautiously. Emphasize effect sizes over statistical thresholds.
7. Randomise supports multivariate regression models. Other software is available for performing Support Vector Regression Lesion-Symptom Mapping (SVR-LSM), such as Lesymap [46] and DeMarco [47].
8. All neuroimaging tools used in this protocol (e.g., FSL, SPM, and BCBtoolkit) assume consistent voxel dimensions, orientation, and spatial alignment across input files. Before running group-level analyses or transformations, confirm the consistency of header information using fslhd or other DICOM/NIfTI viewers. Pay particular attention to the voxel size, image origin, and affine matrix.
9. The guidelines for the correct formatting according to different analysis designs can be found on the official FSL website https://web.mit.edu/fsl_v5.0.10/fsl/doc/wiki/GLM.html.

References

1. Broca P (1861) Remarks on the seat of the faculty of articulated language, following an observation of aphemia (loss of speech). Bull Soc Anat 6:330–357
2. Wernicke C (1881) Lehrbuch der gehirnkrankheiten fur aerzte und studirende. Kassel Theodor Fischer 2:229–242
3. Rorden C, Karnath H-O (2004) Using human brain lesions to infer function: a relic from a past era in the fMRI age?. Nat Rev Neurosci 5:812–819
4. Bates E et al (2003) Voxel-based lesion–symptom mapping. Nat Neurosci 6:448–450
5. Mah Y-H, Husain M, Rees G, Nachev P (2014) Human brain lesion-deficit inference remapped. Brain 137:2522–2531
6. Meynert T (1884) Klinik Der Erkrankungen Des Vorderhirns Begründet Auf Dessen Bau, Leistungen Und Ernährung
7. Geschwind N (1965) Disconnexion syndromes in animals and man. Brain 88:237–237
8. Catani M, Ffytche DH (2005) The rises and falls of disconnection syndromes. Brain 128:2224–2239
9. Doricchi F, Thiebautdeschotten M, Tomaiuolo F, Bartolomeo P (2008) White matter (dis) connections and gray matter (dys)functions in visual neglect: gaining insights into the brain networks of spatial awareness. Cortex 44:983–995
10. Lunven M et al (2015) White matter lesional predictors of chronic visual neglect: a longitudinal study. Brain 138:746–760
11. Thiebaut De Schotten M et al (2014) Damage to white matter pathways in subacute and chronic spatial neglect: a group study and 2 single-case studies with complete virtual 'in vivo' tractography dissection. Cereb Cortex 24:691–706
12. Thiebaut De Schotten M, Forkel SJ (2022) The emergent properties of the connected brain. Science 378:505–510

13. Cohen AL, Ferguson MA, Fox MD (2021) Lesion network mapping predicts post-stroke behavioural deficits and improves localization. Brain 144:e35
14. Foulon C et al (2018) Advanced lesion symptom mapping analyses and implementation as BCBtoolkit. GigaScience 7(3):1–17
15. Griffis JC, Metcalf NV, Corbetta M, Shulman GL (2021) Lesion quantification toolkit: A MATLAB software tool for estimating grey matter damage and white matter disconnections in patients with focal brain lesions. NeuroImage Clin 30:102639
16. Rounis E et al (2024) A preliminary study of white matter disconnections underlying deficits in praxis in left hemisphere stroke patients. Brain Struct Funct 229:2255–2268
17. Rorden C (2025) From MRIcro to MRIcron: the evolution of neuroimaging visualization tools. Neuropsychologia 207:109067
18. Rorden C (2025) MRIcroGL: voxel-based visualization for neuroimaging. Nat Methods 22:1613–1614
19. Jenkinson M, Beckmann CF, Behrens TEJ, Woolrich MW, Smith SM (2012) FSL. NeuroImage 62:782–790
20. Friston KJ et al (1994) Statistical parametric maps in functional imaging: a general linear approach. Hum Brain Mapp 2:189–210
21. Pacella V et al (2020) Anosognosia for theory of mind deficits: a single case study and a review of the literature. Neuropsychologia 148:107641
22. Brett M, Leff AP, Rorden C, Ashburner J (2001) Spatial normalization of brain images with focal lesions using cost function masking. NeuroImage 14:486–500
23. Nachev P, Coulthard E, Jäger HR, Kennard C, Husain M (2008) Enantiomorphic normalization of focally lesioned brains. NeuroImage 39:1215–1226
24. Makris N et al (2006) Decreased volume of left and total anterior insular lobule in schizophrenia. Schizophr Res 83:155–171
25. Tzourio-Mazoyer N et al (2002) Automated anatomical labeling of activations in SPM using a macroscopic anatomical parcellation of the MNI MRI single-subject brain. NeuroImage 15:273–289
26. Rojkova K et al (2016) Atlasing the frontal lobe connections and their variability due to age and education: a spherical deconvolution tractography study. Brain Struct Funct 221:1751–1766
27. Oishi K et al (2008) Human brain white matter atlas: Identification and assignment of common anatomical structures in superficial white matter. NeuroImage 43:447–457
28. Hua K et al (2008) Tract probability maps in stereotaxic spaces: analyses of white matter anatomy and tract-specific quantification. NeuroImage 39:336–347
29. Wakana S et al (2007) Reproducibility of quantitative tractography methods applied to cerebral white matter. NeuroImage 36:630–644
30. De Haan B, Karnath H-O (2018) A hitchhiker's guide to lesion-behaviour mapping. Neuropsychologia 115:5–16
31. Moore MJ, Demeyere N, Rorden C, Mattingley JB (2024) Lesion mapping in neuropsychological research: a practical and conceptual guide. Cortex 170:38–52
32. Shahid H et al (2017) Important considerations in lesion-symptom mapping: Illustrations from studies of word comprehension. Hum Brain Mapp 38:2990–3000
33. Bertagnoli S et al (2022) Disconnections in personal neglect. Brain Struct Funct 227:3161–3171
34. Pacella V, Moro V (2022) Motor awareness: a model based on neurological syndromes. Brain Struct Funct 227:3145–3160
35. Pia L, Neppi-Modona M, Ricci R, Berti A (2004) The anatomy of anosognosia for hemiplegia: a meta-analysis. Cortex 40:367–377
36. Moro V, Pernigo S, Zapparoli P, Cordioli Z, Aglioti SM (2011) Phenomenology and neural correlates of implicit and emergent motor awareness in patients with anosognosia for hemiplegia. Behav Brain Res 225:259–269
37. Jehkonen M, Laihosalo M, Kettunen J (2006) Anosognosia after stroke: assessment, occurrence, subtypes and impact on functional outcome reviewed. Acta Neurol Scand 114:293–306
38. Cocchini G, Beschin N, Cameron A, Fotopoulou A, Della Sala S (2009) Anosognosia for motor impairment following left brain damage. Neuropsychology 23:223–230
39. Pacella V et al (2025) Anarchy in the brain: behavioural and neuroanatomical core of the anarchic hand syndrome. Cortex 182:181–194
40. Dalla Barba G et al (2018) Different patterns of confabulation in left visuo-spatial neglect. Exp Brain Res 236:2037–2046
41. Pacella V et al (2021) The role of white matter disconnection in the symptoms relating to the anarchic hand syndrome: a single case study. Brain Sci 11:632
42. Pacella V et al (2019) Anosognosia for hemiplegia as a tripartite disconnection syndrome. eLife 8:e46075
43. Talozzi L et al (2023) Latent disconnectome prediction of long-term cognitive-behavioural symptoms in stroke. Brain 146:1963–1978
44. Griffis JC, Metcalf NV, Corbetta M, Shulman GL (2019) Structural. Cell Rep 28:2527–2540.e9

45. Nozais V, Forkel SJ, Foulon C, Petit L, Thiebaut De Schotten M (2021) Functionnectome as a framework to analyse the contribution of brain circuits to fMRI. Commun Biol 4:1035
46. Pustina D, Avants B, Faseyitan OK, Medaglia JD, Coslett HB (2018) Improved accuracy of lesion to symptom mapping with multivariate sparse canonical correlations. Neuropsychologia 115:154–166
47. DeMarco AT, Turkeltaub PE (2018) A multivariate lesion symptom mapping toolbox and examination of lesion-volume biases and correction methods in lesion-symptom mapping. Hum Brain Mapp 39:4169–4182

Check for updates

Chapter 12

A Multiscale Methodology for Mapping Human Neuroanatomy

Trang Cao, James C. Pang and Alex Fornito

Abstract

This chapter describes how to use Mode-based Morphometry (MBM) as a new approach to map anatomical variations in the human brain at multiple spatial scales. This approach overcomes the limitations of conventional approaches, such as surface-based morphometry or voxel-based morphometry, which only consider localized and independent effects at specific brain locations. In contrast, MBM analyzes anatomical group differences of a brain structure as a linear combination of its geometric eigenmodes, which represent intrinsic resonant patterns spanning multiple spatial frequencies that are determined by the geometry of the cortex. Assessing the contributions that each mode makes to the pattern of anatomical differences sheds light on the spatial scales at which such differences are preferentially expressed. MBM can reveal the underlying spatial pattern of group differences, enhance reproducibility, and offer insights into the generative mechanisms that shape the observed differences. Here, we provide step-by-step guideline for conducting an MBM analysis with example data. The approach can be applied to a wide range of anatomical data, given a geometric surface or volume of the structure to derive the geometric eigenmodes.

Keywords Geometric eigenmodes, Brain mapping, MRI, Disorder, Morphometry

1 Introduction

A fundamental challenge in neuroscience is establishing robust methods that can quantify and compare differences in brain structure between population groups, such as between patients with a particular disease and healthy controls. Traditionally, such comparisons have been performed by measuring morphometric features of the brain, such as cortical thickness, surface area, or gray matter volume, using T1-weighted magnetic resonance imaging (T1w MRI). These measurements can then be linked to sociodemographic factors, behavior, clinical symptoms, and more[1–4].

Among the most widely used morphometric techniques are voxel-based morphometry [5] (VBM) and surface-based morphometry [6] (SBM). Both involve fitting statistical models at each image voxel (in VBM) or vertex in a triangular mesh representation of the cortical

Antonio Di Ieva and Jennilee M. Davidson (eds.), *Computational Neuroscience*, Methods in Molecular Biology, vol. 3062,
https://doi.org/10.1007/978-1-0716-5440-8_12,

surface (in SBM) to identify regions showing significant group differences. These analyses are not restricted to voxel or vertex level but can be applied to voxels/vertices aggregated across predefined regions of interest (ROIs) using anatomical atlases [7].

The above morphometric approaches have been successful in describing where group differences between a patient group and controls are located in the brain but offer no insights into the fundamental constraints that may have shaped those differences. Moreover, these approaches rest on the assumption that anatomical differences are (1) highly localized, (2) independent across space, and (3) confined to the spatial scale of the measurement (i.e., local collections of voxels/vertices or atlas-based ROIs). These assumptions can limit our ability to detect broader, spatially extended differences and contribute to inconsistencies across studies, especially when different statistical thresholds are applied. This is a particular concern given the poor reproducibility of putative neuroanatomical signatures of some brain disorders [8].

An alternative approach to relying on a coordinate-based representation that considers the data at its finest resolution is suggested by many areas of physics and engineering, in which the structure of a system can be understood through its natural, resonant modes of excitation called eigenmodes (also referred to, in different contexts, as harmonics, eigenvectors, or eigenfunctions) [9, 10]. These eigenmodes form an orthogonal basis that captures patterns across multiple spatial scales, from those that are global to locally varying, allowing for a spectral representation of the data defined on the system's structure, similar to the classical Fourier decomposition of a one-dimensional signal into component frequencies.

Eigenmode-based approaches have recently been adapted in neuroscience. Two types of eigenmodes are particularly relevant: geometric eigenmodes [11], derived from the brain's geometry (e.g., cortical surface), and connectome eigenmodes [12], derived from a graph-based model of the structural connectome (i.e., a wiring diagram) measured with diffusion MRI. Geometric eigenmodes, ψ, and the corresponding eigenvalues, λ, are obtained by solving the eigenvalue problem, $\Delta\psi = -\lambda\psi$, where Δ is the Laplace-Beltrami operator (LBO) of the cortical surface mesh, which captures local vertex-to-vertex spatial relations and curvature. Connectome eigenmodes are obtained by solving the eigenvalue problem, $L'\psi = -\lambda\psi$, where L' is the normalized graph Laplacian, a discrete counterpart of the LBO, which does not directly account for spatial embedding. These modes have been used to reconstruct patterns of brain activity, understand structure–function relationships, and explain spatial distributions of brain atrophy in neurodegenerative and other conditions [13–16]. In each of these cases, empirical spatial brain maps can be viewed as resulting from the preferential involvement, or excitation, of specific resonant modes of brain structure, thus offering insights into the generative physical mechanisms that shape the observed spatial pattern. The specific choice between geometric and connectome eigenmodes depends on the anatomical properties one wishes to prioritize.

Importantly, the spectral perspective offered by eigenmodes provides insight into which scales (spatial frequencies) are most involved in a given pattern [17]. For example, task-based fMRI activations and individual differences in brain structure have been shown to predominantly involve low-frequency (i.e., large-scale) modes, challenging the traditional emphasis on highly localized effects and underscoring the benefits of a spatially informed, multiscale approach [11, 18].

We have proposed a new approach to mapping anatomical variations in the brain called mode-based morphometry (MBM) [19], which leverages geometric eigenmodes to analyze brain anatomy in a spatially multiscale, mechanism-informed manner. Instead of testing each voxel or vertex independently, MBM models group differences as a linear combination of eigenmodes and performs statistical inference on their contributions. This approach preserves the underlying spatial structure, enhances reproducibility, and offers insight into the generative mechanisms shaping observed anatomical patterns (see [19] for details on the performance of MBM).

The MBM workflow is shown in Fig. 1. We begin by extracting morphometric features of the brain, for example, cortical thickness (CT), from T1w MRI for each individual. We then obtain an unthresholded spatial map of statistics quantifying group mean differences in CT from the individual CT maps using a general linear model as usually performed in SBM. The geometric eigenmodes are obtained by solving the eigendecomposition of the Laplace-Beltrami Operator on the cortical surface (Fig. 1b). The resulting eigenmodes are ordered by spatial wavelength according to their eigenvalues. Thus, for instance, the first mode describes a global mean pattern, the next three capture long-wavelength variations along anterior-posterior, dorso-ventral, and medio-lateral axes, and higher-order modes capture fluctuations and progressively finer spatial scales.

With MBM, the unthresholded statistical map is decomposed using geometric eigenmodes (ψ_1) of the brain as a basis set via a second general linear model (Fig. 1c):

$$t = \beta_1\psi_1 + \beta_2\psi_2 + \ldots + \beta_N\psi_N + \epsilon,$$

ϵ is an error vector and N is the number of eigenmodes one chooses to decompose the data (note that the total number of modes in the basis is equivalent to the total number of points on the cortical surface, and in standard analyses is in the order of 10^4-10^5). The set of weights, β_j, comprises the β spectrum, which quantifies the contribution of each eigenmode to the statistical map. In other words, the spectrum encodes the spatial scales that dominate CT differences between the two groups.

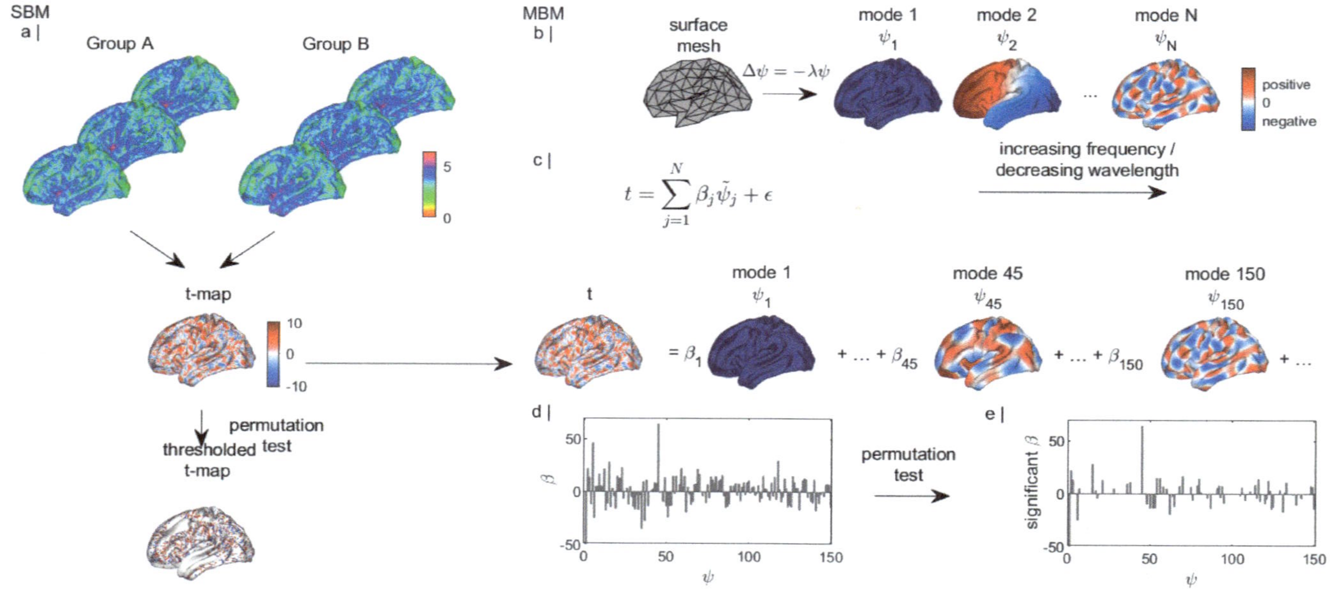

Fig. 1 SBM and MBM analysis pipelines. (**a**) In SBM, a t-statistic is calculated independently at each vertex, quantifying point-wise group differences in CT. A thresholded t-map is derived by comparing the observed t-map and the distribution of null t-maps after permutation testing. (**b**) In MBM, eigenmodes are derived from a cortical surface mesh. The modes are ordered in increasing spatial frequency or decreasing spatial wavelength. Values in each mode are arbitrarily defined, with negative–zero–positive values colored as blue–white–red. (**c**) An empirical t-map can be decomposed as a weighted sum of eigenmodes and errors using a general linear model, with weights given by β_j. The set of β_j is called the β spectrum. (**d**) An example β spectrum showing a large β_{45} that indicates the dominant contribution of mode 45. (**e**) An example of statistically significant βs derived by comparing the observed β spectrum and the distribution of null β spectra after permutation testing. Figure reproduced from [19] with permission under http://creativecommons.org/licenses/by-nc-nd/4.0/

We then perform statistical inference on each β coefficient using permutation tests, wherein group labels are shuffled before repeating the analysis. Repeating this procedure many times yields an empirical null distribution of statistical maps and corresponding β spectra. Observed β coefficients are then compared with their associated null values to estimate the likelihood of obtaining the observed value under the null hypothesis. The resulting set of statistically significant β coefficients can then be used to reconstruct the significant pattern of the group difference. For details on the theory of the framework, see [19]. In this chapter, we present practical instructions to apply the method to analyze cortical thickness (CT) differences between groups of people with mild depression and healthy controls in the left hemisphere, but the method generalizes to other morphometric measures, populations, and eigenmode types.

2 Materials

2.1 Computer Software

1. FreeSurfer [20, 21] (https://surfer.nmr.mgh.harvard.edu/) to preprocess and derive cortical thickness maps from T1w-MRI scans.
2. MATLAB to run the MBM pipeline.
3. MBM package downloaded from https://github.com/NSBLab/MBM.

2.2 Data

1. T1-weighted (T1w) MRI scans in NIFTI format (.nii or.nii.gz). Here, we have 51 subjects with mild depression and 21 healthy controls from the study "Resting state with closed eyes for patients with depression and healthy participants," openly accessible at https://openneuro.org/datasets/ds002748/versions/1.0.5. We have derived the CT of the left hemisphere, which can be downloaded at https://osf.io/6p2m7/. Download RD.zip and extract the folder RD into demo_book_chapter in the MBM package. When downloading the data, make sure the contents are saved following the below folder structure of code and data:

   - MBM
     - data
       - demo_book_chapter
         - RD
           - raw_data
             - sub-01
               - anat
                 - sub-01_T1w.nii.gz

- derivatives
  - freesurfer
    - sub-01
      - surf
        - lh.thickness.fwhm
          - 0.fsaverage.mgh
    - glmfit
      - 6_RestDepress_thick_smooth 0_lh_sex_age
        - y.mgh
    - mbm
      - fsaverage_164k_cortex-lh_ mask.txt
      - fsaverage_164k_midthick-ness-lh.vtk
      - fsaverage_164k_midthick-ness-lh_emode_200.mat
      - fsaverage_164k_midthick-ness-lh_mass_200.mat
      - inputMaps.mat
      - mbmDesignMat.txt
  - participants.tsv
- mbm_main.m
- mbm_demo_bookchapter.m

2. Grouping labels such as diagnosis to run the statistical analysis. For example, here we have the file participants.tsv that shows which participants belong to patients or healthy controls.
3. Demographics such as age and sex to be used as covariates. For example, here we have the file participants.tsv.
4. Cortical surface mesh of a group average to derive the eigenmodes, for example, fsaverage_164k_midthickness-lh.vtk - included in the MBM package.

3 Methods

3.1 Preprocessing Data to Extract Cortical Thickness Maps

We first preprocess the data to extract brain measures and then apply MBM on the measures.

Step 1—Image quality control:

We first exclude scans with poor quality before extracting morphometric features, such as CT, from raw T1w scan by using standard image quality control procedures such as manual visualization for obvious image artifacts and the MRIQC software package (see https://mriqc.readthedocs.io/en/latest/ for details). For example, MRIQC is used to extract 56 different measures quantifying various aspects of image quality, such as the impact of noise, spatial distribution of information, and the impacts of particular artifacts. We apply principal component analysis (PCA) on these measures to extract the first k dimensions accounting for 80% of the variance and discard participants with a normalized score exceeding z= 3.5 (i.e., 99.98 percentile) on any of the k principal components.

Step2—Extracting cortical thickness:
Run "recon-all -i ${BIDS_DIR}/raw_data/${subject_id}/anat/${subject_id}_T1w.nii -s ${subject_id} -all -qcache" command in FreeSurfer (see https://surfer.nmr.mgh.harvard.edu/fswiki/recon-all for details) on the T1w MRI scans. In our data example, BIDS_DIR is the full address containing raw_data directory (which is in MBM/data/demo_book_chapter/RD/). The FreeSurfer pipeline includes volumetric registration to MNI305 atlas; bias field estimation and removal; removal of non-brain tissue using a deformable template model; tissue and broad brain region segmentation using a combination of intensity and atlas-based approaches; generation and refinement of a triangular surface mesh model of the white matter surface by following the intensity gradients between the white and gray matter; generation of a pial surface model by expanding the white matter surface to follow intensity gradients between the gray matter and cerebrospinal fluid (CSF); and then estimation of cortical thickness at each surface vertex as the average distance between the closest points on the white and the pial surfaces. The pial surface is inflated and aligned via spherical registration to the fsaverage atlas (given as default in FreeSurfer), which comprises 163,842 vertices per hemisphere. The extracted cortical thickness measure in the left hemisphere (lh.thickness.fwhm0.fsaverage.mgh) is on the fsaverage atlas, that is, the same atlas of the cortical surface mesh fsaverage_164k_midthickness-lh.vtk that is later used to derive the eigenmodes. Details of the FreeSurfer processing pipeline can be found in [20, 21] and https://surfer.nmr.mgh.harvard.edu/fswiki/FreeSurferAnalysisPipelineOverview. After running "recon-all," manually visualize the surface reconstruction to observe inaccuracies and then discard that subject or manually correct the surfaces and re-extract cortical thickness (see https://surfer.nmr.mgh.harvard.edu/fswiki/FsTutorial/TroubleshootingDataV6.0).

3.2 Running MBM by Command Line

mbm_main.m executes the pipelines in Fig. 1. The inputs to mbm_main.m combined in a Matlab structure named MBM are:

- Paths to anatomical maps to be analyzed and a path to a mask to exclude elements of the maps from the analysis. Anatomical maps are expected as GIFTI, NIFTI, or .mgh files and projected on the same surface template for comparison between maps. In the example given in mbm_demo_book_chapter.m, the left midthickness surface of the fsaverage atlas with 163,842 vertices is used as a template.
- Parameters specifying the statistical test and a design matrix representing effects on subjects. One-sample t-test, two-sample t-test, one-way ANOVA, and ANCOVA (two groups, F-test and Z-test) are supported.
- The eigenmodes (ψ_j in Fig. 1) or a surface mesh (a .vtk file) to calculate the eigenmodes. Eigenmodes should be derived from the same surface template that the maps are projected on.
- Parameters specifying the visualization of the results.

The outputs of mbm_main.m combined in the structure MBM are the statistical map (statistical values at each vertex), p-values of the statistical test at each vertex, thresholded statistical map, β spectrum, p-values of each β, significant β spectrum, the significant pattern which is the reconstructed map of the group difference from the significant β spectrum, and the most influential modes whose β values are the highest and significant. Visualization of the results is provided as in Fig. 2.

Run help mbm_main in the Command Window of MATLAB or open mbm_main.m to see the documentation on all the input and output parameters and their types.

Run example: Run mbm_demo_book_chapter.m for mapping cortical thickness difference in the left hemisphere between groups of mild depression and healthy controls, taking into account covariates of sex and age, that is, ANCOVA with Z-test, on example data. Here, we use the common statistical test applied in FreeSurfer, which uses a GLM to obtain an F-statistical map and then normalizes it to a Z-statistical map (see https://surfer.nmr.mgh.harvard.edu/fswiki/mri_glmfit and https://surfer.nmr.mgh.harvard.edu/fswiki/FsTutorial/GroupAnalysis). The script executes the following steps:

1. Read demographic information in "data/demo_book_chapter/RD/participants.tsv."
2. Make a list file ("data/demo_book_chapter/RD/derivatives/mbm/inputMaps_full_path.txt") of all the paths to the individual CT maps, each path in a line.
3. Make the design matrix file ("data/demo_book_chapter/RD/derivatives/mbm/mbmDesignMat.txt") that includes group labels and covariates, that is, age and sex. For example, with the "ANCOVA_Z" test, the first column has the value "1" or another number (e.g., "2") to present the group effect

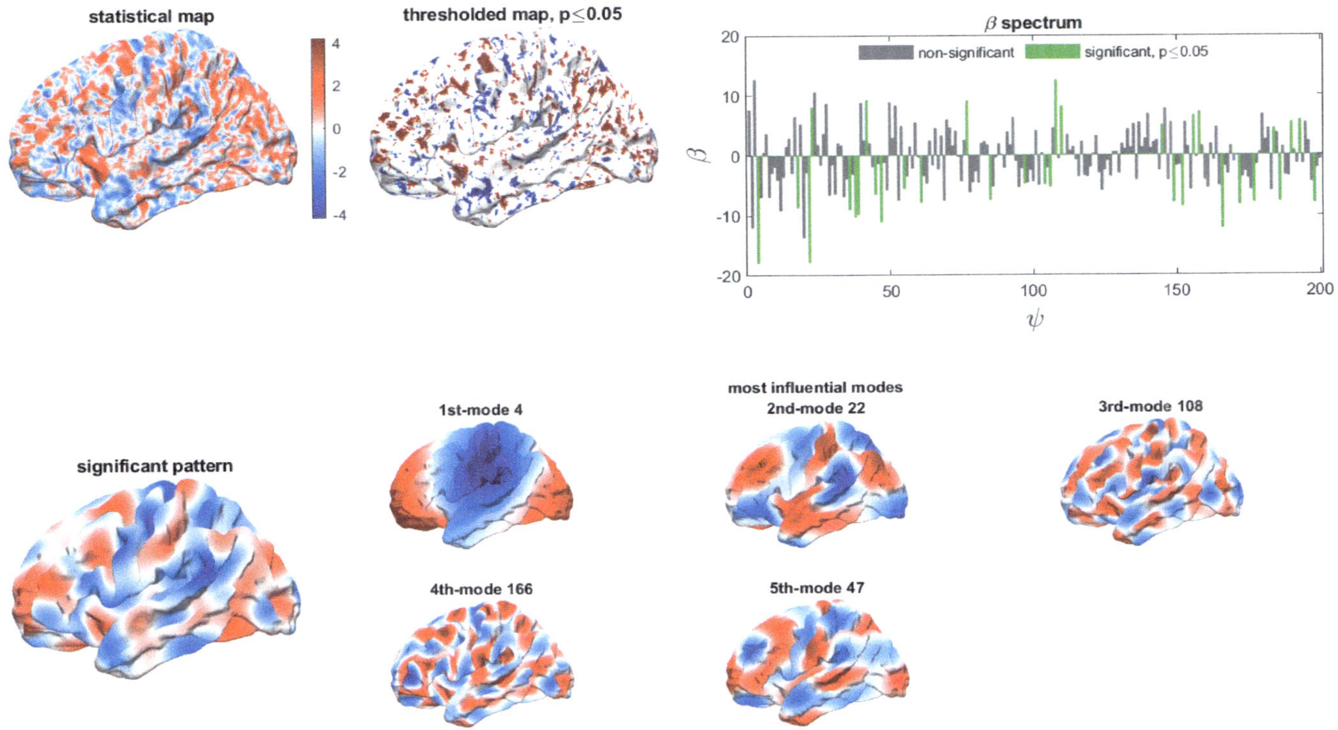

Fig. 2 Visualization of results from MBM analysis of cortical thickness group difference in the example dataset, including the β spectrum, the significant β spectrum, the significant pattern which is the reconstructed map of the group difference from the significant β spectrum, and the most influential modes whose β values are the highest and significant. The statistical map and thresholded map are obtained from SBM

(similar to the input file for mri_glmfit in freesurfer); the second to *k*-th columns are discrete or continuous numbers presenting covariates.

4. Assign parameters of the statistical tests, such as the test name, the number of permutations for the null test, the threshold of *p*-value to consider significant results, the threshold for tail estimation to be used, the indication of False Discovery Rate (FDR) correction (considering FDR correction or not), and the number of modes to be used for the analysis.
5. Read the mask file "data/demo_book_chapter/RD/derivatives/mbm/fsaverage_164k_cortex-lh_mask.txt" containing a binary mask, where values "1" or "0" indicate the vertices of the applied maps to be used or removed. Here, we remove the medial wall.
6. Read the midthickness surface mesh from "data/demo_book_chapter/RD/derivatives/mbm/fsaverage_164k_midthickness-lh.vtk."
7. Assign parameters for visualizing and saving results.
8. Read CT maps.
9. Mask the data.
10. Calculate the statistical map, including permutation tests, and threshold the map.
11. Calculate eigenmodes (*see* **Note 4.1** for more information) or use precalculated eigenmodes in "data/demo_book_chapter/RD/derivatives/mbm/fsaverage_164k_midthickness-lh_emode_200.mat" together with the mass matrix in "data/demo_book_chapter/RD/derivatives/mbm/fsaverage_164k_midthickness-lh_mass_200.mat". The mass matrix is generated in order to calculate the eigenmodes numerically and make the eigenmodes orthogonal.
12. Decompose the statistical map using the eigenmodes to obtain the beta spectrum.
13. Decompose the null maps from the permutation test and determine the significant beta spectrum.
14. Reconstruct the statistical map using the significant beta spectrum.
15. Plot and save the results.

3.3 Running MBM by GUI

To use the GUI in Matlab, open `MBM_toolbox`.mlapp by double-clicking.

The app appears as shown in Fig. 3. On the upper part, the input panel has three tabs: **Maps, Stat**, and **Eigenmodes**. The run panel is on the lower part.

Load the following inputs to run the model:

- In the Maps tab:
  - Map list: a text file comprising the list of paths to the anatomical maps in GIFTI, NIFTI, or .mgh format; or a .mat file containing a matrix whose each row is a map; or a .mgh file containing a 4-D matrix obtained from mri_glmgit output in Freesurfer.
  - Mask: a text file containing a binary mask where values "1" or "0" indicate the vertices of the applied maps to be used or removed.
  - Surface: a vtk file containing a surface used to calculate the eigenmodes and/or plot the results. The surface should be the one from which the eigenmodes are derived and on which the anatomical maps are projected.
  - Hemisphere: to be analyzed, chosen from the drop-down list.
- In the Stat tab:
  - Statistical test: chosen from the drop-down list.

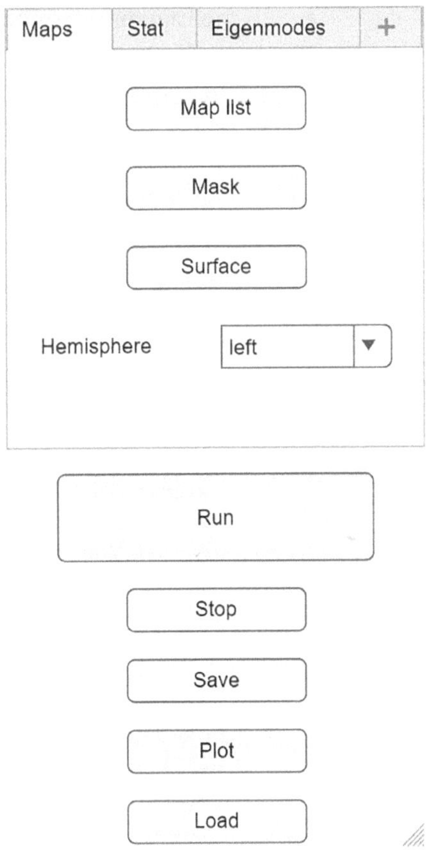

Fig. 3 MBM toolbox interface

  - Design matrix G: a text file containing a design matrix [*m* subjects by *k* effects]. For the design matrix in the statistical test of one sample (one column), two sample (two columns), one-way ANOVA (*k* columns, number of subjects in each group must be equal), "1" or "0" in each column indicates a subject in a group or not. For the design matrix in the statistical test of ANCOVA_F and ANCOVA_Z: "1" or another number (e.g., "2") in the first column indicates the group effect (similar to the input file for mri_glmfit in Freesurfer) and discrete or continuous numbers in the second to *k*-th columns indicate covariates.
  - Permutation: the number of permutations in the statistical test.
  - Pthr: tail approx: the threshold of *p*-values for tail approximation. If the *p*-values are below Pthr, these are refined further using a tail approximation from the Generalised Pareto Distribution (GPD) [22].
  - P threshold: the threshold of *p*-values for being significant.
  - FDR: mark if using FDR correction.

- In the Eigenmodes tab,
  - Eigenmodes: a text file or a .mat file containing eigenmodes in columns. If you do not have pre-calculated eigenmodes, this input can be omitted, but .vtk surface is required. The code will calculate the eigenmodes from the .vtk surface.
  - Mass matrix: a text file or a .mat file containing the mass matrix associated with the eigenmodes to preserve the orthogonality. See https://github.com/Deep-MI/LaPy and the reference papers to understand what a mass matrix is and how eigenmodes are calculated.
  - Number of modes: the number of eigenmodes used for the analysis.
  - Most influential modes: the number of the most influential modes to plot.

- On the lower part:
  - Run: to run the analysis.
  - Save: to save the structure "MBM" containing the parameters and results in a .mat file when the analysis has finished.
  - Plot: to plot the results. The result panel comprises the statistical map, the thresholded statistical map, the beta spectrum, the significant pattern, and the most influential modes.

- Stop: to interrupt the analysis if required. For 5000 permutations, the analysis needs approximately 1 h to run.
- Load: to read all the configuration from the MBM structure in a .mat file. Any of the inputs can be changed after loading the MBM structure.

Run example: load all the inputs in the tabs on the upper part and then hit "Run." For the map list, the text file containing the list of paths to the individual CT maps can be found in "data/demo_book_chapter/RD/derivatives/mbm/inputMaps_full_path.txt" if mbm_demo_book_chapter.m has been run. Otherwise, use "data/demo_book_chapter/RD/derivatives/mbm/inputMaps.mat" or "data/demo_book_chapter/RD/derivatives/freesurfer/glmfit/6_RestDepress_thick_smooth0_lh_sex_age/y.mgh." If mbm_demo_book_chapter.m has been run and an MBM structure has been created and saved in "data/demo_book_chapter/RD/derivatives/mbm/mbm_demo_book_chapter.mat," load the MBM structure by clicking the "Load" button and then run the analysis by clicking "Run."

4 Notes

4.1 Deriving Eigenmodes

We calculate the eigenmodes of the cortical surface represented by a triangular mesh by using utils/modes/calc_geometric_eigenmode.m in the MBM package, based on the LaPy Python library [23, 24]. Here, we use a population-averaged template of the cortical midthickness surface in, for example, fsaverage_164k space (Fig. 1b). We use a population-averaged surface to obtain a common set of eigenmodes for all participants, enabling easier comparisons. Note that individual-specific eigenmodes can also be derived from individual surfaces. However, differences in the geometry of the cortical surface can alter the spatial patterns of the eigenmodes, which makes comparison between individual-specific modes challenging. Here, we perform analyses on the left hemisphere and thus use the modes derived from the left hemisphere.

References

1. DeCasien AR, Guma E, Liu S, Raznahan A (2022) Sex differences in the human brain: a roadmap for more careful analysis and interpretation of a biological reality. Biol Sex Differ 13(1):43
2. Ducharme S, Albaugh MD, Nguyen TV, Hudziak JJ, Mateos-Pérez JM, Labbe A et al (2016) Trajectories of cortical thickness maturation in normal brain development — the importance of quality control procedures. NeuroImage 125:267–279
3. Zhao Y, Zhang Q, Shah C, Li Q, Sweeney JA, Li F et al (2022) cortical thickness abnormalities at different stages of the illness course in Schizophrenia: a systematic review and meta-analysis. JAMA Psychiatry 79(6):560
4. Douaud G, Lee S, Alfaro-Almagro F, Arthofer C, Wang C, McCarthy P et al (2022) SARS-CoV-2

is associated with changes in brain structure in UK Biobank. Nature 604(7907):697–707
5. Ashburner J, Friston KJ (2000) Voxel-based morphometry–the methods. NeuroImage 11(6 Pt 1):805–821
6. Fischl B, Dale AM (2000) Measuring the thickness of the human cerebral cortex from magnetic resonance images. Proc Natl Acad Sci U S A 97(20):11050–11055
7. Desikan RS, Ségonne F, Fischl B, Quinn BT, Dickerson BC, Blacker D et al (2006) An automated labeling system for subdividing the human cerebral cortex on MRI scans into gyral based regions of interest. NeuroImage 31(3):968–980
8. Cao T (2025) The cross-site reproducibility of MRI morphometric phenotypes in psychiatric disorders. Available from: https://www.medrxiv.org/content/10.1101/2025.07.09.25331220v2.full
9. Beards C (1996) Structural vibration: analysis and damping [Internet]. Elsevier Science. Available from: https://books.google.com.au/books?id=OQvoRTt8H0QC
10. Felippa CA, Park KC, Farhat C (2001) Partitioned analysis of coupled mechanical systems. Comput Methods Appl Mech Eng 190(24–25):3247–3270
11. Pang JC, Aquino KM, Oldehinkel M, Robinson PA, Fulcher BD, Breakspear M et al (2023) Geometric constraints on human brain function. Nature 618(7965):566–574
12. Atasoy S, Donnelly I, Pearson J (2016) Human brain networks function in connectome-specific harmonic waves. Nat Commun 7(1):10340
13. Robinson PA, Henderson JA, Gabay NC, Aquino KM, Babaie-Janvier T, Gao X (2022) Determination of dynamic brain connectivity via spectral analysis. Front Hum Neurosci 15:655576 [cited 2022 Nov 6] Available from: https://www.frontiersin.org/articles/10.3389/fnhum.2021.655576
14. Griffa A, Amico E, Liégeois R, Van De Ville D, Preti MG (2022) Brain structure-function coupling provides signatures for task decoding and individual fingerprinting. NeuroImage 250:118970
15. Abdelnour F, Mueller S, Raj A (2015) Relating cortical atrophy in temporal lobe epilepsy with graph diffusion-based network models. PLOS Comput Biol 11(10):e1004564
16. Wang MB, Owen JP, Mukherjee P, Raj A (2017) Brain network eigenmodes provide a robust and compact representation of the structural connectome in health and disease. PLOS Comput Biol 13(6):e1005550
17. Robinson PA, Zhao X, Aquino KM, Griffiths JD, Sarkar S, Mehta-Pandejee G (2016) Eigenmodes of brain activity: neural field theory predictions and comparison with experiment. NeuroImage 142:79–98
18. Chen YC, Arnatkevičiūtė A, McTavish E, Pang JC, Chopra S, Suo C et al (2022) The individuality of shape asymmetries of the human cerebral cortex. eLife 11:e75056
19. Cao T, Pang JC, Segal A, Chen Y, Aquino KM, Breakspear M et al (2024) Mode-based morphometry: a multiscale approach to mapping human neuroanatomy. Hum Brain Mapp 45(4):e26640
20. Dale AM, Fischl B, Sereno MI (1999) Cortical surface-based analysis: I. Segmentation and surface reconstruction. NeuroImage 9(2):179–194
21. Fischl B, Sereno MI, Dale AM (1999) Cortical surface-based analysis: II: inflation, flattening, and a surface-based coordinate system. NeuroImage. 9(2):195–207
22. Winkler AM, Ridgway GR, Douaud G, Nichols TE, Smith SM (2016) Faster permutation inference in brain imaging. NeuroImage 141:502–516
23. Reuter M, Wolter FE, Peinecke N (2006) Laplace–Beltrami spectra as 'Shape-DNA' of surfaces and solids. Comput-Aided Des 38(4):342–366
24. Wachinger C, Golland P, Kremen W, Fischl B, Reuter M (2015) Alzheimer's Disease Neuroimaging Initiative. BrainPrint: a discriminative characterization of brain morphology. NeuroImage 109:232–248

Check for updates

Chapter 13

Federated Learning Pipeline for Whole-Slide Image Classification in Digital Neuropathology

Cong Cong, Antonio Di Ieva and Sidong Liu

Abstract

Federated learning (FL) has emerged as a powerful paradigm for privacy-preserving, collaborative training of machine learning models, particularly valuable in fields like digital pathology, where data sharing is constrained by institutional policies and regulatory concerns. This chapter presents a reproducible computational pipeline for whole-slide image (WSI) classification in digital neuropathology using a federated learning framework. In addition to outlining the end-to-end implementation, including preprocessing, model training, and deployment, we highlight key challenges specific to applying FL in WSI analysis, such as data heterogeneity and communication efficiency. Detailed code snippets, implementation guidance, and deployment recommendations are provided to support real-world adoption. We hope this chapter serves as a valuable resource for neuropathologists, researchers, and machine learning practitioners aiming to bridge technology and medical science in the evolving landscape of digital neuropathology.

Keywords Federated learning, Digital neuropathology, Image classification

1 Introduction

Digital pathology has transformed the way histopathological assessments are conducted in clinical and research settings. By digitizing conventional glass slides into high-resolution whole-slide images (WSIs), it has enabled the integration of computational algorithms into pathology workflows [1, 2]. These advancements have opened new frontiers in diagnostics, enabling AI-driven approaches to improve accuracy, consistency, and speed of interpretation.

Among computational techniques, deep learning-based methods, especially those leveraging weakly supervised learning and multiple instance learning (MIL), have become essential in WSI classification [3–6]. However, training robust models requires access to large and diverse datasets, which individual institutions often lack. Data sharing across hospitals is hindered by privacy regulations, ethical concerns, and logistical challenges, particularly in neuropathology where datasets are small and case diversity is limited.

Antonio Di Ieva and Jennilee M. Davidson (eds.), *Computational Neuroscience*, Methods in Molecular Biology, vol. 3062,
https://doi.org/10.1007/978-1-0716-5440-8_13,

Federated learning (FL) provides a promising framework to overcome these barriers by allowing institutions to collaboratively train machine learning models without centralizing patient data [7–10]. Instead, each participating site trains a model locally and shares only model updates or abstracted information with a central server, thereby preserving data privacy. Despite its potential, applying FL to WSI classification introduces several domain-specific challenges:

1. **Client Heterogeneity**: Medical institutions often operate under diverse computational environments, employing different hardware configurations, software pipelines, feature extractors, and model architectures tailored to their clinical needs. This diversity poses a significant challenge for standard federated learning methods, which typically assume a uniform model architecture and identical resource availability across clients. Such assumptions are unrealistic in real-world clinical settings, where institutions may prefer convolutional neural networks (CNNs), Vision Trans-formers (ViTs), or distinct MILs classifiers. As a result, standard FL frameworks struggle to accommodate this variability, often leading to degraded performance, unstable convergence, or even failure to train a unified global model effectively.
2. **Communication Overhead**: Standard FL protocols involve the frequent exchange of entire model weights or gradients between clients and a central server. This leads to substantial communication overhead. In digital pathology, where WSIs require complex models with millions of parameters, such overhead becomes impractical for institutions with limited network bandwidth or computational infrastructure. Smaller hospitals or resource-constrained clinics may face delays, connection instability, or outright inaccessibility of FL participation due to these demands, impeding widespread adoption.
3. **Privacy and Security**: Although FL reduces privacy risks by avoiding the transfer of raw data, standard FL approaches still require sharing model weights or gradients that can potentially leak sensitive patient information. Recent studies have demonstrated that adversaries can reconstruct input data or infer sensitive attributes from these shared updates. Standard FL lacks built-in mechanisms to guard against such inversion attacks or leakage pathways, thereby raising concerns about its adequacy for high-stakes applications like digital neuropathology, where patient confidentiality is paramount.

To address the above challenges, personalized federated learning (pFL) methods [10–15] have been proposed, allowing each client to train a customized model. These approaches have shown promise in mitigating client heterogeneity by enabling local adaptation and have introduced mechanisms to improve communication efficiency and robustness. However, most existing pFL methods only partially address the three key challenges outlined above. For instance, some rely on model decoupling or synthetic data layers [10–12, 14], which still require substantial model parameter exchange and assume consistent feature extractors across clients. Others depend on class abstraction sharing [13, 16, 17], which may not generalize well in real-world WSI scenarios with diverse MIL architectures.

This chapter provides a practical guide to implementing a novel pFL pipeline for WSI classification in digital neuropathology, with a specific focus on simultaneously alleviating the three key challenges: client heterogeneity, communication inefficiency, and privacy risk. We begin by introducing the datasets and computational setup required to reproduce our pipeline. Next, we walk through the foundational algorithms, including standard FL, MIL, and their integration in a federated setting. We then describe how to set up a simulated federated environment and introduce our proposed method, FedDFP [18], and provide implementation details for each of its key components, feature extraction, deep feature prompting, communication protocols, and model aggregation, so readers can replicate and adapt the approach to their own settings. Finally, we demonstrate the effectiveness of FedDFP through experimental results on an exemplary WSI dataset.

2 Materials

2.1 Dataset

To demonstrate the efficacy of our proposed personalized federated learning approach in digital neuropathology, we systematically explore the entire WSI analysis pipeline using the TCGA-IDH datasets [19]. The dataset contains 200 randomly selected patient samples from The Cancer Genome Atlas (TCGA) data portal, including 100 glioblastoma (GBM) patients from the TCGA GBM project and another 100 astrocytoma and oligodendroglioma cases from the TCGA low-grade glioma (LGG) project. We provide the following steps to download and process the dataset, or you may use the link (https://drive.google.com/drive/folders/1VIzLFelixKAlNzPt9t9bhg7-fX-2c2n5?usp=drive) to download the processed dataset.

2.1.1 Dataset Retrieval Pipeline

We obtained glioma cases from the TCGA-GBM and TCGA-LGG projects using the Genomic Data Commons (GDC) Data Portal[1]. The following steps were performed:

1. **Case Selection**: We selected all cases under the TCGA-GBM (glioblastoma) and TCGA-LGG (lower-grade glioma) projects in the GDC "Exploration" interface.

[1] https://portal.gdc.cancer.gov/

2. **Whole-Slide Images**: Under the Files tab, we filtered for Data Category: Biospecimen and Data Type: Diagnostic Slide Image (.svs). All resulting WSIs were added to the cart, and a manifest file was downloaded.
3. **Data Download**: Files were retrieved programmatically using the GDC Data Transfer Tool:

   The dataset used for demonstration comprises contributions from eight centers, each providing more than 40 WSIs, resulting in a total of 968 slides. Each slide is annotated as either IDH wildtype (WT) or mutant (MU) based on immunohistochemistry and/or genetic sequencing results. The center-wise distribution is as follows: C1: 292 WT / 119 MU, C2: 52 WT / 24 MU, C3: 8 WT / 40 MU, C4: 57 WT / 61 MU, C5: 38 WT / 130 MU, C6: 3 WT / 45 MU, and C7: 84 WT / 15 MU. For each center, WSIs were randomly split into 80% for training and 20% for testing.

```
./gdc-client download -m gdc_manifest.json
```

2.2 Dataset Preprocessing

The downloaded WSIs were provided in the .svs format. We preprocessed these slides following a standard whole-slide image analysis workflow provided in the CLAM [20] code repository. The preprocessing was implemented in Python, using the OpenSlide library for efficient WSI reading and tiling, and a MIL framework for subsequent slide-level classification tasks. The main steps are summarized below.

2.2.1 Slide Format and Metadata Extraction

We used the OpenSlide Python package to load the slides and extract associated metadata, including magnification levels, image dimensions, and downsampling factors.

```
import openslide
# Extract metadata
dimensions = slide.dimensions                         # (width, height) in pixels
downsampling_factors = slide.level_downsamples  # list of down sample factors
level_dimensions = slide.level_dimensions          # dimensions at each level
# Print metadata
print("Base dimensions:", dimensions) print("Down sampling factors:",
downsampling_factors) print("Level dimensions:", level_dimensions)
```

2.2.2 Patch Extraction Strategy

WSIs are extremely large (often exceeding 100,000 × 100,000 pixels) and cannot be processed directly by deep learning models due to GPU memory limitations. Therefore, we adopted a patch-based strategy to divide each WSI into manageable tiles. For each slide, tissue regions were first identified using a tissue-detection mask (Otsu [21] on a downsampled image). Patches were then sampled at the 10× magnification level as nonoverlapping tiles of size 1024×1024 pixels. Instead of saving each tile as an image file (e.g.,.png or.jpg), we stored the patch coordinates, slide identifier, and magnification level in a compact.h5 file. This approach substantially reduces storage requirements while preserving all spatial and hierarchical information required for downstream MIL frameworks. As a result, each slide is associated with a single.h5 file containing the complete set of patch metadata, which can be used for on-the-fly feature extraction and training. The above process can be done by running the following code:

```
python3 create_patches_fp.py \
--source $DATA_DIRECTORY \
--save_dir $RESULTS_DIRECTORY \
--patch_size 256 \
--patch_level 1 \
--patch --seg --stitch
```

The above command will segment every slide in DATA DIRECTORY using default parameters, extract all patches (each of size 256×256) within the segmented tissue regions, create a stitched reconstruction for each slide using its extracted patches, and generate the.h5 files at the specified RESULTS DIRECTORY.

2.2.3 Patch Feature Extraction

By default, the extracted patches are stored in the patches subdirectory within the specified RESULTS DIRECTORY. Each slide is represented by a single.h5 file containing an array of patch coordinates, where each entry specifies the top-left corner of a 256×256 pixel patch in the original WSI. In accordance with standard MIL workflows, these raw patches must be transformed into compact feature embeddings before training. This is accomplished by reading each patch and passing it through a pretrained feature extractor (e.g., a convolutional neural network (ResNet50 [22]) trained on ImageNet [23]), thereby producing a lower-dimensional feature representation suitable for efficient downstream slide-level classification. The entire feature extraction process can be reproduced by running the following command:

```
python3 extract_patches_fp.py \
--model_name 'resnet50' \
--data_h5_dir $RESULTS_DIRECTORY \
--data_slide_dir $DATA_DIRECTORY \
--feat_dir $FEATURES_DIRECTORY \
--batch_size 512
```

The above command reads the patch coordinates from each.h5 file in RESULTS DIRECTORY and converts the corresponding image patches into feature embeddings. The resulting embeddings are saved in the FEATURES DIRECTORY as individual.pt files. Each file contains the complete set of patch-level embeddings, which are subsequently used as inputs for model training.

The above pipeline was executed independently at each center, ensuring that both training and testing slides were processed locally. Each slide was stored as a separate.pt file containing its patch-level embeddings.

2.3 Hardware and Training Protocol

All experiments were conducted using a single NVIDIA V100 GPU with 32 GB of memory and 64 GB of system RAM. To simulate the federated learning setup, training was performed sequentially: at each communication round, local training was conducted on one center at a time. Once all centers completed their local updates, the information designated for sharing (e.g., model weights, gradients, or task-specific representations) was sent to the server, aggregated, and then redistributed to all centers for the next training round. In our framework, this shared information corresponds to the Deep Feature Prompts (DFPs).

2.4 Software and Dependencies

All code was implemented in Python 3.9 using PyTorch 1.13.1. Experiments were conducted on a Linux-based system (CentOS 8, kernel version 4.18.0-553.47.1.el8.nci.x86 64) hosted on the Gadi supercomputing cluster. The full implementation, including environment setup and instructions, is available on GitHub at: https://github.com/thomascong121/FedDFP.

3 Methods

3.1 Federated WSI Classification with FedDFP

This section provides a step-by-step guide for implementing a pFL pipeline for WSI classification using the FedDFP framework. We provide a visual demonstration of the FedDFP pipeline in Fig. 1. The goal is to train distributed MIL classifiers across multiple medical centers without sharing raw data or full model parameters.

Step 1: Feature Extraction and Local Preprocessing
Each participating center starts by preprocessing its WSIs using the pipeline described in Sect. 2.2. A pretrained feature extractor is used to convert tissue-containing patches into fixed-length embeddings. These patch-level features are stored locally in pt files and serve as input to the downstream MIL classifier.

Step 2: Federated Learning Workflow
At each global epoch, the server initializes (or updates) the global prompt and distributes it to all participating clients. Each client

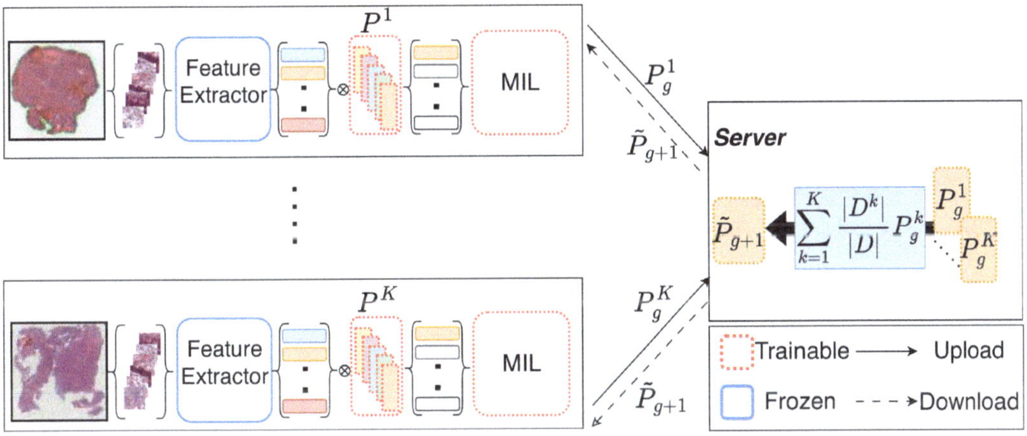

Fig. 1 Overview of FedDFP: Each site trains a local MIL classifier on patch features modified by a learnable Deep Feature Prompt (DFP). Only the DFPs are shared with the server, aggregated, and redistributed in each communication round

performs local training on its private data, updates the local model and prompt parameters, and returns the updated weights and evaluation metrics. The server then aggregates the local prompts to update the global prompt. The process repeats for multiple global epochs until convergence, and the best-performing model and prompt states are saved for downstream evaluation.

```
for epoch in range(global_epochs):
    local_prompt = []
    for idx in n_clients:
        prompt_w = clients[idx].local_train(global_prompt)
        local_prompt.append(deepcopy(prompt_w))
```

In the subsequent sections, we detail the client-side procedure and describe how the DFP is integrated into the local MIL training updates.

Step 3: Initialize Local Models and Prompts

At each center k, the following components are first initialized:

1. A MIL classifier, denoted as M^k, was instantiated independently at each site to accommodate local computational infrastructure and performance preferences. We provide example implementations for several MIL model initializations in utils/Get model.py. For example, when using CLAM [20] as the local MIL classifier, the model can be initialized by executing the following code:

```
def define_clam(args):

    model_dict = {"dropout": args.drop_out, 'n_classes': args.n_classes}

    if args.mil_method != 'mil':

        if 'ResNet50' in args.ft_model:

            model_size = 'small'

        elif 'PLIP' in args.ft_model:

            model_size = 'medium'

        elif 'ViT_S_16' in args.ft_model:
            model_size = 'ultra_small'

        elif 'ViT_T_16' in args.ft_model:
            model_size = 'tiny'

        else:

            raise NotImplementedError
        model_dict.update({"size_arg": model_size})

    if args.subtyping:
        model_dict.update({'subtyping': True})

    if args.B > 0:
        model_dict.update({'k_sample':args.B})

    if args.inst_loss == 'svm':

        from topk.svm import SmoothTop1SVM

        instance_loss_fn = SmoothTop1SVM(n_classes=2)
        instance_loss_fn = instance_loss_fn.cuda()

    else:

        instance_loss_fn = nn.CrossEntropyLoss()

    if args.mil_method == 'CLAM_SB':

        model = CLAM_SB(**model_dict, instance_loss_fn=instance_loss_fn)

    else:

        raise NotImplementedError

        model.relocate()

    return model
```

- By default, ResNet50 is used as the feature. The number of top-selected patches is set to 1 because some bags in our exemplary dataset contain only a single patch. Following the original CLAM configuration, we specify SVM as the instance-level loss function and use CLAM_SB as the default MIL method.
- A DFP $P^k = \{p^k\}_{t=1}^{T}$, consisting of T learnable vectors with the same dimensionality as the patch embeddings:

```
Prompt_embeddings = nn.Parameter(torch.zeros(1, num_tokens, prompt_size).to(device))
nn.init.uniform(prompt_embeddings.data, 0.0, 1.0)
```

A full list of DFP configurations is provided in our code base, and guidance on configuration selection can be found in **Note 2**.

Step 4: Apply Deep Feature Prompt

FedDFP introduces DFPs that are applied directly to the patch-level feature embeddings:

$$e_i^{k,P} = \frac{1}{T} \sum_{t=1}^{T} e_i^k \odot p_t^k$$

Here, e_i^k is the patch embedding for patch i at site k, and $\odot$ denotes element-wise multiplication, and this is implemented in the forward function of the Prompt object:

```
class Prompter(nn.Module):
    def __init__(self, args):
        super(Prompter, self).__init__()
        self.prompt_embeddings, self.prompt_norm = prompt_init(args)
        self.prompt_aggregation = args['prompt_aggregation']
    def forward(self, e):
        B = e.size(0)
        prompt = self.prompt_norm(self.prompt_embeddings).expand(B, -1, -1)
        prompt = torch.permute(prompt, (1, 0, 2)).to(device)
        e = e.expand(prompt.size(0), -1, -1) * prompt
        e = torch.mean(e, dim=0)
        return e
```

Step 5: Local Training

During local client training, MIL model is first instantiated:

```
model = define_clam(args)
```

We then configure the optimizer to jointly update the parameters of both the MIL classifier and the DFP:

```
prompter_params.append({'params': prompt_embeddings.parameters(), 'lr': prompt_lr})
prompter_params.append({'params': filter(lambda p: p.requires_grad, model.parameters()), 'lr': model_lr})
optimizer = optim.Adam(prompt_params)
```

The prompted embeddings $e_i^{k,P}$ are then fed into the MIL to produce the prediction:

```
logits = model(data)
```

Here, data denotes the set of prompted patch embeddings, and logits represents the unnormalized class-wise outputs. The slide-level classification loss is computed as:

$$\mathcal{L}^k = \mathrm{E}_{(x,y)\in D^k}\mathcal{L}\left(M^k, P^k, x, y\right)$$

where D^k is the local dataset at site k. Notably, the feature extractor is kept frozen during training to improve computational efficiency. In our implementation, the loss is instantiated using cross-entropy.

Based on the computed loss, we update both the model and the DFP parameters via gradient backpropagation:

```
loss = nn.CrossEntropyLoss()(logits, label)
```

```
optimizer.zero_grad()
loss.backward()
optimizer.step()
```

Step 6: Communication with Server

After completing local training, each center uploads only its updated prompt vectors P^k to the central server. The MIL classifier M^k and raw feature representations remain strictly local, thereby preserving data privacy and reducing communication overhead. The server then aggregates the received prompt vectors from all participating centers to obtain the global prompt:

$$\widetilde{P}_{t+1} = \sum_{k=1}^{K} \frac{|D^k|}{|D|} P_t^k$$

In practice, this aggregation is implemented as follows:

```
for client_idx, client_prompt in enumerate(local_prompt):
    if client_idx == 0:
        global_prompt = client_prompt
    else:
        global_prompt += client_prompt
```

The resulting global prompt $\widetilde{P}_{t+1}$ is then broadcast back to each participating site for the next communication round.

Step 7: Iterative Optimization

Steps 3–6 are repeated for T global communication rounds. Over time, the local model benefits from shared feature prompting knowledge.

Step 8: Inference

After training, each site performs WSI inference using its local M^k in combination with the final shared prompt $\widetilde{P}_T$. Because the prompts are lightweight and model-agnostic, the framework naturally supports heterogeneity across sites in both feature extractors and classifiers. The inference process can be implemented as follows:

```
model.eval()
total_acc = 0.0
with torch.no_grad():
    for batch_idx, (data_ft, label) in enumerate(test_loader):
        prompted_data = prompt(data_ft)
        logits = model(prompted_data)
        acc = logits.softmax(dim=1).eq(label).float().mean().item()
        total_acc += acc
final_acc = total_acc/len(test_loader)
```

3.2 FedDFP for Collaborative IDH Biomarker Prediction

Having introduced the design and training pipeline of FedDFP, we now demonstrate its application to collaborative IDH biomarker prediction across eight participating centers (C1–C8) using the TCGA-IDH dataset. The quantitative results are summarized in Table 1, comparing Local training, FedHisto, and our proposed FedDFP. To encourage reproducibility, we provide the full implementation details of FedDFP in Sect. 3 and share an example training/testing command below:

Table 1

Performance comparison of selected FL methods on TCGA-IDH (IDH), using ResNet50 as the patch-level feature extractor and CLAM [20] as the MIL classifier. The table reports center-wise accuracy, p-values, and communication cost per round (#Com). Best-performing results are highlighted in bold

Center	Local	FedHisto	FedDyn	MOON	FedDFP
C1	89.5±2.2	89.8±2.1	84.0±2.4	67.9±1.6	**94.5±1.6**
C2	64.6±3.0	66.7±2.9	87.5±5.1	66.7±5.6	**93.8±5.1**
C3	80.0±1.6	70.0±1.4	86.7±9.4	76.7±2.5	**86.7±4.7**
C4	68.3±2.4	78.3±6.2	80.0±7.1	83.3±4.7	**91.7±2.4**
C5	66.7±1.7	86.1±1.4	90.3±1.1	86.7±4.7	**91.7±3.4**
C6	74.5±7.8	72.6±3.7	89.2±3.7	80.4±5.5	**91.2±2.4**
C7	50.0±1.6	50.0±1.6	86.7±4.7	45.8±6.8	**90.0±8.2**
C8	71.7±9.4	80.0±1.5	93.3±2.4	85.0±0.0	**96.7±2.4**
p-value	$6.67e^{-08}$	$1.03e^{-04}$	$5.59e^{-04}$	$2.81e^{-06}$	–
# Com	–	$791K$	$791K$	$791K$	**1.02*k***

```
Python3 main.py \
--feature_type $FEATURE_TYPE \
--ft_model $MODEL_NAME \
--mil_method $MODEL_NAME_MIL \
--fed_method $MODEL_NAME_FED
```

We set FEATURE_TYPE=R50 features and use MODEL_NAME=ResNet50 as the ImageNet-pretrained feature extractor. For the MIL model, we select MODEL_NAME_MIL=CLAM SB, and for the federated method, we use MODEL_NAME_FED=FedDFP. We also implemented several baseline federated learning approaches, and for all baseline experiments, we used the same experimental setup (*see* **Note 1**):

- FedHisto [7]: a histopathology-specific adaptation of FedAvg.
- FedDyn [24]: introduces dynamic regularization to mitigate client drift.
- MOON [25]: uses contrastive learning to align local and global model representations.

  For personalized FL, where clients maintain customized local models, we include:
- FedProto [13]: shares class prototypes to guide local adaptation.

- DESA [26]: transfers knowledge across clients via synthetic data generation.
- SGPT [14]: applies spatial visual prompts for personalized learning.
- pFedPG [27]: uses gradient-based prompt tuning for efficient local adaptation.

In addition to the above experiments, we further evaluated FedDFP under heterogeneous client settings to simulate more realistic federated learning scenarios. To simulate this realistic scenario, we define a heterogeneous FL setup in which each medical center independently adopts a classifier randomly selected from a pool of widely used MIL methods: CLAM [20], TransMIL [28], and ABMIL [29]. The quantitative results are summarized in Table 2.

3.2.1 Interpretation of the Results

Per-center classification accuracies in Table 1 show that federated learning consistently improves performance across nearly all centers. For example, in the TCGA-IDH dataset, local models performed poorly on centers with limited data (e.g., C7: 50.0%), whereas FL methods substantially improved performance (FedDyn: 86.7%, MOON: 67.9%). These results demonstrate that federated training mitigates performance degradation caused by data scarcity and institutional heterogeneity.

Although standard FL methods such as FedHisto, FedDyn, and MOON improve performance, they require transmitting large model updates, on the order of hundreds of thousands of parameters per communication round, posing challenges for bandwidth-limited clinical settings. FedDFP addresses this by exchanging only a

Table 2

Comparison of FedDFP with personalized FL methods on TCGA-IDH (IDH). Each row reports the MIL method used by the center and the corresponding accuracy. Best results are highlighted in bold

Methods			DESA	SGPT	pFedPG	FedDFP
IDH	C1	TransMIL	84.4±5.7	75.8±2.2	70.0±5.7	**86.9±2.1**
	C2	ABMIL	89.6±10.6	83.3±3.0	83.3±3.0	**93.8±4.7**
	C3	CLAM	90.0±0.4	90.0±0.0	90.0±8.2	**90.0±0.0**
	C4	TransMIL	86.7±4.8	76.7±4.7	80.0±7.1	**86.7±2.4**
	C5	ABMIL	72.2±10.9	76.4±5.2	62.5±3.4	**91.7±5.9**
	C6	CLAM	87.3±5.6	80.4±3.6	85.3±8.7	**87.3±5.0**
	C7	TransMIL	83.3±4.7	86.7±4.7	93.3±4.7	**96.7±3.2**
	C8	ABMIL	91.7±6.3	83.3±8.5	93.3±6.2	**93.3±4.7**
	p-value	–	$6.67e^{-03}$	$1.29e^{-04}$	$4.04e^{-05}$	–

compact prompt vector, reducing communication cost from ~791 K parameters to just 1.02 K (over 99.9% reduction), while achieving superior accuracy. This efficiency makes FedDFP particularly attractive for real-world deployments where communication resources are constrained.

As shown in Table 2, FedDFP consistently outperforms all baselines, highlighting its effectiveness in real-world heterogeneous settings. By applying prompts at the feature-embedding level, FedDFP enables architecture-agnostic collaboration: each center can retain its preferred MIL design while still benefiting from shared knowledge. This flexibility, combined with superior accuracy, positions FedDFP as a practical and scalable solution for federated digital pathology, where institutional variability is the norm rather than the exception. We also note that FedDFP has several limitations and potential extensions, which are discussed in **Note 3**.

4 Notes

1. To ensure reproducibility, all experiments were conducted with fixed random seeds and frozen pretrained feature extractors. Each federated round included 50 local training epochs per client. We recommend tracking client-specific training logs to monitor prompt convergence over time. For all experiments, the DFP length was set to 1, the learning rate for DFP updates was 1e-3, and the optimizer used was Adam with a weight decay of 1e-4. Clients used a batch size of 1 slide per update (bag-level learning), and training was performed for ten communication rounds unless otherwise specified.
2. Convergence issues may arise if the feature extractors are not well-aligned across clients. We recommend using standardized or pretrained backbones (e.g., ResNet50 trained on ImageNet or pathology foundation models). If prompts fail to converge, reducing the learning rate can help. Additionally, avoid prompt overfitting by regularizing with dropout or prompt decay mechanisms.
3. FedDFP currently assumes fixed feature extractors, which may limit adaptability in domains with large domain shifts. Moreover, the prompt representation size is fixed across clients, which may not optimally reflect each institution's data complexity. Extending the framework to include adaptive or hierarchical prompts is an area for future research.

References

1. Jahn SW, Plass M, Moinfar F (2020) Digital pathology: advantages, limitations and emerging perspectives. J Clin Med 9:3697
2. Baxi V, Edwards R, Montalto M, Saha S (2022) Digital pathology and artificial intelligence in translational medicine and clinical practice. Mod Pathol 35:23–32
3. Cong C, Liu S, Di Ieva A et al (2022) Colour adaptive generative networks for stain normalisation of histopathology images. Med Image Anal 82:102580
4. Waqas A, Bui MM, Glassy EF et al (2023) Revolutionizing digital pathology with the power of generative artificial intelligence and foundation models. Lab invest 103:100255
5. Jose L, Liu S, Russo C et al (2023) Artificial Intelligence–assisted classification of gliomas using whole slide images. Arch Pathol Lab Med 147:916–924
6. Campanella G, Hanna MG, Geneslaw L et al (2019) Clinical-grade computational pathology using weakly supervised deep learning on whole slide images. Nat Med 25:1301–1309
7. Lu MY, Chen RJ, Kong D et al (2022) Federated learning for computational pathology on gigapixel whole slide images. Med Image Anal 76:102298
8. Huang L, Shao L, Bao M et al (2024) Federated learning with comparative learning-based dynamic parameter updating on glioma whole slide images. Eng Appl Artif Intell 137:109233
9. du Terrail J, Leopold A, Joly C et al (2023) Federated learning for predicting histological response to neoadjuvant chemotherapy in triple-negative breast cancer. Nat Med 29:135–146
10. Jin H, Liu S, Cong C et al (2025) FedWSIDD: federated whole slide image classification via dataset distillation. Preprint at https://arxiv.org/abs/2506.15365
11. Arivazhagan MG, Aggarwal V, Singh AK, Choudhary S (2019) Federated learning with personalization layers. Preprint at https://arxiv.org/abs/1912.00818
12. Collins L, Hassani H, Mokhtari A, Shakkottai S (2021) Exploiting shared representations for personalized federated learning. In: Proceedings of the 38th International Conference on Machine Learning, vol 139. PMLR, Cambridge, MA, pp 2089–2099.
13. Tan Y, Long G, Liu L et al (2022) FedProto: federated prototype learning across heterogeneous clients. In: Proceedings of the AAAI Conference on Artificial Intelligence, vol 36, no 8. AAAI Press, Palo Alto, CA, pp 8432–8440
14. Jia Y, Vahidian S, Sun J et al (2025) Unlocking the potential of federated learning: the symphony of dataset distillation via deep generative latents. In: Leonardis A, Ricci E, Roth S et al (eds) Computer Vision—ECCV 2024. Lecture Notes in Computer Science, vol 15136. Springer, Cham, pp 18–33
15. Jing L, Cong C, Chen Y, Cao Y (2026) Federated distillation for whole slide image via Gaussian-mixture feature alignment and curriculum integration. Preprint at https://arxiv.org/abs/2605.00578
16. Jeong E, Oh S, Kim H et al (2018) Communication-efficient on-device machine learning: Federated distillation and augmentation under non-iid private data. Preprint at https://arxiv.org/abs/1811.11479
17. Chan YH, Ngai ECH (2021) FedHE: heterogeneous models and communication-efficient federated learning. In: Proceedings of the 17th International Conference on Mobility, Sensing and Networking (MSN). IEEE, Piscataway, NJ, pp 207–214
18. Liu S, Shah Z, Sav A et al (2020) Isocitrate dehydrogenase (IDH) status prediction in histopathology images of gliomas using deep learning. Sci Rep 10:1–11
19. Lu MY, Williamson DFK, Chen TY et al (2021) Data-efficient and weakly supervised computational pathology on whole-slide images. Nat Biomed Eng 5:555–570
20. Cong C, Song Y, Di Ieva A et al (2026) Flexible and scalable federated learning with deep feature prompts for digital pathology. npj Digit Med. https://doi.org/10.1038/s41746-026-02710-6
21. Otsu N (1979) A threshold selection method from gray-level histograms. IEEE Trans Syst Man Cybern 9:62–66
22. He K, Zhang X, Ren S, Sun J (2016) Deep residual learning for image recognition. In: Proceedings of the IEEE Conference on Computer Vision and Pattern Recognition. IEEE Computer Society, Los Alamitos, CA, pp 770–778
23. Deng J, Dong W, Socher R et al (2009) ImageNet: a large-scale hierarchical image database. In: Proceedings of the IEEE Conference on Computer Vision and Pattern Recognition. IEEE, Piscataway, NJ, pp 248–255
24. Acar DAE, Zhao Y, Navarro RM et al (2021) Federated learning based on dynamic regularization. Preprint at https://arxiv.org/abs/2111.04263
25. Li Q, He B, Song D (2021) Model-contrastive federated learning. In: Proceedings of the IEEE/CVF Conference on Computer Vision and Pattern Recognition. IEEE Computer Society, Los Alamitos, CA, pp 10713–10722

26. Huang C-Y, Srinivas K, Zhang X, Li X (2024) Overcoming data and model heterogeneities in decentralized federated learning via synthetic anchors. Preprint at https://arxiv.org/abs/2405.11525
27. Yang FE, Wang CY, Wang YCF (2023) Efficient model personalization in federated learning via client-specific prompt generation. In: Proceedings of the IEEE/CVF International Conference on Computer Vision. IEEE Computer Society, Los Alamitos, CA, pp 19159–19168
28. Shao Z, Bian H, Chen Y et al (2021) Transmil: transformer based correlated multiple instance learning for whole slide image classification. Adv Neural Inf Process Syst 34:2136–2147
29. Ilse M, Tomczak J, Welling M (2018) Attention-based deep multiple instance learning. In: Proceedings of the 35th International Conference on Machine Learning, vol 80. PMLR, Cambridge, MA, pp 2127–2136

Check for updates

Chapter 14

Using Large Language Model Embeddings to Analyze Neural Activity During Language Processing

Pengli Shao, Griffin A. Shimamoto and Jing Cai

Abstract

Human language emerges from the dynamic interplay of many interconnected elements, including meaning, context, and grammar. This richness makes it difficult to study the neural processes that respond to language. A major challenge is the absence of a precise, comprehensive numerical representation of language that can serve as a bridge between language and neural activity. Recent advances in large language models (LLMs) offer a potential solution by translating language into numerical representations that preserve meaning, grammar, and context. In this paper, we provide a general protocol for using LLM embeddings in conjunction with electrophysiological recordings during language tasks to reveal specific neural patterns in response to language activity. Our approach enables the identification of neural units selectively involved in language production and comprehension by correlating LLM embeddings with neural signals precisely aligned to word onset timestamps. The workflow consists of four stages: (i) extraction and preprocessing of numerical language features representing meaning and context obtained from pretrained language models, (ii) preprocessing and temporal alignment of neural recordings with word-onset timestamps, (iii) correlation and specificity analysis, and (iv) visualization of significantly responding neural units.

Keywords Human language, Large language model, Electrophysiological recordings, Contextual word embeddings, Neurolinguistics

1 Introduction

Human language is a complex process, encompassing structure (syntax), meaning (semantics), context, and intonation (prosody), all of which interact fluidly in real time [1, 2]. This places tight computational demands on the neural substrates of the brain, which must rapidly transform acoustic input into meaning and, conversely, generate speech from intended messages through coordinated computations [1, 2]. Due to the rich information conveyed during language activity, identifying the neural mechanisms underlying language in its natural form is difficult, requiring a theoretical framework that can handle complex inputs, distributed representations (information encoded

Antonio Di Ieva and Jennilee M. Davidson (eds.), *Computational Neuroscience*, Methods in Molecular Biology, vol. 3062,
https://doi.org/10.1007/978-1-0716-5440-8_14,

across many dimensions rather than in single neurons or units), and rapidly evolving dynamics.

Recent advances in LLMs have provided a structured, numerical format, called embeddings, for representing language that preserves many of its core relationships [3]. These models assign each word or subset of a word to a point in a high-dimensional space, which forms a compositional representation that captures both the meaning of the word and contextual information [4, 5]. In addition, LLMs have demonstrated good performance on a variety of different language tasks [6–8], suggesting that the internal representations—numerical vectors that encode each word in the context of surrounding words—within these models preserve much of the information required to comprehend and produce language. Supporting this, recent work has shown that embeddings from these models can predict patterns of brain activity during passive listening [9, 10]. As such, LLMs offer a promising bridge between the symbolic and abstract forms of language and the distributed, fine-grained neural processes that support it in the brain.

In this paper, we introduce a protocol for mapping language embeddings from pre-trained LLMs onto neural recordings collected during language-related tasks. The protocol can be applied to almost all LLMs, and is compatible with a variety of neural recording modalities (e.g., EEG, ECoG, MEG) and experimental designs (e.g., natural conversation, structured sentence reading, narrative comprehension, and visual language presentation). The pipeline consists of four main stages:

1. **Language embedding extraction:** Transcribed language data are passed through pre-trained LLMs. Embeddings are extracted from the model for each word.
2. **Neural recording preprocessing and alignment:** Neural signals are processed, and then precisely aligned to language events (e.g., word or sentence onsets) using a task-specific window (e.g., pre- and post-onset periods).
3. **Correlation and specificity analysis:** Embeddings are compared to neural features via machine learning models or related techniques to identify areas of significant encoding.
4. **Visualization:** Results are aggregated across time and subjects to visualize the brain regions, channels, or temporal dynamics most selectively responsive to linguistic features.

This framework enables researchers to localize language-related responses across brain regions, and to investigate how neural signals relate to specific levels of linguistic abstraction. While we apply the method here to a particular recording method (sEEG) involving natural speech, the approach is extensible to a wide range of language tasks and recording types.

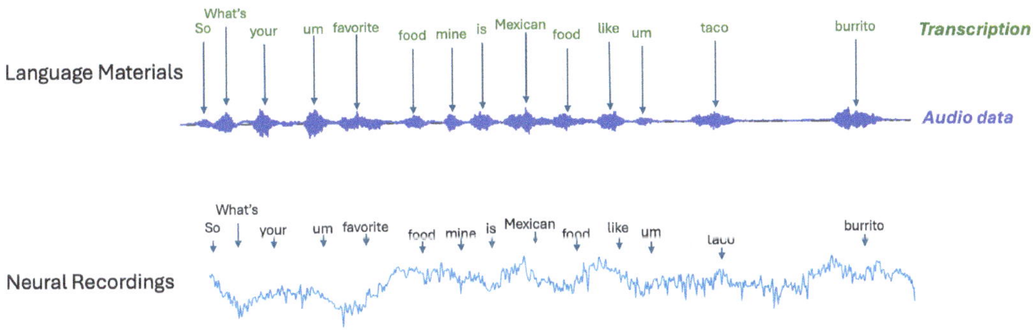

Fig. 1 Illustration of the materials. The conversation is recorded and transcribed with each word's onset/offset time. Neural recordings are aligned to the audio using the same word-level onsets/offsets. This schematic was created by the authors and uses no real participant data

2 Materials

To study neural responding to language, we need to have simultaneous recording of neural activity and language task (Fig. 1). Specifically, neural recordings for each participant's electrodes are in.npy or.mat format, audio recordings in.wav, and a transcription of the conversation or speech in.csv format. Transcriptions should also include precise timing information for each word onset.

2.1 Language Data

1. **Language tasks.**
   Using LLM embeddings to study neural activity can be applied to a wide range of language tasks performed alongside neural recordings. These may include passive listening, narrative comprehension, language production, or natural conversation. The approach can be applied to both carefully controlled experimental designs and naturalistic settings. The pipeline can also be adapted for visual language presentation, not limited to auditory input.
2. **Audio Data**
   For language tasks involving auditory inputs or speech production, audio recordings from the language task are collected and retained. These recordings will be used to generate transcriptions and extract timing information for each individual word.

2.2 Neural Recording and Preprocessing

1. Neural recordings are preprocessed into discrete time series of signals, corresponding to each recording unit. This is done by the standard preprocessing pipeline applied to the specific neural recording method. In this paper, we used sEEG as an example, but the same logic applies to other recording techniques. For sEEG, bipolar referenced voltages are band-pass filtered into 5 frequency bands (e.g., α: 8–13 Hz, β: 13–30 Hz, low-γ: 30–55 Hz, high-γ_1: 70–110 Hz, high-γ_2: 130–170 Hz) to obtain the envelopes for each channel and frequency, and the extreme values were treated (*see* **Note 1**). The resulting envelopes for

each channel are concatenated into an $\mathbb{R}^{T \times b}$ matrix, where T represents time points, b represents the number of frequency bands.

2.3 Software and Packages

1. **Data preprocessing**
   We primarily use MATLAB for initial signal processing routines and data conversion. Raw recordings are loaded, and standard procedures are performed, such as AC noise filtering (60 Hz in the U.S.). The processed data is then output as a.mat file that can be read by Python.
2. **Modeling and analysis**
   The modeling and analysis are performed in Python ($\geq 3.9, \leq 3.12$). Python is used for most computational and visualization tasks, including feature extraction from deep learning models, neural signal preprocessing, alignment, and statistical analysis.
   Below is a complete list of all packages used:
   `Torch (≥ ver. 2.1)`
   `pytorch_transformers (ver. 1.2.0)`
   `NumPy (ver.1.23.5)`
   `SciPy (ver. 1.10.1)`
   `Scikit-learn (ver. 1.3)`
   `Pandas (ver. 1.5.3)`
   `Matplotlib (ver. 3.7.1)`
   `Seaborn (ver. 0.12.2)`
   `NiBabel (ver. 5.2.0)`
   `Nilearn (ver. 0.10.3)`
   `whisperx (ver. 3.3.4)`

2.4 LLM Models

Large language models (LLMs) like Bidirectional Encoder Representations from Transformers (BERT) [8] and Generative Pre-trained Transformer 2 (GPT-2) [11] learn to represent language numerically, mapping words to vectors in a high-dimensional space, preserving linguistic information like meaning and context. These numerical features, known as embeddings, change with their surrounding words, providing context-dependent representations for language.

As shown in Fig. 2, to process language, LLMs first convert raw text into a numerical form through a process called `tokenization.A token` may be a word or a part of a word—for example, "going" may be broken into "go" and "ing"—each of which is assigned a unique identifier [11, 12]. These token IDs are then mapped to high-dimensional numerical vectors from a learned embedding matrix. These initial embeddings capture the identity of

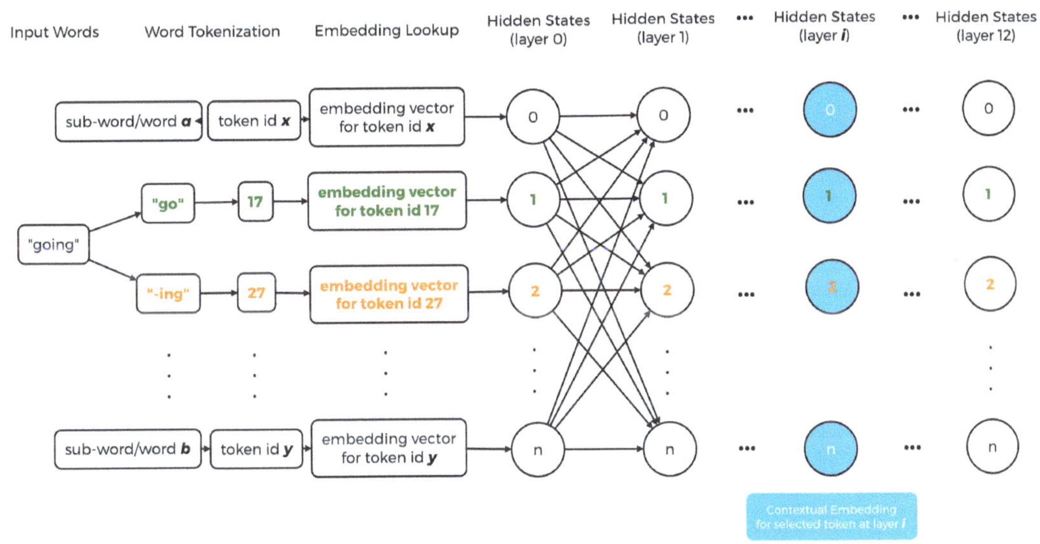

Fig. 2 Illustration of a transformer LLM (GPT-2 small). Input words (e.g., "going") are split into subword tokens (e.g., "go", "-ing") by BPE tokenizer and are converted via an embedding lookup into d-dimensional vectors (one per token). These token vectors feed a stack of transformer blocks ("hidden layers")—shown here as three for clarity, but in practice, there are many (e.g., GPT-2 small has 12) [11]. Each block outputs a hidden state for every token with the same dimensionality as its input. These hidden states are like the artificial "neurons" that are used to compare to the human neural activity. This schematic was created by the authors and uses no real participant data

each token before any contextual information is added. They are then fed into the model's multiple hidden layers that iteratively refine the representation based on surrounding context (*see* **Note 2**).

As the input embeddings pass through multiple transformer layers (e.g., 12 layers in GPT-2 [11]), each layer updates the representation of every token based on other tokens. This process produces a series of hidden states—vector representations of tokens at different layers—where each hidden state reflects the token's compositional linguistic features after incorporating progressively more context. Early layers typically capture low-level lexical or syntactic information, whereas later layers encode more abstract, high-level features [5]. These layer-wise hidden states form the basis for linking language model representations to neural activity.

The contextual word embeddings were extracted from a widely used large language model. We used the GPT-2 small (12 layers, 768 hidden units, 117 M parameters [11], *see* **Note 3**). The model was loaded with frozen weights to obtain the activations of the embedding lookup layer (layer 0) plus all 12 transformer layers, 13 layers in total.

While we primarily demonstrate this protocol with GPT-2 small for its accessibility and widespread usage, the pipeline is not limited to this model. In principle, any language model that provides access to hidden states can be integrated, including GPT-2 Medium/XL, RoBERTa, or more recent models like Large Language Model Meta AI (LLaMA) [13] or DeepSeek [14]. Models vary in the following ways:

1. **Model architecture and tokenizer:** Different models may use different tokenization schemes and vary in number of layers, hidden size, and context length. The context length will affect tokenization and the model architecture will affect the dimensionality of the embeddings (*see* **Note 4**).
2. **Embedding dimension and computational cost:** Larger models often produce higher-dimensional embeddings and require more memory to compute. Dimensionality reduction [e.g., principal component analysis (PCA) or layer selection] may be necessary to reduce computational demands.
3. **Context handling and alignment implications:** LLMs differ in how they incorporate context into embeddings. Autoregressive models (e.g., GPT-2) compute each token's representation from preceding tokens only, approximating left-to-right processing; bidirectional models (e.g., BERT) condition on both preceding and following tokens [8, 11]. This distinction matters for interpreting correlations between neural signals and embeddings: for a mid-sentence word, a BERT embedding may already reflect information from later words, whereas a GPT-2 embedding does not.

In summary, the current pipeline offers flexibility for experimentation with diverse models, provided that their outputs are formatted into a consistent tensor (tokens/words × hidden state size × layers).

3 Methods

3.1 Process Language Data

3.1.1 Deidentification

To ensure participant privacy, all information that may reveal personal identity was rigorously deidentified prior to analysis. This includes, but is not limited to:

- **Participant identifiers** originally assigned by clinical staff (e.g., hospital medical IDs) were replaced with anonymized codes that bear no link to the original record.
- **Transcription content** was scanned for personally identifiable information, including names, addresses, ages, schools, and regions of residence. These entities were replaced using a consistent substitution scheme: e.g., all place names were mapped to "Boston", all school names to "Harvard."

3.1.2 Auto Transcription

To compare neural activity with the high-level linguistic features, the raw audio waveform needs to be transcribed into words together with their start and end timestamp. A cost-efficient alternative to manual or crowdsourced transcription is to use WhisperX [15] (*see* **Note 5**). The overall transcription and alignment pipeline consists of three steps:

1. **Speech-to-text transcription**: Convert continuous audio into raw textual output using Whisper Automatic Speech Recognition (ASR).
2. **Word-level onset and offset timestamp alignment**: Apply forced alignment to assign onset and offset times to each word in the transcript.
3. **Speaker identification and labeling**: Optionally tag each word or segment with a unique speaker identifier, either via diarization tools or manual annotation.

The following Python code shows how automatic speech-to-text was performed and timestamps to the word level were refined with WhisperX:

----

a. Import the dependencies for ASR and force alignment.


```
import whisperx
import torch
import pandas as pd
import json, os, re
```

----

b. Setup the WhisperX speech-to-text model. It converts audio into text and gives rough timestamps for speech segments. We enable voice activity detection (VAD) to separate speech from silence and use hyperparameters/settings tuned for casual conversation (*see* **Note 6**)


```
device = "cuda" if torch.cuda.is_available() else
"cpu"
model = whisperx.load_model(
"large-v3",
device=device,
vad_method = 'pyannote',
compute_type="float32",
asr_options = {
"temperatures": [0.0, 0.2],
"beam_size": 10,
"best_of": 10,
```

```
"patience": 2.0,
"compression_ratio_threshold": 3.5,
"length_penalty": 1.0,
"log_prob_threshold": -4.0,
"no_speech_threshold": 0.01,
"suppress_numerals": True,
"suppress_tokens": [-1],
"initial_prompt":"...",
},
vad_options = {
"chunk_size": 15,
"vad_onset": 0.1,
"vad_offset": 0.1,
})
```

----

c. Load the alignment model to refine timestamps at the word level (start and end time for each word). We use a HuBERT-based aligner for English.

```
align_model, metadata = whisperx.
load_align_model(
language_code='en',
model_name="HUBERT_ASR_XLARGE",
device=device)
```

----

d. Transcribe the input audio file (.wav) to text with coarse segment timestamps.

```
audio_wav = "your.wav audio file directory"
audio = whisperx.load_audio(audio_wav)
result = model.transcribe(audio,
language='en',
verbose=True,)
```

----

e. Refine the timestamps to the word level by using the alignment model. We also request character-level alignments.

```
aligned = whisperx.align(result["segments"],
align_model,
metadata,
audio,
device,
interpolate_method = 'linear',
return_char_alignments=True)
--------------------------------------------------------------
```

The variable `aligned` is a JSON-style Python dictionary. If a different format is needed, the output can be saved as JSON, or converted to CSV/TXT for downstream use.

By using this pipeline, WhisperX reduces the manual effort required for large-scale audio dataset preprocessing. However, it should be noted that timestamp accuracy may not always meet the precision required (*see* **Note 7**).

3.1.3 Manual Checking and Alignment of Auto Transcription

If the automatic transcription or its word-level timing is not sufficiently accurate for the experiment's needs, manual correction and alignment should be performed.

We use Audacity [16] to load the audio together with a word-level label track produced by WhisperX and to interactively verify word onsets/offsets.

1. **Prepare files**. Export the task audio as `.wav` (keep the original sample rate; e.g., 16 kHz or 44.1 kHz). Convert the WhisperX word timestamps into an Audacity label file with one row per word in the format `<start time> <end time> <text>` (tab-separated, times in seconds). E.g., 0.532 0.812 going.
2. **Open the project**. Launch Audacity and set Project Rate (bottom-left) to match the audio's sample rate. In the Selection Toolbar, set the time display to hh:mm:ss + milliseconds.
3. **Import audio and labels**. Use `File`→Import → audio_wav... to load the.`wav`. Then `File→ Import→Labels...` to load the label`.txt`. Audacity creates a *Label Track* aligned beneath the waveform; each label shows the word text with adjustable start/end handles.
4. **Inspect and correct**. Click a label to select it. Drag its left/right edges to align onset/offset with the visible acoustic boundary (e.g., plosive release or voicing onset).
5. **Export corrected labels**. When finished, choose `File→ Export→ Export Labels...` to save a corrected.`txt`. Convert this file back to the project's CSV schema (`word, start_time, end_time, sid`) and ensure times remain absolute seconds from file start. Use these corrected timestamps in the downstream alignment and analysis.

For each word, its word-level onset and offset timestamps fall within an acceptable error margin—e.g. within ±30 milliseconds, although this threshold may be adjusted based on the specific needs of the study.

3.2 Tokenization and Embeddings

Because GPT-2 is an autoregressive language model, it processes tokens sequentially and computes each embedding based only on the tokens present within its fixed-size context window. To maximize the contextual information available for each extracted embedding, we process the transcription using a large sliding context window (e.g., $c_n = 600$ words) and, from each window, retain embeddings for only the final k^t words (e.g., $k^t = 10$). The window is then advanced by k^t words, ensuring that each extracted embedding is computed with substantial preceding context while avoiding redundant processing.

3.2.1 Tokenize Text

For GPT-2 tokenization, the conversation transcription S for each participant is processed sequentially in time order. We use a sliding context window c of up to c_n preceding words for each step (*see* **Note 8**). For each context window c, the words are tokenized using GPT-2Tokenizer [byte-level (BPE)], yielding a token ID sequence

$$T_c = BPE(c) \in \mathbb{Z}^{L_c}$$

where L_c represents the number of integer tokens in the current context window.

The following code demonstrates the minimal Python code to tokenize the current window c with GPT-2's byte-level BPE.

---

a. Import and initialize GPT-2's byte-level BPE tokenizer, and Pandas.

```
from pytorch_transformers import GPT2Tokenizer
import pandas as pd
tokenizer = GPT2Tokenizer.from_pretrained('gpt2')
```

---

b. Load a `transcript` CSV file with time-ordered "word" column.

```
transcription_df = pd.read_csv('your/
transcription.csv')
```

---

c. Define the current context window c by indices [start, end) (cap at cn words, here we define cn = 600 only for demonstration).

```
# the maximum number of words in window c
window_size = 600 # correspond to c_n

# exclusive end index of the window [start, end)
end = min(len(transcription_df), window_size)
# inclusive start index of the window
start = max(0, end - window_size)
```

d. Build the window text and tokenize to integer IDs (`Tc = BPE(c))`.

```
words =" ".join(transcription_df.word.
iloc[start:end].tolist())
token_ids = tokenizer.encode(words)
```

Here, `window_size` denotes as c_n in formulas, replace `transcription.csv` with the transcription directory, `start/end` with the bounds of the current window; `words` is the space-joined text of c. The output `token_ids` is the tokens ID sequence Tc for current window c.

For the current window c, we obtain the token ID sequence T_c as `token_ids` with length L_c=`len (token_ids)`.

3.2.2 Extract the Last Token Indices for Each Word

Within each context window c, we tokenize it to obtain T_c. For words that correspond to more than one token, we represent the word by its last token (*see* **Note 9**). We then locate in T_c the positions of the final token for each of the last k^t words w_j. These positions form an index vector $I_c \in \mathbb{Z}^{k^t}$.

The following code demonstrates the Python code to compute the final-subtoken positions for the last $k^t = 10$ words in the current window (forming I_c).

a. Allocate the index vector Ic for current window c.

```
import numpy as np
```

```
# Number of most recent words per step (= k^t)
last_words = 10
# Sliding-window step index;
# each step advances by k_t words
window_idx = 0

# final token position of each word
token_ind = np.zeros(last_words, dtype=int)
```

---

b. Populate each entry with the position of the final token of the j-th most recent word wj in Tc.

```
for cur_word in range(last_words):
# Absolute word index of the k-th word in current
window
word_idx = window_idx*last_words + cur_word

if word_idx < window_size:
words = get_words(transcription_df, 0, word_idx+
1)
else:
words = get_words(
transcription_df,
window_idx * last_words - window_size,
word_idx + 1)
token_ind[cur_word] = len(tokenizer.
encode(words)) - 1
```

Here, `words` is the window text up to (and including) wj, `get_words` is a helper function that concatenates the transcript tokens from the word column over the half-open index range `[start, end]` into a single string. This block runs inside the sliding-window loop, each `token_ind[cur_word]`≡ `Ic[j]` represents the final token position of the j-th most recent word wj, combining all Ic[j] yields an index vector Ic. Advancing the sliding-window loop by kt words creates a new context window c (and tokenization Tc) and therefore a new index vector Ic for that window.

We form the index vector Ic as token_ind, where each entry stores the position of the final token of wj.

3.2.3 Pass Tokenized Text Through Model

Load the GPT-2 model with the chosen parameters (*see* **Note 10**). For each context window c, the token sequence T_c is passed through GPT-2, producing hidden states

$$H_c \in \mathbb{R}^{l \times L_c \times d},$$

where l is the number of layers, d is the hidden embedding dimension, and L_c is the number of tokens in the current window.

The following code runs GPT-2 and retrieves the hidden states for the current window.

----

a. Import and load GPT-2 configured to return hidden states.

```
from pytorch_transformers import GPT2Model
model = GPT2Model.from_pretrained(
'gpt2',
output_hidden_states=True,
output_attentions=False)

model.eval()
```

----

b. Forward the current token sequence *Tc* to obtain *Hc*.

```
import torch
input_ids = torch.tensor([token_ids])
hidden_states = model(input_ids)[2]
```

For the current window c, we obtain the layer-wise hidden states H_c as hidden_states.

3.2.4 Extract Embeddings for Words

For each c_n-word context window c, the embeddings for the k^t most recent words are obtained by selecting the hidden states at positions given by I_c, where $I_c[j]$ specifies the position of the final token of the j-th word (*see* **Note 11**). Stacking across all k^t words in the window yields

$$E_c \in \mathbb{R}^{k^t \times d \times l}.$$

The following code computes the per-window embeddings $E_c \in \mathbb{R}^{k^t \times d \times l}$ for the last $k^t = 10$ words.

---

a. Allocate a container for the current window's embeddings *(kt × d × l)*.

```
emb_size = 768 # hidden embedding dimension
layers = 13 # number of layers
E_c = np.zeros((last_words, emb_size, layers))
```

---

b. Fill *Ec* by selecting, for each layer *L* and word index *k*, the hidden state at position *Ic[k]* in *Tc*.

```
for layer in range(layers):
for cur_word in range(last_words):
idx = int(token_ind[cur_word])
E_c[cur_word,:, layer] = (hidden_states[layer][0,
idx,:].detach().numpy())
```

Here, `layer` corresponds to each of the $l = 13$ layers (including 12 layers of transformer module and 1 input layer), `cur_word` corresponds to each of the $k^t = 10$ most recent words in the current window, and each `idx` is `Ic[j]`, the position (in *Tc*) of the final token for word *wj*.

For the current window c, we obtain the window-level embeddings Ec, ordered along the first axis (most recent k^t words).

3.2.5 Concatenate Embeddings Across Words

Window-level tensors $E_c \in \mathbb{R}^{k^t \times d \times l}$ are concatenated along the word axis to form a transcript-level array

$$E = concat_c(E_c) \in \mathbb{R}^{N \times d \times l}$$

where N is the total number of words across the transcript.

In our implementation, we pre-allocate E (named output_emb) of size (N, d, l) and write into contiguous rows as the sliding window advances.

---

a. Preallocate and write all complete windows into contiguous rows.

```
# total number of words across the transcript
num_words = transcription_df.shape[0]

output_emb = np.zeros([num_words, emb_size,
layers])
output_emb[:] = np.nan

window_idx = 0 # window index (0, 1, 2,...)
while window_idx < num_words // last_words:
token_ind = np.zeros(last_words, dtype = int)
for cur_word in range(last_words):
# absolute word index
word_idx = window_idx * last_words + cur_word
if word_idx < window_size:
words = get_words(transcription_df, 0, word_idx
+ 1)
else:
words = get_words(
transcription_df,
window_idx * last_words - window_size,
word_idx + 1)
# last token
token_ind[cur_word] = (len(tokenizer.
encode(words)) - 1)

# One forward pass for this window's context
input_ids = torch.tensor([tokenizer.
encode(words)])
hidden_states = model(input_ids)[2]

# Write k_t rows into the transcript-level array
for layer in range(layers):
for cur_word in range(last_words):
output_emb[window_idx * last_words + cur_word,:,
layer] = (
hidden_states[layer][0,token_ind[cur_word],:].
detach().numpy())
window_idx += 1
```

---

b. After processing all complete windows, the following block handles the tail case (fewer than kt remaining words) by indexing the last token in Tc.

```
word_idx = window_idx * last_words
while word_idx < num_words:
if word_idx < window_size:
words = get_words(transcription_df, 0, word_idx
+ 1)
else:
words = get_words(transcription_df, word_idx-
window_size, word_idx+1)

input_ids = torch.tensor([tokenizer.
encode(words)])
hidden_states = model(input_ids)[2]

for layer in range(layers):
output_emb[word_idx,:, layer] = (
hidden_states[layer][0, -1,:].detach().numpy())
word_idx += 1
```

Concatenating all window blocks yields the transcript-level embedding array $E \equiv \text{output\_emb} \in \mathbb{R}^{N \times d \times l}$.

3.2.6 Split Embeddings by Speaker Identity

This step is only for conversational or multispeaker tasks where the transcript includes speaker label. E is then split by speaker identity into two embedding sets, corresponding to (i) words spoken by the participant and (ii) words spoken by the other speaker(s)

$$E_s \in \mathbb{R}^{n_s \times d \times l},\ E_o \in \mathbb{R}^{n_o \times d \times l}.$$

Here, E_s denotes embeddings for speak (self-speaking) words (i.e., words produced by the participant), and E_o denotes embeddings for listen (other-speaking) words (i.e., words produced by other speaker(s) and perceived by the participant).

If your study is pure language comprehension with a single speaker (no self-speech), you may skip this step and simply let $E_o \leftarrow E$ and E_s be empty; use only E_o in Sects. 3.3 and 3.4.

The following minimal code builds boolean masks from the transcript and applies them to the transcript-level embedding array ($E \equiv$ `output_emb`).

a. Build speaker masks from the transcript's sid column (0 = participant/self; nonzero = other).

```
sid = transcription_df["sid"].to_numpy()
self_mask = (sid == 0)
other_mask = (sid!= 0)
```

b. Split into self/other arrays.

```
E_s = output_emb[self_mask,:,:]
E_o = output_emb[other_mask,:,:]
```

E_s contains the contextual embeddings for produced words spoken by the participant, whereas E_0 contains the embeddings for perceived words spoken by other speakers.

3.3 Channel-Level Neural Data Processing

All neural data analyses are performed separately for each electrode (or recording channel). For a given electrode, the aligned neural data are extracted and processed to generate envelope power for words spoken and heard, respectively. These matrices are later paired with the shared GPT-2 embeddings during regression analysis. The neural recordings are continuous time-series and are further processed at high frequency (e.g., 1000 Hz), while GPT-2 embeddings are computed for each word in the transcription. To enable direct comparison between the two datasets, we align the neural signals to the onset of each word and summarize the activity within a fixed-length window around it. This transforms continuous neural data into word-level features, consistent with the granularity of the embeddings.

3.3.1 Extract Neural Signal Windows Aligned to Word Onsets

The envelopes of the neural data were time-aligned to word onsets using a fixed-length window ($\pm w$ samples), producing word-level neural features that were then paired with the corresponding word-level embeddings. The choice of window size can be adjusted based on the desired temporal resolution.

For each band $i \in \{1, \ldots, n_b\}$, we align the amplitude envelope to each word onset and extract a fixed-length window of $\pm w$ samples:

$$T_s^{(i)} \in \mathbb{R}^{n_s \times (2w)}, \quad T_o^{(i)} \in \mathbb{R}^{n_o \times (2w)}.$$

Here, n_s and n_o are the counts of self- and other-produced words, respectively. Windows are centered on the onset (e.g., $2w = 8000$ samples); edge samples are mean-filled per channel to maintain length.

The following code shows how we compute these word-aligned windows for each frequency band.

---

a. Apply the helper for each frequency band.

```
T_s_list, T_o_list = [], []
for band in range(num_bands):
T_s_i, T_o_i = align_to_word(
transcription_df, # transcript
ampl[:, band].reshape(-1, 1)) # single-band
envelope
T_s_list.append(T_s_i)
T_o_list.append(T_o_i)
```

Here, we defined a helper function align_to_word, which takes the transcript transcription_df (with word onset times and speaker labels) and a single-band envelope ampl[:, i].reshape(-1, 1) as input, and outputs word-aligned neural matrices for self-produced ($T^{(i)}s$) and other-produced ($T^{(i)}o$) words. The detailed implementation of this helper function is available in the code repository [17].

Each row in T_s_i or T_o_i corresponds to an 8-second (8000 sample) window ($[-4000, +4000)$ms) around a word onset at 1 kHz sampling, aligned to either self-spoken or other-spoken words.

3.3.2 Compute Pre- and Post-onset Power Envelope Average

We average the envelope over two different time intervals aligned to word onset for produced/perceived words (δ ms before or after word onset), respectively:

$$A_s^{pre} \in \mathbb{R}^{n_s \times b}, \quad A_o^{post} \in \mathbb{R}^{n_o \times b},$$

where b indexes the frequency bands for electrophysiological neural recordings (*see* **Note 12**).

```
num_bands = len(T_s_list) # n_b
num_self = T_s_list[0].shape[0] # n_s
num_other = T_o_list[0].shape[0] # n_o

A_s_pre = np.zeros((num_self, num_bands))
A_o_post = np.zeros((num_other, num_bands))
pre_idx = slice(3500, 4000) # [-500,0) ms to word
onset
post_idx = slice(4000, 4500) # [0,+500) ms to word
onset

for band in range(num_bands):
T_s_i, T_o_i = T_s_list[band], T_o_list[band]
A_s_pre[:, band] = T_s_i[:, pre_idx].mean(axis=
1)
A_o_post[:, band] = T_o_i[:, post_idx].
mean(axis=1)
```

Here, `A_s_pre[:, band]` is neural activity obtained before the onset of spoken words, `A_o_post[:, band]` is the neural activity obtained after onset of listened words.

3.3.3 Remove Autocorrelation

Many statistical analyses, such as linear regression, require that the residuals be approximately uncorrelated and have roughly constant variance across observations. However, time-series neural recordings often exhibit serial dependence, which can induce autocorrelation in the residuals. In our data, both the neural activity and the embedding series show strong lag-1 autocorrelation, so we remove this first-order dependence before regression.

Specifically, every 1-D series in A_s^{pre}, A_o^{post}, E_s, and E_o are processed by removing lag-1 autocorrelation, which refers to the correlation between a sample x_t and its immediately previous value x_{t-1}. Mathematically, this is done by estimating the lag-1 correlation coefficient ρ:

$$\rho = \mathrm{corr}(x_t, x_{t-1}), \quad x_t' = x_t - \rho \; x_{t-1} \quad (t = 2, \ldots, T),$$

where T is the length (number of samples) of the original 1-D series $x_{1:T}$.

For every $t \geq 2$, we subtract ρ times the previous sample x_{t-1} from the current sample x_t; the first sample is dropped to keep alignment. If the original sequence exhibits strong first-order autoregression and other orders of autoregressions are low, the resulting series x' has almost no autocorrelation. Then, we treated the extreme values:

for every x', compute the first quartile Q_1, the third quartile Q_3, and the IQR ($IQR = Q_3 - Q_1$); any sample falling below $Q_1 - 1.5 \cdot IQR$ or above $Q_3 + 1.5 \cdot IQR$ is set to the corresponding threshold.

Below is the minimal code for this step.

----

a. Lag-1 removal (estimate ρ and form x′).

```
rho = np.corrcoef(x[1:], x[:-1])[1, 0]
x_de = x[1:] - rho * x[:-1]
```

----

b. IQR clipping (suppress extremes in the deautocorrelated series).

```
q1, q3 = np.percentile(x_de, [25,75]); iqr = q3
- q1
lower_bound, upper_bound = q1 - 1.5*iqr, q3 +
1.5*iqr
x_de = np.clip(x_de, lower_bound, upper_bound)
```

3.4 Linear Regression Model to Correlate Neural Activity and GPT-2 Embeddings

The goal of this analysis is to identify neural recording channels that exhibit significant correlations with the contextual embeddings extracted from a large language model (LLM). If a brain channel's activity (envelope in specific frequency bands) is significantly correlated to GPT-2 hidden states, it suggests that the electrode is responsive to language representations, hence involved in language processing. In this way, these "language-responsive" channels are selected, and then used in downstream interpretation and visualization.

3.4.1 Prepare Input Variables

Let $A_s^{pre} \in \mathbb{R}^{n_s \times b}$ and $E_s \in \mathbb{R}^{n_s \times d \times l}$ represent the processed neural power signals and GPT-2 embeddings for participant-spoken words, respectively. Likewise, A_o^{post} and E_o are defined for words spoken by others. Where:

- n_s, n_o: number of self/other word samples
- b: number of frequency bands
- d: embedding dimensions
- l: number of layers

Each 1-D signal is preprocessed by de-autocorrelating and IQR-based clipping (*see* Sect. 3.3.3).

3.4.2 Linear Regression

For a given frequency band b^*, embedding unit d^*, and layer l^*, fit

$$\text{NeuralActivity}_{b^*}^{i} = \alpha + \beta \cdot \textit{Embedding}_{d^*,l^*}^{i} + \varepsilon$$

where i indexes words. The regression is computed using `scipy.stats.linregress`, which returns slope (β), intercept (α), correlation coefficient (r), p-value (p), and standard error (stderr).

3.4.3 Repeat Across All Feature Combinations

To fully characterize the relationship between neural activity and model representations, we repeat the regression for all combinations of:

- Frequency bands $b^* \in \{1, \ldots, b\}$
- Embedding units $d^* \in \{1, \ldots, d\}$
- LLM layers $l^* \in \{0, \ldots, l-1\}$

For each (b^*, d^*, l^*) combination, a linear regression is independently fit between the neural response vector $A_{s/o}^{pre/post}[:, b^*]$and the GPT-2 embedding vector $E_{s/o}[:, d^*, l^*]$.

The following minimal code sweeps all combinations for (i) produced words and (ii) perceived words, storing the five regression statistics along the last axis.

---

a. Produced spoken words (A_s^{pre} vs. *Es)*.

```
from scipy import stats

num_bands = 5 # Band number
emb_size = 768 # hidden embedding dimension
layers = 13 # number of layers

# initialize matrix that stores stats outputs
sigs = np.zeros([num_bands, emb_size, layers, 5])
sigs[:] = np.nan

for band in range(num_bands):
# de-autocorrelation and IQR clipping

A_s_treated = deauto_confineOutlier(A_s_pre[:,
band])

for emb in range(emb_size):
for layer in range(layers):
# de-autocorrelation and IQR clipping
```

```
E_treated = deauto_confineOutlier(E_s[:, emb,
layer])


res = stats.linregress(A_s_treated, E_treated)
sigs[band, emb, layer,:] = [
res.slope,
res.intercept,
res.rvalue,
res.pvalue,
res.stderr]
```

---

b. Perceived words (A_o^{post} vs. Eo).

For perceived words, we simply apply the same process by replacing sigs with sigo, A_s_treated with `A_o_treated`, `A_s_pre` with `A_o_post`. Here, the index mapping is `band`$\equiv b^*$, `emb`$\equiv d^*$ (embedding unit), `layer`$\equiv l^*$. The last axis of `sigs` and `sigo` stores {slope, intercept, r, p, stderr}, `deauto_confineOutlier` is a helper function that first removes lag-1 autocorrelation and then suppresses extremes by `IQR` clipping, check code availability for more details [17].

3.4.4 Stack Regression Results Across Electrodes

The previous steps (from Sects. 3.4.1–3.4.3) are repeated for all electrodes of all participants, using each electrode's envelope-derived word-level features against the shared embeddings. This produces per-electrode regression outputs stored as 4D tensors

$$S_s^{(p,e)} \equiv \text{sigs} \in \mathbb{R}^{b\times d\times l\times 5},\ S_o^{(p,e)} \equiv \text{sigo} \in \mathbb{R}^{b\times d\times l\times 5}$$

If computational constraints are a concern, PCA can be used to reduce the dimensionality of the embeddings (*see* **Note 13**).

For each participant p with n_p electrodes, we first extract the p -values (index 3 along the last axis) from each per-electrode tensor $S_s^{(p,e)}[:,:,:,3]$, $S_o^{(p,e)}[:,:,:,3]$, and stack them to obtain

$$P_s^{(p)} \in \mathbb{R}^{n_p\times b\times d\times l},\ P_o^{(p)} \in \mathbb{R}^{n_p\times b\times d\times l}$$

where $P_s^{(p)}$ represents the stack of pre-onset p-values for self-spoken words, and $P_o^{(p)}$ represents the stack of post-onset p-values for other-spoken words.

Finally, concatenating participants along the electrode axis yields the group-level arrays

$$P_s, P_o \in \mathbb{R}^{n\times b\times d\times l}$$

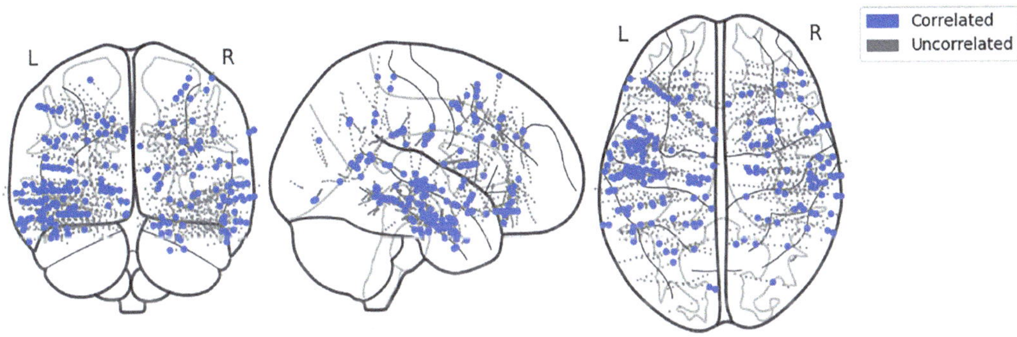

Fig. 3 Electrode location visualization. The locations of electrodes whose activity was significantly correlated with GPT-2 embeddings during speech production were shown in both hemispheres. Only neural activity occurring before word onset was analyzed, and electrodes were considered significant if their *p*-value fell below the Bonferroni-corrected threshold. Colors indicate correlation status: blue for electrodes significantly correlated, and gray for electrodes not significantly correlated. This schematic was created by the authors and uses no real participant data

Here, n denotes the total number of electrodes across all participants, i.e., $n = \sum_p n_p$.

3.4.5 Identify Significant Channels

To determine whether an electrode is significantly correlated with LLM embeddings, we apply a Bonferroni-corrected significance threshold of

$$p \leq \frac{0.05}{d \times l \times b}$$

Electrodes were marked as significant if any *p*-value across all embedding dimensions, hidden layers, or power bands fell below this threshold. This significance threshold was used to create two binary maps showing the locations of electrodes that were significant during speech production and comprehension.

The following minimal code demonstrates how we compute the Bonferroni-corrected threshold and identify significant electrodes.

---

a. Compute Bonferroni threshold and generate binary masks.

```
P_s = np.load("pre_onset_p_values.npy")
P_o = np.load("post_onset_p_values.npy")

alpha = 0.05
threshold = alpha / (emb_size * layers *
num_bands)
```

```
sig_mask_s = (P_s <= threshold).any(axis=(1,2,3))
sig_mask_o = (P_o <= threshold).any(axis=(1,2,3))
```

Here, `sig_mask_s` and `sig_mask_o` are boolean arrays that mark electrodes that are significantly correlated to LLM embeddings for spoken and listened words. Each entry corresponds to one electrode across all participants.

4 Visualization: Spatial Plot of Significant Electrodes

To visualize brain regions involved in speech comprehension and production, we plotted electrodes whose activity was highly correlated with the LLM embeddings.

The Montreal Neurological Institute (MNI) positions of all electrodes were loaded into an array

$$pos \in \mathbb{R}^{n\times 3}$$

where each row corresponds to an electrode.

As shown in Fig. 3, we created a spatial plot with two conditions: correlated and uncorrelated. We used the binary maps obtained in Sect. 3.4.5 to determine which electrodes had at least one significant *p*-value (To reduce visual clutter and improve anatomical interpretability, we limited plots to brain regions with sufficient electrodes; *see* **Note 14**.). The 3D positions of the significant electrodes were plotted as large blue dots, and the positions of the non-significant electrodes were plotted as small gray dots. All plotting was done using nilearn (*see* **Note 15**).

The following code loads the results and visualizes significant versus non-significant electrodes in the MNI space.

a. Load in necessary libraries and MNI positions.

```
import matplotlib.pyplot as plt
from nilearn import plotting
import matplotlib.patches as mpatches
from matplotlib.colors import ListedColormap

# pre-onset p-values for self-spoken, shape
(n,b,d,1)
P_s = np.load("pre_onset_p_values.npy")
```

```
# MNI position data, shape (n,3)
pos = np.load("MNI_position.npy")
```

----

b. Create significance mask


```
# reuse threshold from Sec. 3.4.5
sig_mask_s = (P_s <= threshold).any(axis=(1, 2,
3))
```

----

c. Visualization


```
inactive_mask = ~sig_mask_s
active_mask = sig_mask_s

display = plotting.plot_markers(
values=np.zeros(np.sum(inactive_mask)),
coords=pos[inactive_mask],
node_size=0.5,
node_cmap=ListedColormap(['grey']),
colorbar=False)

display.add_markers(coords=pos[active_mask],
marker_color='b',
marker_size=10)

handles = [mpatches.Patch(color='blue', label=
'Correlated'),
mpatches.Patch(color='grey', label=
'Uncorrelated')]

plt.gca().legend(handles=handles,
loc='upper left',
bbox_to_anchor=(1.05, 1),
frameon=True)
plt.show()
```


Here, we create significance by the method of computing Bonferroni threshold, we mentioned in Sect. 3.4.5 to get `sig_mask_s`, `P_s` denotes the group-level tensor of pre-onset *p*-values for self-spoken words, pos denotes the MNI position of each electrode.

Further, if a researcher is interested in visualizing how the neural pattern evolves across time, the same procedures can be repeated in successive time windows (e.g., pre-onset, post-onset, or sliding Δt bins). This will render one spatial map per window using the corresponding binary masks from Sect. 3.4.5. In this way, it is possible to study how language-responsive electrodes evolve over time, or whether the neural pattern progression depends on brain area.

6 Notes

1. Bipolar signals were calculated by subtracting the signal from adjacent contacts along each electrode. All bipolar time series were stored in participant-specific folders. Each channel was band-pass filtered into five frequency bands using a 4th-order Chebyshev Type II filter (stop-band attenuation = 40 dB). For each band, the analytic signal was computed using the Hilbert transform, and the amplitude envelope was extracted by taking its magnitude. Outlier samples exceeding $\pm 1.5 \times IQR$ of the envelope distribution were suppressed.
2. Tokenization splits text into smaller units like words or subwords. Each token is assigned a numerical ID used to retrieve its corresponding vector from a learned embedding matrix. For example, for the word "waiting," "wait" might map to token 614, "ing" to 172, etc. So "waiting" becomes [614, 172]. These `initial embeddings` are identical for the same token regardless of context before being passed through the model's hidden layers—for instance, the token "duck" has the same initial embedding whether in "the duck swam" or "duck and cover." Contextualization occurs only after passing through the model's hidden layers.
3. In this study, we used the small GPT-2 model with $l = 13$ (one embedding + 12 transformer layers) and $d = 768$, which balances model expressivity and computational feasibility. Increasing either l or d (e.g., in GPT-2 medium or large) will substantially increase the memory usage and runtime during extraction.

   - **Layers**. As text flows through the model, each layer refines the vector that represents a token: early layers tend to capture simpler patterns (e.g., local word forms), while later layers capture more abstract, context-dependent information (e.g., meaning in a sentence). GPT-2 small has 12 such layers.
   - **Hidden units (hidden size)**. This is the length of the vector used to represent each token at a given layer. In GPT-2 small, each token is represented by a vector of 768

numbers per layer. Larger hidden size means a higher-dimensional representation.

- **Parameters**. These are the trainable numbers (weights) the model learns from text during training. They determine how inputs are transformed across layers. GPT-2 small has about 117 million parameters [11]. More parameters generally increase capacity but also memory and compute demands.

4. BERT and GPT-2 small differ in both tokenization and architecture. BERT uses WordPiece tokenization and processes text bidirectionally, often one sentence at a time [8]. GPT-2 uses byte-pair encoding (BPE) and processes text in fixed-length left-to-right windows, and each window does not necessarily correspond to a single sentence [11]. The design differences influence how tokens are segmented.
5. WhisperX is an open-source automatic speech recognition (ASR) and force alignment framework built upon OpenAI's Whisper model. While Whisper is a large-scale encoder-decoder speech model trained to perform speech-to-text tasks, WhisperX extends its capabilities by incorporating additional alignment and voice activity detection components to produce word-level timestamps. Users can choose from a range of Whisper models such as `small`, `medium`, `large-v2`, or optimized variants like `faster-whisper`, depending on desired trade-offs between speed, accuracy, and resource usage.
6. General settings:

- `vad_method`: VAD backend to split speech versus silence. There are two VAD methods that WhisperX provides. `pyannote` is more accurate, but it consumes more computational resources. As an alternative, you can use `silero`.
- `temperatures`: The model first uses deterministic decoding (0.0), then a slightly non-deterministic pass (0.2) if needed. If transcripts look stuck/repetitive, add a higher value (e.g., 0.4).
- `beam_size`: How many hypotheses are explored in beam search (when temperature is 0). Large beams can improve accuracy but are slower.
- `best_of`: How many sampled candidates to keep when temperature is larger than 0.
- `patience`: Beam-search patience (stops earlier if no gains); 1.0≈ standard beam search. Lower the value toward 1.0 when decoding is slow with little gain.
- `compression_ratio_threshold`: Detects "garbled" text, higher means more tolerant. Raise if it falsely flags good text, lower if you see junk output.

- `length_penalty`: Penalizes very long outputs; 1.0= neutral normalization.
- `log_prob_threshold`: Treat a segment as failed if average log-probability is below this.
- `no_speech_threshold`: If the probability of "no speech" is above this and decoding failed, mark as silence. If the background noise is too loud or the speech is soft, set this value to a lower level.
- `suppress_numerals`: Avoid digits/currency in output, because many aligners handle them poorly. If you need exact numbers in the transcript, set it to `False`.
- `suppress_tokens`: IDs of tokens to suppress; [−1] is a shorthand for suppress most special characters.
- `initial_prompt`: Give the model context about the conversation style/domain. Helps with colloquial speech, fillers, repetitions.
- `chunk_size`: Maximum seconds to merge into one chunk.
- `vad_onset`: Threshold to start speech; lower = more sensitive (starts earlier)
- `vad_offset`: Threshold to end speech; lower = more tolerant (ends later)

7. The alignment precision in WhisperX is largely influenced by the choice of the alignment model. By default, WhisperX uses the Wav2Vec2 model for forced alignment [15], but more recent alternatives such as the HuBERT model—particularly its "extra large" architecture—have also been adopted for improved robustness [18]. However, based on our experience, most off-the-shelf alignment models cannot meet the stringent timing precision required for experimental neural analysis without manual parameter adjustment. Typical alignment errors range from 30 ms to 100 ms pre-tuning. Parameter tuning for speech-to-text model and force alignment model is highly dependent on the quality of the audio input and the desired level of conversational fidelity—e.g., whether alignment should focus solely on close-microphone speakers or also include background speech. For detailed guidance on alignment tuning, refer to the official documentation for Wav2Vec2 [19, 20] and HuBERT [18, 20].
8. We use a finite window of up to c_n preceding words to bound the amount of context input to the model at each step. In early steps, when fewer than c_n words are available, the window size is set to the current number of available words. At each step, we tokenize the current window, run the model once; the window then advances by k^t words and the procedure repeats

to the end of the transcript. This is because GPT-2 (and many LLMs) use sub-word tokenization, c_n words typically expand to L_c tokens ($L_c \geq c_n$). Each model has a maximum input length; thus, we limit the number of words to be tokenized and input into LLMs to keep the number of tokens within the maximum input length. As an example, GPT-2 small has a maximum input token limit of 1024 tokens, a window size of $c_n = 600$ reduces the chance of exceeding the maximum input length (1024 tokens for GPT-2 small). Moreover, increasing the number of most recent words k^t covered by window reduces the total number of model evaluations (lower overall computation) but increases the number of embeddings processed per step, which may raise memory usage. Therefore, we set $k^t = 10$ and $c_n = 600$.

9. Byte-level BPE tokenization maps each word w_j in the current window to one or more token indices $[t_1]$ or $[t_1, \ldots, t_n]$. Assume the current word is *"doing,"* it might be tokenized into *"do"* and *"-ing,"* where the token index of *"do"* is $t_1 = 14$, *"-ing"* is $t_2 = 52$.
10. Load the GPT-2 small model with `output_hidden_states = True` and switch to evaluation mode by `model.eval()`.
11. For models like GPT-2 small, the hidden state for the *last token* of a word integrates information from all preceding tokens in the window, including the full sequence of that word's tokens. Earlier tokens have not yet incorporated information from later tokens of the same word, so they provide a less complete word-level representation. Using the last token, therefore, yields a single, context-complete vector per word.
12. In our implementation, we set $w = 4000$ ms and extracted an 8-s window around each word onset (centered at word onset) with a sampling rate of 1 kHz. Although the statistical analysis in this study uses only a subset of neural activity to compute pre- and post-onset averages, we extract longer segments of neural signal for each word to ensure flexibility for alternative analyses.
13. In the case where the output embeddings from your model result in prohibitively high-dimensional tensor, we provide an optional step to reduce dimensionality using PCA, which is done by reducing the dimension of hidden units d. For each word-level embedding tensor $E_o \in \mathbb{R}^{n_o \times d \times l}$, we first loop over all layers, resulting in a 2D matrix $\mathbb{R}^{n_o \times d}$, then apply PCA to retain a target number of components(e.g., $n^{PCA} = 100$). The resulting embeddings are then used in regression analysis as described in Sect. 3.4. This step is especially recommended when computational constraints (e.g., memory usage or overfitting risk) arise from full embedding dimensionality.

The PCA decomposition can be performed using `sklearn.decomposition.PCA` [21] with standard preprocessing (mean centering and optional whitening).

14. To reduce visual clutter and improve anatomical interpretability, we limited spatial plots to brain regions with at least 25 electrodes recorded across all participants. This threshold retained nine regions of interest, including middle temporal, hippocampus, inferior temporal, inferior parietal, precentral, postcentral, insula, fusiform, and rostral anterior cingulate.
15. Electrodes can be colored either by brain area or by the strength of their correlation with the LLM embeddings. We used `nilearn.plotting.plot_markers` [22], which takes a color array followed by a corresponding array of 3D electrode positions.

Acknowledgments

The authors are grateful to Alex Hadjinicolaou, Angelique Paulk, Ziv Williams, and Sydney Cash for their assistance with original data collection. We also acknowledge the usage of OpenAI's GPT-5 Thinking to help polish language and fix minor mistakes (e.g., typos and variable names).

Code Availability

All analysis code and scripts used in this study are available at GitHub [17].

References

1. Friederici AD (2012) The cortical language circuit: from auditory perception to sentence comprehension. Trends Cogn Sci 16(5):262–268. doi:10.1016/j.tics.2012.04.001
2. Hickok G, Poeppel D (2007) The cortical organization of speech processing. Nat Rev Neurosci 8(5):393–402. doi:10.1038/nrn2113
3. Mikolov T, Chen K, Corrado G, Dean J (2013) Efficient estimation of word representations in vector space. arXiv preprint arXiv:1301.3781
4. Peters ME, Neumann M, Iyyer M, Gardner M, Clark C, Lee K, Zettlemoyer L (2018) Deep contextualized word representations. Proc NAACL-HLT 2018:2227–2237
5. Tenney I, Das D, Pavlick E (2019) BERT rediscovers the classical NLP pipeline. In: Proceedings of the 57th Annual Meeting of the Association for Computational Linguistics, pp 4593-4601. doi:10.18653/v1/P19-1452
6. Brown T, Mann B, Ryder N et al (2020) Language models are few-shot learners. Adv Neural Inf Process Syst (NeurIPS) 33:1877–1901
7. Wang A, Pruksachatkun Y, Nangia N, Singh A, Michael J, Hill F, Levy O, Bowman SR (2019) SuperGLUE: A stickier benchmark for general-purpose language understanding systems. Adv Neural Inf Process Syst 32:3261-3275
8. Devlin J, Chang MW, Lee K, Toutanova K (2019) BERT: pre-training of deep bidirectional transformers for language understanding. In: Proceedings of the 2019 Conference of the North American Chapter of the Association for Computational Linguistics: Human Language Technologies, Volume 1 (Long and Short

Papers), pp 4171-4186. doi:10.18653/v1/N19-1423
9. Goldstein A, Zada Z, Buchnik E (2022) Shared computational principles for language processing in humans and deep language models. Nat Neurosci 25:369–380. doi:10.1038/s41593-022-01026-4
10. Caucheteux C, King J-R (2022) Brains and algorithms partially converge in natural language processing. Commun Biol 5:134. https://doi.org/10.1038/s42003-022-03036-1
11. Radford A, Wu J, Child R, Luan D, Amodei D, Sutskever I (2019) Language models are unsupervised multitask learners. OpenAI Blog 1(8):9
12. Sennrich R, Haddow B, Birch A (2016) Neural machine translation of rare words with subword units. In: Proceedings of the 54th Annual Meeting of the Association for Computational Linguistics, Volume 1: Long Papers, pp 1715-1725. doi:10.18653/v1/P16-1162
13. Touvron H, Lavril T, Izacard G et al (2023) LLaMA: Open and efficient foundation language models. arXiv preprint arXiv:2302.13971
14. DeepSeek-AI (2024) DeepSeek-V2: A strong, economical, and efficient mixture-of-experts language model. arXiv preprint arXiv:2405.04434
15. Bain M, Huh J, Han T, Zisserman A (2023) WhisperX: time-accurate speech transcription of long-form audio. Interspeech 2023, pp 4489-4493. doi:10.21437/Interspeech.2023-78
16. Audacity Team (2025) Audacity: free, open-source, cross-platform audio software (Version 3.7.4). https://www.audacityteam.org/. Accessed 14 Aug 2025
17. Cai J (2024) sEEG_conversation [Source code]. *GitHub*. https://github.com/jill620/sEEG_conversation. Accessed 14 Aug 2025
18. Hsu W-N, Bolte B, Tsai Y-HH et al (2021) HuBERT: self-supervised speech representation learning by masked prediction of hidden units. IEEE/ACM Trans Audio Speech Lang Process 29:3451-3460. doi:10.1109/TASLP.2021.3122291
19. Baevski A, Zhou Y, Mohamed A, Auli M (2020) wav2vec 2.0: a framework for self-supervised learning of speech representations. Adv Neural Inf Process Syst (NeurIPS) 33: 12449-12460
20. PyTorch Audio Team (2024) torchaudio: HUBERT_ASR_XLARGE pretrained ASR model (bundle). https://pytorch.org/audio/stable/generated/torchaudio.pipelines.HUBERT_ASR_XLARGE.html. Accessed 15 Aug 2025
21. Pedregosa F, Varoquaux G, Gramfort A et al (2011) Scikit-learn: machine learning in python. J Mach Learn Res 12:2825–2830
22. Abraham A, Pedregosa F, Eickenberg M et al (2014) Machine learning for neuroimaging with scikit-learn. Front Neuroinform 8:14

Check for updates

Chapter 15

Leveraging Identity-by-Descent for Disease Gene Discovery

Lyndal Henden

Abstract

Genomic regions inherited from a common ancestor are said to be identical-by-descent (IBD). These IBD regions, or segments, can be leveraged to aid in the discovery of disease-associated genetic variation in the absence of large cohorts or pedigrees. In this approach, genetic relatedness is established between affected individuals through the detection of IBD segments. Shared segments that segregate with the disease across individuals can then be prioritized as candidate regions for causal or risk variant discovery, offering a targeted strategy for investigating the genetic basis of diseases. This chapter presents a practical workflow for performing IBD analysis for disease mapping in small-to-medium-sized cohorts using a simulated genomic dataset.

Keywords Identical-by-descent, Disease gene mapping, Genomics, Single nucleotide polymorphisms, Bioinformatics

1 Introduction

Calculating genetic relatedness by comparing alleles between individuals is a fundamental genetics methodology for the identification of genes linked to disease. Alleles are defined as identical-by-state (IBS) if they share an identical nucleotide sequence, for example, two individuals have an "A" allele at a specific single nucleotide polymorphism (SNP) site in the genome. A subset of IBS alleles can be more specifically defined as identical-by-descent (IBD) if they are also inherited from a common ancestor. Since IBD alleles arise from shared ancestry, they are only found in individuals who are genetically related. Closely related individuals typically share a large proportion of their genome IBD. As relationships become more distant, the portions of genome that are inherited from a common ancestor, called IBD segments, become smaller and fewer as recombination during meiosis gradually breaks them into smaller fragments over successive generations (Fig. 1). The proportion of genome shared IBD between two individuals can be estimated using the number of meioses (m) separating them, with the average proportion calculated

Antonio Di Ieva and Jennilee M. Davidson (eds.), *Computational Neuroscience*, Methods in Molecular Biology, vol. 3062, https://doi.org/10.1007/978-1-0716-5440-8_15,

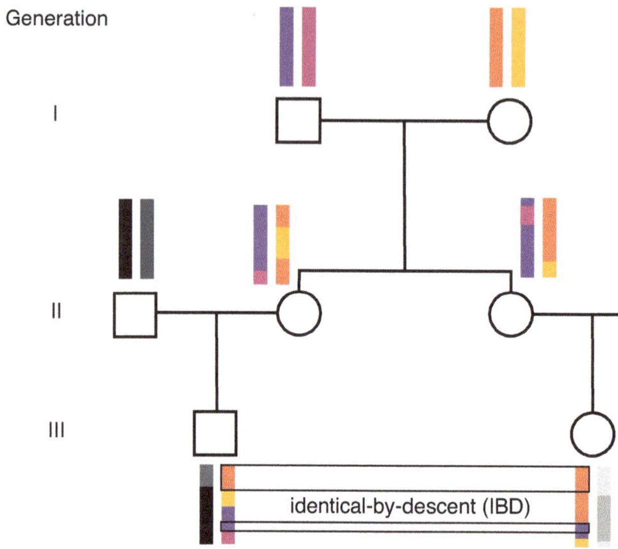

Fig. 1 A three-generation pedigree highlighting two IBD segments that are shared between first cousins, separated by 4 meioses. The orange IBD segment has been inherited from their grandmother and the purple IBD segment from their grandfather. Males are represented by squares and females by circles

as 2^{-m+1}, and the average segment length as $100/m$ centiMorgans (cM) [1]. A centiMorgan is a commonly used measure of genetic distance that represents the likelihood of recombination between two loci on a chromosome during meiosis. A 1% recombination frequency corresponds to 1 cM of map distance, which is equivalent to $\frac{1}{100}$ of a Morgan. For instance, second cousins, separated by six meioses, share about 3.1% of their genome IBD, with segments averaging 16.7 cM. After many meioses, the likelihood of inheriting IBD segments is low, though small segments may persist due to linkage disequilibrium (LD), which reflects ancient population structure. This chapter focuses on detecting recent ancestry, limiting analysis to within 25 generations (50 meioses) where the smallest detectable IBD segment is expected to span 2 cM.

IBD segments can be leveraged to aid in the discovery of disease-associated genetic variation through a method known as disease gene mapping [2, 3]. In this approach, genetic relatedness is established—whether known or previously unrecognized—between affected individuals through the detection of IBD segments. In disease gene mapping, shared segments that segregate with the disease across individuals can then be prioritized as putative candidate regions for causal or risk variant discovery, offering a targeted strategy for investigating the genetic basis of diseases without the need for genealogical pedigrees or population-scale data.

IBD inference is a valuable strategy for disease mapping in rare complex diseases characterized by both phenotypic and genetic heterogeneity and has been leveraged in gene discovery efforts of neurological and neurodegenerative disorders. Coppieters et al. applied IBD-based disease mapping to identify critical genomic regions for variant interrogation in 26 consanguineous families affected by retinal dystrophy [4]. Subsequent genetic analysis by exome sequencing within shared IBD regions successfully identified causal mutations in 20 out of the 26 families. Tuncay et al. employed IBD inference to identify candidate critical regions in a cohort of 22 families affected by autism spectrum disorder [5]. Within IBD-segregating segments, potentially pathogenic variants in known neurodevelopmental disorder genes were identified in five families. Henden et al. used IBD inference to refine the critical region in four families with familial adult myoclonic epilepsy [6], with Corbett et al. subsequently identifying the causal mutation within the IBD-refined region [7].

Bioinformatic-based IBD inference is performed using SNP genotype data. Advances in sequencing technologies, including genotyping arrays and next-generation sequencing, have enabled the generation of high-resolution datasets containing hundreds of thousands to millions of SNPs. These dense genomic datasets improve the resolution of IBD detection; however, the increased SNP density comes at a computational cost as analysis runtime typically scales with both the number of SNPs and the size of the cohort [2, 3]. The underlying algorithm of the IBD inference method also affects the runtime, so careful consideration should be taken when selecting a bioinformatic tool for performing IBD analysis.

IBD algorithms are generally categorized into probabilistic and non-probabilistic approaches [2, 3] . Probabilistic methods, such as PLINK [8], Beagle IBD [9], and XIBD [10], rely on Hidden Markov Models (HMMs) to estimate the posterior probability that two individuals share an allele IBD at each SNP, given their observed genotypes and allele frequencies in the population. These models account for the probabilistic nature of inheritance and recombination by allowing for transitions between IBD and non-IBD states across the genome. Probabilistic methods are highly accurate in detecting IBD segments but are computationally intensive in large-scale cohorts. These approaches also typically require additional inputs, such as population allele frequencies from a control cohort and genetic map positions (in Morgans) for each SNP, which can add complexity to the analysis. Non-probabilistic methods identify long stretches of genome that are IBS and classify these stretches as IBD if they are longer than a predefined threshold. One of the most popular non-probabilistic methods is GERMLINE [11], which scans phased genomes using sliding windows of a predefined length, generating hash codes for each window to identify exact haplotype matches. Individuals sharing the same hash code are considered candidate IBD pairs, and the algorithm then attempts to extend these matches beyond the window to identify longer IBD segments. GERMLINE can perform analyses on thousands of individuals using dense phased haplotype data in a matter of seconds, making it a very popular method for population-based cohort studies. Given its efficiency, the hash-based technique has formed the basis of newer IBD tools, including fastSMC [12] and iLASH [13]. The main drawback of these methods, however, is their reliance on phased genomic data. This introduces limitations, including the need for appropriate reference panels to estimate haplotype frequencies and the use of additional phasing tools. For a more detailed review of the current tools for IBD analysis, see [3].

A primary entry-point challenge for IBD analysis is tool selection. There are more than 20 tools available that differ in algorithms, input data requirements (such as reference cohorts, data formats, pedigree specification, and phasing), scalability to large cohorts, programming languages, and overall ease of use. This chapter outlines a straightforward workflow for performing IBD analysis for disease gene mapping of small-to-medium-sized cohorts. PLINK [8] is used for quality control processing of genotype data, and XIBD [10] is employed for IBD segment detection. Although earlier versions of PLINK (up to v1.9) supported IBD inference, this functionality has since been retired. Nonetheless, PLINK remains a widely used and reliable tool for genomic data processing. XIBD is an R-based [14] tool that uses unphased genotype data from either SNP microarrays or next-generation sequencing to infer IBD segments using an HMM similar to that previously implemented in PLINK. It is one of the only tools that incorporates analysis of the X chromosome, making it invaluable for disease studies with suspected X-linked inheritance. As R continues to gain popularity due to its flexibility in statistical analysis and data visualization, tools like XIBD offer an accessible and integrated approach for post-hoc exploration of IBD results.

2 Materials

Coding scripts accompanying this workflow are available for download from https://gitlab.com/mq-mnd/grp_williams/ibd_simulation_2025.git.

2.1 Software

PLINK: Download and install PLINK 1.9 from https://www.cog-genomics.org/plink/1.9/. Although PLINK 2.0 is available, PLINK 1.9 is preferred for this workflow due to its full support for reading and writing PED/MAP file formats.

- **R and RStudio:** Ensure that both R (https://cran.rstudio.com/) and RStudio (https://posit.co/download/

rstudio-desktop/) are installed and updated to their latest versions to ensure compatibility with all packages and functions.

- **XIBD:** The XIBD package imports the "identity" R package, which is no longer available on the Comprehensive R Archive Network (CRAN) and must be installed manually. To install both R packages, run the following code in RStudio.

```
[language=R, numbers=none]# Download the identity
package tarball from CRAN archive
url <- "https://cran.r-project.org/src/contrib/
Archive/identity/identity_0.2-1.tar.gz"
pkgFile <- "identity_0.2-1.tar.gz"
download.file(url = url, destfile = pkgFile)
# Install identity package
install.packages(pkgs=pkgFile, type="source",
repos=NULL)
# Delete package tarball
unlink(pkgFile)
# Install XIBD
install.packages("devtools")
devtools::install_github("bahlolab/XIBD")
```

2.2 Unphased Genotype Datasets

Data are available for download from https://doi.org/10.5281/zenodo.16519091.

2.2.1 Simulated Family

To demonstrate the IBD inference workflow, simulated datasets are used. SNP genotype data has been simulated for eight individuals from a hypothetical five-generation pedigree (Family 1), with IBD segments segregating along the disease-implicated lineage (Fig. 2). Such comprehensive sampling is uncommon in real-world studies, where genetic data are typically available for only a few individuals from more recent generations. Nonetheless, this dataset offers a valuable demonstration of the IBD inference workflow.

The dataset includes 141,212 SNPs, spanning both autosomes and the X chromosome. Simulation outputs are provided in PLINK PED/MAP format (*see* **Note 1**) and are named family.ped and family.map.

2.2.2 Simulated Reference

Genotype data for a matched reference cohort of 500 unrelated individuals has been simulated using the same set of 141,212 SNPs as in Family 1. The reference cohort data are provided in PLINK PED/MAP format as reference.ped and reference.map.

2.2.3 HapMap Phase II and III

Genomic data from HapMap Phase II and III populations have been included for quality control purposes [15]. These cohorts comprise individuals from 11 distinct populations. The HapMap data have been filtered and harmonized to include only SNPs present in the simulated dataset, resulting in 100,846 overlapping SNPs out of

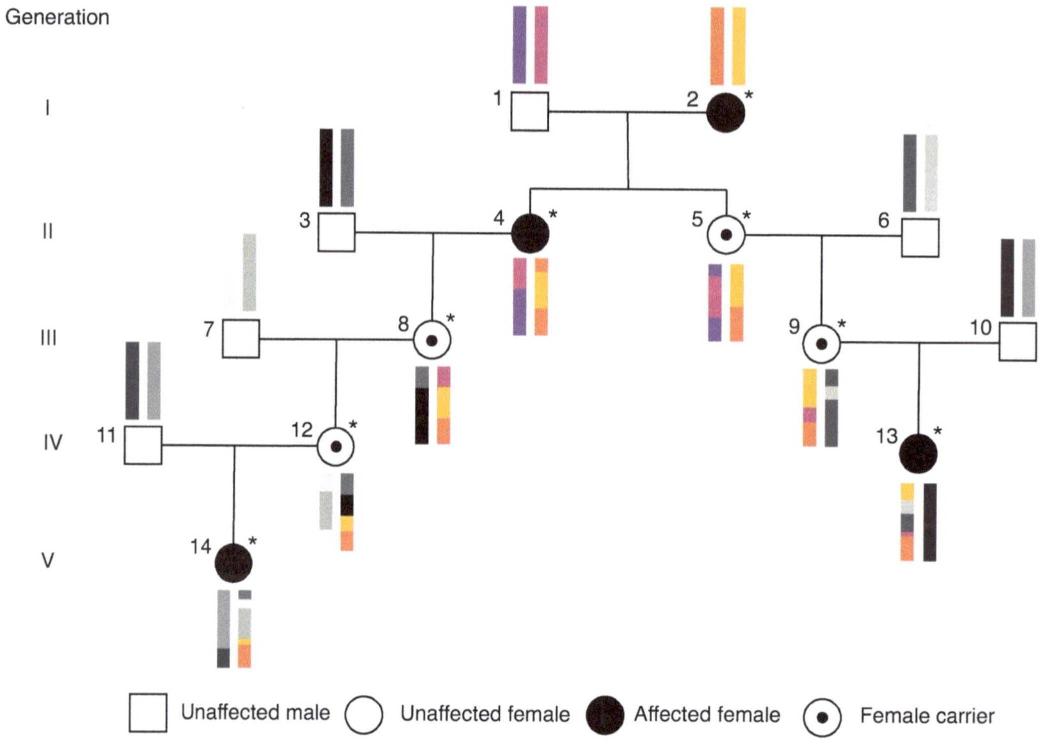

Fig. 2 A hypothetical five-generation pedigree (Family 1) illustrating the segregation of an inherited disease. Genome-wide genotype data for 141,212 SNPs has been simulated and provided for the eight individuals marked with an asterisk

the original 141,212. The data are provided in PLINK PED/MAP format as hapmap.ped and hapmap.map, along with a sample manifest (hapmap_pedigree.txt) indicating the population assignment for each individual.

3 Methods

3.1 Data Quality Control Processing Using PLINK

Performing data quality control processing prior to IBD analysis is recommended. While quality control can be conducted in R, large datasets, such as those with hundreds of thousands of SNPs and hundreds to thousands of individuals, may exceed the memory limitations of R. To mitigate this, datasets can be pruned during the quality control process using criteria such as LD and minor allele frequency (MAF) thresholds in PLINK. A reasonable data processing strategy that includes SNP pruning is provided here.

3.1.1 SNP and Sample Filtering

SNPs and individuals with high rates of missing genotype calls should be excluded from analysis to ensure data quality. Remove SNPs and samples with missing genotype call rates exceeding 10%. Use the --mind argument to filter individuals based on missingness, and the --geno argument to filter SNPs. It is recommended that SNP and sample filtering be applied to both the family and reference datasets.

```
[language=bash, numbers=none]# Remove SNPs and
samples with high missing call rates
plink --ped family.ped \
--map family.map \
--mind 0.1 \
--geno 0.1 \
--make-bed \
--out family.noMiss
plink --ped reference.ped \
--map reference.map \
--mind 0.1 \
--geno 0.1 \
--make-bed \
--out reference.noMiss
```

3.1.2 Sex Check

Validate the reported sex in the input dataset by comparing it to the sex imputed from the X chromosome heterozygosity. Use the --check-sex argument to perform this test. An output file with the suffix .sexcheck will be generated. Users should manually inspect this file for sex discrepancies in their cohort and update accordingly.

```
[language=bash, numbers=none]# Check the sex in the
input dataset with the X chromosome
plink --bfile family.noMiss \
--check-sex \
--out family.noMiss
plink --bfile reference.noMiss \
--check-sex \
--out reference.noMiss
```

3.1.3 Prune SNPs in High LD

Several approaches are available to prune the dataset based on LD, helping to reduce redundancy due to correlated SNPs. Here, the argument --indep 50 5 2 is used. Within a 50 SNP window, the variance inflation factor (VIF) is calculated, and SNPs with a VIF greater than 2 are removed. The window then shifts by 5 SNPs, and the process is repeated. These parameters are adjustable. In practice, a VIF threshold between 1.5 and 2 is recommended. However, in smaller samples, setting the threshold too low or using a large window size may result in excessive SNP removal. LD statistics should be calculated using the reference dataset, and SNPs identified as being in high LD should then be removed from both the family and reference datasets.

```
[language=bash, numbers=none]# Generate a list of
SNPs to be prune based on LD
plink --bfile reference.noMiss \
--indep 50 5 2 \
--out reference.noMiss.LD
# Prune SNPs in high LD
plink --bfile reference.noMiss \
--extract reference.noMiss.LD.prune.in \
--make-bed \
--out reference.noMiss.LDpruned
plink --bfile family.noMiss \
--extract reference.noMiss.LD.prune.in \
--make-bed \
--out family.noMiss.LDpruned
```

3.1.4 Population Stratification

IBD analysis using XIBD requires that all individuals (family and reference) belong to a single, genetically homogeneous population. PLINK provides multiple approaches for assessing population stratification, including principal component analysis (PCA), multidimensional scaling, and complete linkage clustering, each helping to identify underlying population structure and outliers. Incorporating external reference panels such as the HapMap Phase II and III datasets can provide a more comprehensive assessment of population structure and ancestry. Here, PCA is performed on a merged dataset that includes the family and reference individuals, along with the HapMap samples that have been filtered and harmonized to retain only SNPs present in the simulated dataset.

First generate a list of SNP IDs common to the pruned simulated dataset and the HapMap dataset.

```
[language=bash, numbers=none] # List of SNPs common
to pruned simulated and hapmap
comm -12 <(cut -f2 hapmap.map | sort) <(cut -f2
reference.noMiss.LDpruned.bim | sort) > common_
snps.txt
```

Extract the SNPs shared between the simulated and HapMap datasets, and merge the datasets into a single, harmonized dataset. PCA analysis is then performed on the merged data.

```
[language=bash, numbers=none] # Merge family and
reference data extract common SNPs
plink --bfile family.noMiss.LDpruned \
--bmerge reference.noMiss.LDpruned.bed \
reference.noMiss.LDpruned.bim \
reference.noMiss.LDpruned.fam \
--make-bed \
--extract common_snps.txt \
--out merge.family.reference
# Merge simulated with HapMap data and extract
common SNPs
plink --bfile merge.family.reference \
```

```
--merge hapmap.ped hapmap.map \
--make-bed \
--extract common_snps.txt \
--out merge.family.reference.hapmap
# PCA analysis
plink --bfile merge.family.reference.hapmap \
--pca \
--mind 0.1 \
--out merge.family.reference.hapmap
```

PLINK returns the first 20 principal components (PCs), which can be used to assess population substructure. Interpreting substructure and identifying outlier samples is best achieved through visualization of PCs rather than by examining raw numerical output. While PLINK does not provide built-in tools for visualizing PCA results, this can be done in RStudio.

Plotting the first two PCs allows for manual inspection of clustering patterns and outliers. In this case, the simulated dataset originates from a genetically homogeneous population that clusters with the CEU/TSI populations in the HapMap reference panel (Fig. 3). For datasets requiring the removal of outlier samples, PLINK's --exclude argument can be used to filter out specified individuals.

```
[language=R, numbers=none]# In RStudio:
# Install colour package
install.packages("RColorBrewer")
# Load packages and select HapMap colour palette
library("RColorBrewer")
getPalette <- colorRampPalette(brewer.pal(12,
"Accent"))
# Load PLINK PCA results
plink_pca <- read.table("merge.family.reference.
hapmap.eigenvec", header = FALSE)
# Load HapMap populations
hapmap_samples <- read.table("hapmap_pedigree.txt",
header = TRUE)
# Assign colours to HapMap populations
plot_cols <- rep("black", nrow(plink_pca))
pops <- c("ASW", "CEU", "CHB", "CHD", "GIH", "JPT",
"LWK", "MEX", "MKK", "TSI", "YRI")

for (i in 1:length(pops)) {
pedigree_pop <- hapmap_samples[hapmap_samples[,
"pop"] == pops[i],]
which_pop <- which(plink_pca[, 1] %in% pedigree_
pop[, 2])
plot_cols[which_pop] <- getPalette(11)[i]
}
# Plot HapMap samples
plot(plink_pca[plink_pca$V2 %in% hapmap_
samples$V2, "V3"], plink_pca[plink_pca$V2
%in% hapmap_samples$V2, "V4"], pch = 1, col =
```

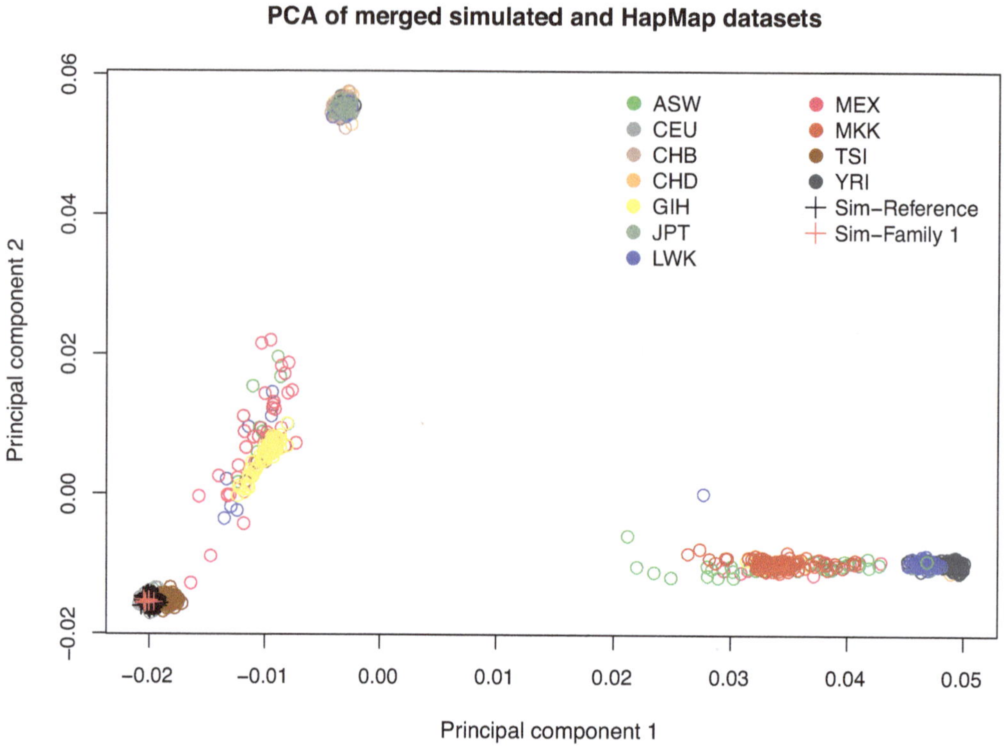

Fig. 3 PCA results produced using PLINK for the simulated datasets (Family 1 and reference) merged with the HapMap Phase II and III populations. The simulated data clusters with the HapMap CEU and TSI populations. The HapMap populations are defined as: ASW—African Ancestry in SW, USA; CEU—CEPH/Utah Collection; CHB—Han Chinese in Beijing, China; CHD—Chinese in Metropolitan Denver, CO, USA; GIH—Gujarati Indians in Houston, TX, USA; JPT—Japanese in Tokyo, Japan; LWK—Luhya in Webuye, Kenya; MEX—Mexican Ancestry in LA, CA, USA; MKK—Maasai in Kinyawa, Kenya; TSI—Tuscany in Italy; and YRI—Yoruba in Ibadan, Nigeria

```
rgb(col2rgb(plot_cols)[1,], col2rgb(plot_cols)[2,],
col2rgb(plot_cols)[3,], alpha = 150, maxColorValue
= 255),

xlab = "Principal component 1",
ylab = "Principal component 2",
main = "PCA of merged simulated and HapMap
datasets",
cex = 1.2)
# Add reference samples

points(plink_pca[plink_pca$V1 == "ref", "V3"],
plink_pca[plink_pca$V1 == "ref", "V4"], pch = 3,
col = 1, cex = 1.2)
# Add family samples

points(plink_pca[plink_pca$V1 == "family1", "V3"],
plink_pca[plink_pca$V1 == "family1", "V4"], pch =
3, col = 2, cex = 1.2)
```

```
# Add figure legend
legend("topright",
pch = c(rep(16, 11), 3, 3),
legend = c(unique(pops), "Sim-Reference", "Sim-
Family 1"),
col = c(getPalette(11), 1, 2),
bty = "n",
pt.cex - 1.5,
ncol = 2)
# Identify homogeneous cohort as necessary
```

3.1.5 Export Pruned PED/MAP and LD Files for XIBD

The current pruned datasets are in binary format (.bed, .bim, and. fam); however, XIBD requires input data in PED/MAP format. To export the final quality-controlled datasets in PED/MAP format, use the --recode argument in PLINK. If the reference population is assumed to be genetically homogeneous, an optional LD file containing squared correlation values based on genotypic allele counts can also be generated. This file is used by XIBD to support its advanced LD-HMM framework.

```
[language=bash, numbers=none]# Extract pruned
simulated PED/MAP files for XIBD
plink --bfile family.noMiss.LDpruned
--recode \
--out family.noMiss.LDpruned
plink --bfile reference.noMiss.LDpruned \
--recode \
--out reference.noMiss.LDpruned
# Generate LD file
plink --bfile reference.noMiss.LDpruned \
--r2 \
--out reference.noMiss.LDpruned.R2
```

3.2 IBD Segment Detection with XIBD in RStudio

3.2.1 Find IBD Segments

Since genotype files can be large, a useful function that loads large files quickly in R is fread from the data.table package. Install this package, then load all required libraries and quality-controlled input datasets into RStudio.

```
[language=R, numbers=none]# Install data.table
package
install.packages("data.table")
# Load libraries
library("XIBD")
library("data.table")
# Load PED/MAP data
family_ped <- fread("family.noMiss.LDpruned.ped",
data.table = FALSE)
family_map <- fread("family.noMiss.LDpruned.map",
data.table = FALSE)
```

```
reference_ped <- fread("reference.noMiss.LDpruned.
ped", data.table = FALSE)

reference_map <- fread("reference.noMiss.LDpruned.
map", data.table = FALSE)
# Load the LD file
reference_ld <- fread("reference.noMiss.LDpruned.
R2.ld", data.table = FALSE)
```

XIBD can implement two HMMs for IBD inference: Model 1 and Model 2. Model 1 replicates the algorithm used in PLINK, with the added capability of analyzing the X chromosome. It assumes that SNPs are in approximate linkage equilibrium, so LD pruning should be performed beforehand. Model 2 is a more advanced algorithm that accounts for some LD between SNPs. It requires an additional input file (reference.noMiss.LDpruned.R2.ld) containing LD squared correlation estimates, which were generated using PLINK. Given its ability to incorporate LD information, Model 2 is preferred for this analysis, although it is more computationally intensive than Model 1.

The PED/MAP datasets should be reformatted before IBD segment detection using the getGenotypes function from XIBD. It is important to define the input.map.distance and reference.map.distance arguments to indicate whether the genetic map positions in the MAP files are in centiMorgans (cM) or Morgans (M, *see* **Note 2**). The simulated datasets use genetic positions measured in Morgans.

```
[language=R, numbers=none]# Combine PED/MAP into a
list
fam_pedmap <- list(family_ped, family_map)
ref_pedmap <- list(reference_ped, reference_map)
# Format genotype calls for XIBD
xibd_geno <- getGenotypes(ped.map = fam_pedmap,
reference.ped.map = ref_pedmap, snp.ld = reference_
ld, model = 2, input.map.distance = "M", reference.
map.distance = "M")
```

The HMM requires prior probability estimates of sharing 0, 1, or 2 alleles IBD for each pair of individuals. These probabilities are computed using the function getIBDparameters, followed by IBD segment detection using the function getIBDsegments.

```
[language=R, numbers=none]# Estimate prior
probabilities
xibd_param <- getIBDparameters(ped.genotypes =
xibd_geno)
# Infer IBD segments
```

```
xibd_ibd <- getIBDsegments(ped.genotypes = xibd_
geno, parameters = xibd_param)
```

3.2.2 Examine Shared IBD Segments

The hypothetical disease in Family 1 exhibits moderate penetrance. This suggests that a causal variant has likely been inherited IBD among affected individuals. Thus, IBD segments shared by multiple affected individuals are strong candidate regions for identifying disease-causing genetic variation. In small cohorts such as this family example, genome-wide visualization of IBD segments allows identification of overlapping regions among affected individuals, which may indicate the location of a causal variant. XIBD includes the function plotIBDsegments to generate such visualizations, enabling intuitive inspection of shared IBD regions across the genome.

```
[language=R, numbers=none]# Plot IBD segments
genome-wide
plotIBDsegments(ibd.segments = xibd_ibd, ped.
genotypes = xibd_geno, add.fid.name = FALSE, add.
iid.name = TRUE, plot.title = "IBD segments in
Family 1")
```

IBD segments inferred between all pairwise combinations of individuals in Family 1 are visualized in Fig. 4. The most distantly related affected individuals in this pedigree are individuals 13 and 14, who are second cousins once removed. This pair shares three IBD segments: two on chromosome 2 and one on chromosome 4. Notably, other members of Family 1 also share overlapping IBD segments on chromosome 2. To assess whether specific loci exhibit elevated IBD sharing, potentially highlighting regions of interest for candidate disease gene discovery, the plotIBDproportions function can be used to visualize the proportion of IBD sharing genome-wide in a linkage-style plot (Fig. 5).

```
[language=R, numbers=none]# Get IBD proportions
genome-wide
locus_matrix <- getLocusMatrix(ped.genotypes =
xibd_geno, ibd.segments = xibd_ibd)

locus_prop <- getLocusProportion(ped.genotypes =
xibd_geno, locus.matrix = locus_matrix)
# Plot IBD proportions genome-wide

plotIBDproportions(locus.proportions = locus_prop,
plot.title = "Distribution of IBD sharing in Family
1", add.rug = FALSE)
```

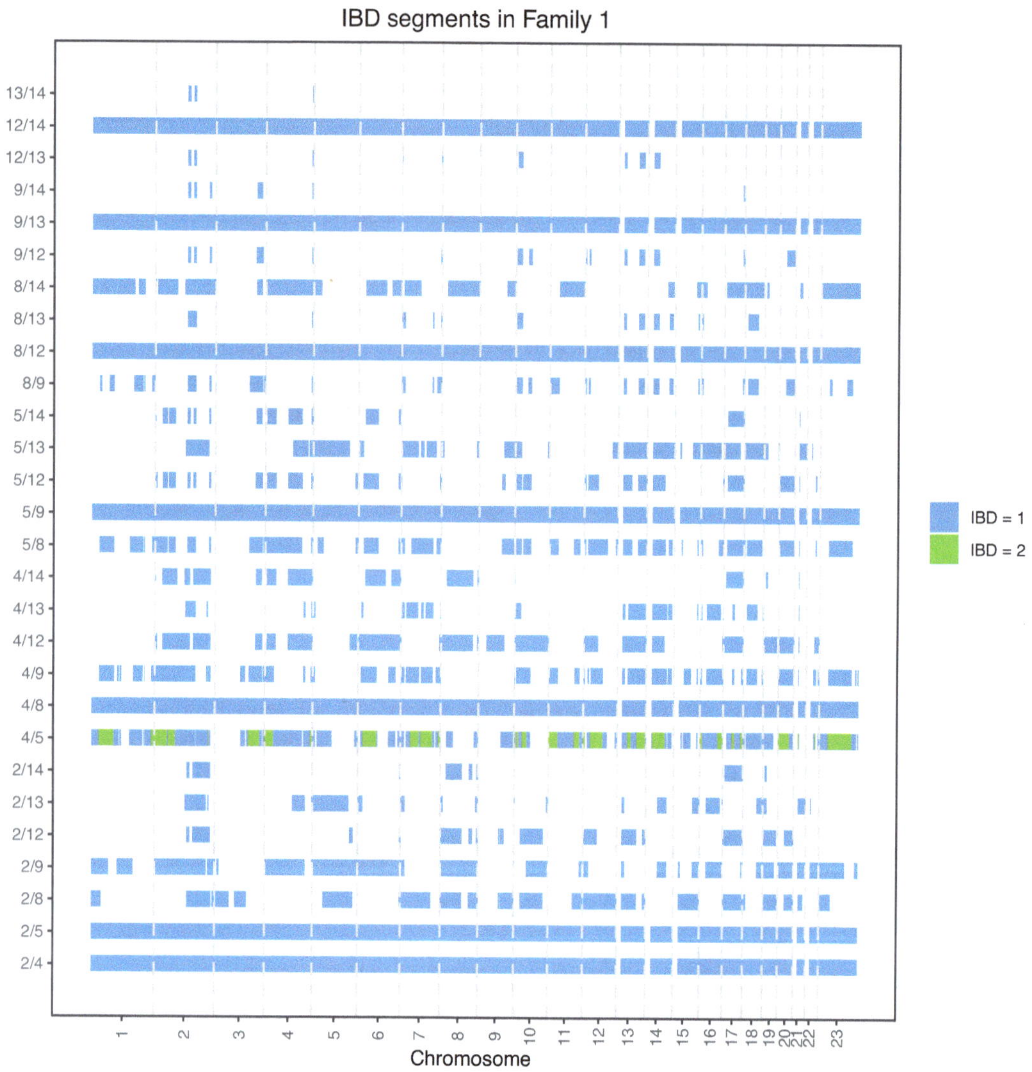

Fig. 4 IBD segments inferred by XIBD among the eight genotyped members of Family 1. Pedigree identifiers are on the y-axis. The most distantly related affected individuals, IDs 13 and 14, are second cousins once removed and share three IBD segments across the genome. Segments labeled IBD = 1 indicate sharing of one allele IBD; IBD = 2 indicates sharing of both alleles, typically observed in siblings or consanguineous relationships

All pairwise comparisons within Family 1 reveal one shared IBD locus on chromosome 2. This region can be considered a candidate disease region in affected individuals. The region's approximate boundaries can be inferred from the IBD segment output file (xibd_ibd). Subsequent candidate disease variant identification and prioritization can now be performed in this family using genomic technologies and bioinformatic workflows.

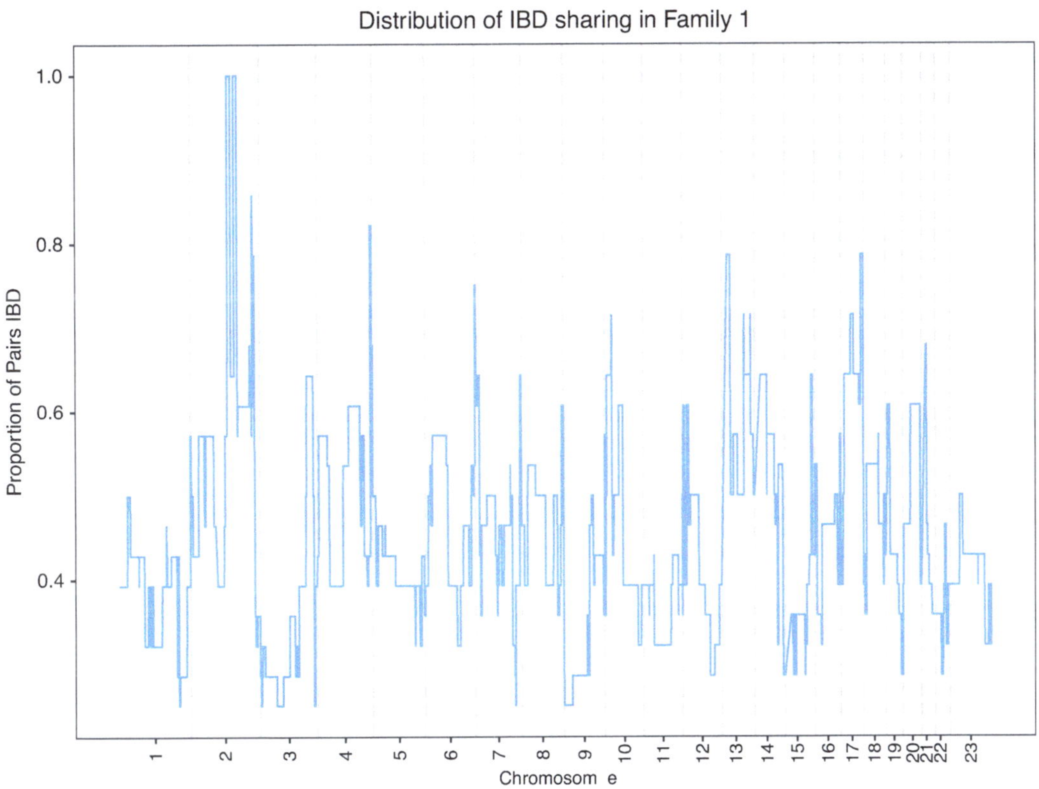

Fig. 5 Genome-wide plot showing the proportion of pairs in Family 1 sharing IBD at each SNP. A locus on chromosome 2 is shared IBD among all genotyped members of Family 1, highlighting it as a high-priority region for candidate gene discovery

4 Notes

1. PED/MAP are a standard and widely used format for genetic analysis, including IBD detection. These file types can be generated using various tools depending on the raw data type, such as PLINK, bcftools [16] and vcftools [17] for next-generation sequencing data, and GenomeStudio or Axion Analysis Suite Software for microarry data.
2. If genetic map positions are unavailable for the selected SNPs, base pair (bp) positions can be used as an approximation, assuming 1 cM = 1,000,000 bp (or 1 Mb).

References

1. Thomas A, Skolnick MH, Lewis CM (1994) Genomic mismatch scanning in pedigrees. Math Med Biol 11(1):16
2. Browning SR, Browning BL (2012) Identity by descent between distant relatives: detection and applications. Annu Rev Genet 46(617):633

3. Sticca EL, Belbin GM, Gignoux CR (2021) Current developments in detection of identity-by-descent methods and applications. Front Genet 12:722602
4. Coppieters F, Van Schil K, Bauwens M et al. (2014) Identity-by-descentâ€“guided mutation analysis and exome sequencing in consanguineous families reveals unusual clinical and molecular findings in retinal dystrophy. Genet Med 16:671–680
5. Tuncay IO, Parmalee NL, Khalil R et al (2022) Analysis of recent shared ancestry in a familial cohort identifies coding and noncoding autism spectrum disorder variants. npj Genom Med 7(1):13. https://doi.org/10.1038/s41525-022-00284-2
6. Henden L, Freytag S, Afawi Z et al (2016) Identity by descent fine mapping of familial adult myoclonus epilepsy (FAME) to 2p11.2--2q11.2. Hum Genet 135:1117–1125
7. Corbett MA, Kroes T, Veneziano L et al (2019) Intronic ATTTC repeat expansions in STARD7 in familial adult myoclonic epilepsy linked to chromosome 2. Nat Commun 10(1):4920. https://doi.org/10.1038/s41467-019-12671-y
8. Purcell S, Neale B, Todd-Brown K et al. (2007) PLINK: a tool set for whole-genome association and population-based linkage analyses. Am J Hum Genet 81(559):575
9. Browning SR, Browning BL (2010) High-resolution detection of identity by descent in unrelated individuals. Am J Hum Genet 86(526):539
10. Henden L, Wakeham D, Bahlo M (2016) XIBD: software for inferring pairwise identity by descent on the X chromosome. Bioinform 32(2389):2391
11. Gusev A, Lowe JK, Stoffel M et al. (2009) Whole population, genome-wide mapping of hidden relatedness. Genome Res 19(318):326
12. Nait Saada J, Kalantzis G, Shyr D et al (2020) Identity-by-descent detection across 487,409 British samples reveals fine scale population structure and ultra-rare variant associations. Nat Commun 11(1):6130. https://doi.org/10.1038/s41467-020-19588-x
13. Shemirani R, Belbin GM, Avery CL et al (2021) Rapid detection of identity-by-descent tracts for mega-scale datasets. Nat Commun 12(1):3546. https://doi.org/10.1038/s41467-021-22910-w
14. R Core Team (2025) R: a language and environment for statistical computing. R Foundation for Statistical Computing. https://www.R-project.org/
15. International HapMap Consortium (2003) The International HapMap project. Nat 426(789):796
16. Danecek P, Bonfield JK, Liddle J et al (2021) Twelve years of SAMtools and BCFtools. GigaScience 10(2):giab008. https://doi.org/10.1093/gigascience/giab008
17. Danecek P, Auton A, Abecasis G et al. (2011) The variant call format and VCFtools. Bioinform 27(2156):2158

Kolibri 360 GmbH
Ferdinand-Jühlke-Straße 7
99095 Erfurt, Deutschland
produktsicherheit@kolibri360.de